Plant Cell Culture Protocols

METHODS IN MOLECULAR BIOLOGY™

John M. Walker, SERIES EDITOR

132. **Bioinformatics Methods and Protocols,** edited by *Stephen Misener and Stephen A. Krawetz, 1999*
131. **Flavoprotein Protocols,** edited by *S. K. Chapman and G. A. Reid, 1999*
130. **Transcription Factor Protocols,** edited by *Martin J. Tymms, 1999*
129. **Integrin Protocols,** edited by *Anthony Howlett, 1999*
128. **NMDA Protocols,** edited by *Min Li, 1999*
127. **Molecular Methods in Developmental Biology:** Xenopus *and Zebrafish*, edited by *Matt Guille, 1999*
126. **Developmental Biology Protocols, Vol. II,** edited by *Rocky S. Tuan, 1999*
125. **Developmental Biology Protocols, Vol. I,** edited by *Rocky S. Tuan, 1999*
124. **Protein Kinase Protocols,** edited by *Alastair D. Reith, 1999*
123. **In Situ Hybridization Protocols, 2nd. ed.,** edited by *Ian A. Darby, 1999*
122. **Confocal Microscopy Methods and Protocols,** edited by *Stephen W. Paddock, 1999*
121. **Natural Killer Cell Protocols:** *Cellular and Molecular Methods,* edited by *Kerry S. Campbell and Marco Colonna, 1999*
120. **Eicosanoid Protocols,** edited by *Elias A. Lianos, 1999*
119. **Chromatin Protocols,** edited by *Peter B. Becker, 1999*
118. **RNA–Protein Interaction Protocols,** edited by *Susan R. Haynes, 1999*
117. **Electron Microscopy Methods and Protocols,** edited by *Nasser Hajibagheri, 1999*
116. **Protein Lipidation Protocols,** edited by *Michael H. Gelb, 1999*
115. **Immunocytochemical Methods and Protocols (2nd ed.),** edited by *Lorette C. Javois, 1999*
114. **Calcium Signaling Protocols,** edited by *David Lambert, 1999*
113. **DNA Repair Protocols:** *Eukaryotic Systems,* edited by *Daryl S. Henderson, 1999*
112. **2-D Proteome Analysis Protocols,** edited by *Andrew J. Link 1999*
111. **Plant Cell Culture Protocols,** edited by *Robert Hall, 1999*
110. **Lipoprotein Protocols,** edited by *Jose M. Ordovas, 1998*
109. **Lipase and Phospholipase Protocols,** edited by *Mark H. Doolittle and Karen Reue, 1999*
108. **Free Radical and Antioxidant Protocols,** edited by *Donald Armstrong, 1998*
107. **Cytochrome P450 Protocols,** edited by *Ian R. Phillips and Elizabeth A. Shephard, 1998*
106. **Receptor Binding Techniques,** edited by *Mary Keen, 1999*
105. **Phospholipid Signaling Protocols,** edited by *Ian Bird, 1998*
104. **Mycoplasma Protocols,** edited by *Roger J. Miles and Robin A. J. Nicholas, 1998*
103. ***Pichia* Protocols,** edited by *David R. Higgins and James Cregg, 1998*
102. **Bioluminescence Methods and Protocols,** edited by *Robert A. LaRossa, 1998*
101. **Myobacteria Protocols,** edited by *Tanya Parish and Neil G. Stoker, 1998*
100. **Nitric Oxide Protocols,** edited by *M. A. Titheradge, 1997*
99. **Human Cytokines and Cytokine Receptors,** edited by *Reno Debets, 1999*
98. **DNA Profiling Protocols,** edited by *James M. Thomson, 1997*
97. **Molecular Embryology:** *Methods and Protocols,* edited by *Paul T. Sharpe and Ivor Mason, 1999*
96. **Adhesion Proteins Protocols,** edited by *Elisabetta Dejana, 1999*
95. **Protocols in DNA Topology and Topoisomerases,** *Part II: Enzymology and Drugs,* edited by *Mary-Ann Bjornsti and Neil Osheroff, 1999*
94. **Protocols in DNA Topology and Topoisomerases,** *Part I: DNA Topology and Enzymes,* edited by *Mary-Ann Bjornsti and Neil Osheroff, 1999*
93. **Protein Phosphatase Protocols,** edited by *John W. Ludlow, 1997*
92. **PCR in Bioanalysis,** edited by *Stephen Meltzer, 1997*
91. **Flow Cytometry Protocols,** edited by *Mark J. Jaroszeski, 1998*
90. **Drug–DNA Interactions:** *Methods, Case Studies, and Protocols,* edited by *Keith R. Fox, 1997*
89. **Retinoid Protocols,** edited by *Christopher Redfern, 1997*
88. **Protein Targeting Protocols,** edited by *Roger A. Clegg, 1997*
87. **Combinatorial Peptide Library Protocols,** edited by *Shmuel Cabilly, 1997*
86. **RNA Isolation and Characterization Protocols,** edited by *Ralph Rapley, 1997*
85. **Differential Display Methods and Protocols,** edited by *Peng Liang and Arthur B. Pardee, 1997*
84. **Transmembrane Signaling Protocols,** edited by *Dafna Bar-Sagi, 1997*
83. **Receptor Signal Transduction Protocols,** edited by *R. A. J. Challiss, 1997*
82. ***Arabidopsis* Protocols,** edited by *José M Martinez-Zapater and Julio Salinas, 1998*
81. **Plant Virology Protocols,** edited by *Gary D. Foster, 1998*
80. **Immunochemical Protocols,** *SECOND EDITION,* edited by *John Pound, 1998*
79. **Polyamine Protocols,** edited by *David M. L. Morgan, 1998*
78. **Antibacterial Peptide Protocols,** edited by *William M. Shafer, 1997*
77. **Protein Synthesis:** *Methods and Protocols,* edited by *Robin Martin, 1998*
76. **Glycoanalysis Protocols,** edited by *Elizabeth F. Hounsel, 1998*
75. **Basic Cell Culture Protocols,** edited by *Jeffrey W. Pollard and John M. Walker, 1997*
74. **Ribozyme Protocols,** edited by *Philip C. Turner, 1997*
73. **Neuropeptide Protocols,** edited by *G. Brent Irvine and Carvell H. Williams, 1997*
72. **Neurotransmitter Methods,** edited by *Richard C. Rayne, 1997*
71. **PRINS and *In Situ* PCR Protocols,** edited by *John R. Gosden, 1997*
70. **Sequence Data Analysis Guidebook,** edited by *Simon R. Swindell, 1997*
69. **cDNA Library Protocols,** edited by *Ian G. Cowell and Caroline A. Austin, 1997*
68. **Gene Isolation and Mapping Protocols,** edited by *Jacqueline Boultwood, 1997*

METHODS IN MOLECULAR BIOLOGY™

Plant Cell Culture Protocols

Edited by

Robert D. Hall

CPRO-DLO, Wageningen, The Netherlands

Humana Press **Totowa, New Jersey**

999 Riverview Drive, Suite 208
Totowa, New Jersey 07512

This publication is printed on acid-free paper. ∞
ANSI Z39.48-1984 (American Standards Institute)
Permanence of Paper for Printed Library Materials.

Cover illustration: Figure 1B from Chapter 17, "Protoplast Isolation, Culture, and Plant Regeneration from *Passiflora*," by Paul Anthony, Wagner Otoni, J. Brian Power, Kenneth C. Lowe, and Michael R. Davey.

Cover design by Patricia F. Cleary.

For additional copies, pricing for bulk purchases, and/or information about other Humana titles, contact Humana at the above address or at any of the following numbers: Tel.: 973-256-1699; Fax: 973-256-8341; E-mail: humana@humanapr.com; or visit our Website: http://humanapress.com

Printed in the United States of America. 10 9 8 7 6 5 4 3 2 1

Library of Congress Cataloging in Publication Data

Main entry under title:

Methods in molecular biology™.

Plant cell culture protocols / edited by Robert D. Hall.
p. cm. -- (Methods in molecular biology™ ; v. 111)
Includes bibliographic references and index.
ISBN 0-89603-549-2 (alk. paper)
1. Plant cell culture—Laboratory manuals. 2. Plant tissue culture—Laboratory manuals.
I. Hall, Robert D. (Robert David), 1958– . II. Series: Methods in molecular biology (Totowa, NJ) ; 111.
QK725.P5535 1999
571.6'382—dc21

98-48664
CIP

Preface

Plant cell culture technology has come of age over the last 5–10 years. Many once-novel culture techniques are now standard laboratory practice and several are already under full-scale industrial application. This proves not only the value of the methodology to science and to world agriculture, but also that it can be an economically viable alternative/extension to existing procedures. For example, micropropagation protocols, as one of the earliest technologies to be developed, now account for the production of nearly 100,000,000 plants a year in The Netherlands alone. Furthermore, transgenic plants with additional, valuable agronomic traits—introduced either via *Agrobacterium*-mediated gene integration or by one of the alternative, direct DNA transfer strategies—are now being grown commercially for a number of crops, such as cotton, maize, and soybean. Many more new transgenic varieties of a wide range of agricultural, horticultural, and ornamental species are at the seed multiplication or registration stages in preparation for imminent commercial release.

The aim of *Plant Cell Culture Protocols* is to provide, in a single volume, a step-by-step guide to the most frequently used and broadly applicable techniques for plant cell and tissue culture, producing a volume that should, on the one hand, be comprehensive and on the other, be complementary to the companion volumes in the *Methods in Molecular Biology* series on other plant-related topics. In addition, a number of specialized protocols have been included to illustrate the diversity of the techniques available and their widespread application. Many of these protocols already form the basis of industrial projects aimed at producing improved varieties of some our most important crop plants that will increase their economic viability and reduce the strain on the environment.

The scope of *Plant Cell Culture Protocols* covers a wide variety of protocols on various aspects of culture initiation, maintenance, manipulation, application, and long-term storage. Particular emphasis has been given to techniques for micropropagation and genetic modification, which represent the two key areas in which plant tissue culture methods have so far had greatest impact. In addition, a number of protocols have been included to indicate how this technology is continually becoming more sophisticated, thus enabling, for example, the isolation of protoplasts from specific cell types (gametes, guard cells), or the transformation of crop plants through such alternative strategies as tissue electroporation or silicon whisker-mediated DNA transfer. The style of presentation is a well-tried and -tested one that minimizes ambiguity and maximizes success. Care has been taken to select examples that will not only be readily reproducible, but also easily modified to suit the particular needs of the reader's own plant material and aims.

To conclude, I would like to thank all those who have assisted me with the compilation of *Plant Cell Culture Protocols* and in particular the authors of the individual chapters. Their usually rapid responses to the >100 letters and approximately 500 E-mail requests (demands) that I have had to send out has made it possible to bring the project to fruition. Finally, I would also like to thank the DLO Centre for Plant Breeding and Reproduction Research (CPRO-DLO) for making their facilities available to me to perform this task.

Robert D. Hall

Contents

PART III. PLANT PROPAGATION IN VITRO

Contributors

PAUL ANTHONY • *Department of Life Sciences, University of Nottingham, University Park, Nottingham, UK*

KASIMALAI AZHAKANANDAM • *Department of Life Sciences, University of Nottingham, University Park, Nottingham, UK*

PILAR BARCELO • *Biochemistry and Physiology Department, IACR-Rothamsted, Harpenden, Hertfordshire, UK*

GEORGE W. BATES • *Department of Biological Sciences, Florida State University, Tallahassee, FL*

ERICA E. BENSON • *School of Molecular and Life Sciences, University of Abertay Dundee, Dundee, Scotland*

NIGEL W. BLACKHALL • *Department of Life Sciences, University of Nottingham, University Park, Nottingham, UK*

JAN BLAAS • *DLO-Centre for Plant Breeding and Reproduction Research, CPRO-DLO, Wageningen, The Netherlands*

ELS BONNE • *Plant Genetic Systems, Gent, Belgium*

GRAHAM BOORSE • *Department of Biology, Willamette University, Salem, OR*

MARTIEN BOSSUT • *Plant Genetic Systems, Gent, Belgium*

PHILLIPE BOXUS • *Biotechnology Department, Agricultural Research Centre, Gembloux, Belgium*

ALAN C. CASSELLS • *Department of Plant Science, University College Cork, Cork, Ireland*

ROSARIO F. CURRY • *Department of Plant Science, University College Cork, Cork, Ireland*

EDWARD C. COCKING • *Department of Life Sciences, University of Nottingham, University Park, Nottingham, UK*

MICHAEL R. DAVEY • *Department of Life Sciences, University of Nottingham, University Park, Nottingham UK*

KARABI DATTA • *Plant Breeding, Genetics and Biochemistry Division, IRRI, Manila, The Philippines*

SWAPAN K. DATTA • *Plant Breeding, Genetics and Biochemistry Division, IRRI, Manila, The Philippines*

KATHLEEN D'HALLUIN • *Plant Genetic Systems, Gent, Belgium*
PAUL DIJKHUIS • *Department of Developmental Biology, DLO-Centre for Plant Breeding and Reproduction Research, CPRO-DLO, Wageningen, The Netherlands*
PHILIP J. DIX • *Department of Biology, Saint Patrick's College, Maynooth, Co. Kildare, Ireland*
JIM M. DUNWELL • *Department of Agricultural Botany, University of Reading, Whiteknights, Reading, UK*
ULRIKA EGERTSDOTTER • *Norwegian Forest Research Institute, Ås, Norway*
JAN FAHLESON • *Department of Physiological Botany, Uppsala University, Uppsala, Sweden*
MICHAEL F. FAY • *Royal Botanic Gardens, Kew, Richmond, Surrey, UK*
MARIANNE FOLLING • *Department of Agricultural Sciences, Plant Breeding and Biotechnology, Royal Veterinary and Agricultural University, Copenhagen, Denmark*
MATTHEW V. FORD • *Royal Botanic Gardens, Kew, Richmond, Surrey, UK*
HIDEYUKI FUNATSUKI • *Plant Genetic Resources Laboratory, Hokkaido National Agricultural Experiment Station, Shinsei, Memuro, Kasai, Hokkaido, Japan*
KRISTINA GLIMELIUS • *Department of Plant Breeding, Swedish University of Agricultural Sciences, Uppsala, Sweden*
ALI GOLDMIRZAIE • *Genetic Resources Department, International Potato Centre (CIP), Lima, Peru*
JILL GRATTON • *Royal Botanic Gardens, Kew, Richmond, Surrey, UK*
BRIAN W. W. GROUT • *Consumers Association, London, UK*
MARC J. M. HAGENDOORN • *Department of Plant Physiology, Wageningen Agricultural University, Wageningen, The Netherlands*
ROBERT D. HALL • *DLO-Centre for Plant Breeding and Reproduction Research, CPRO-DLO, Wageningen, The Netherlands*
INDRA S. HARRY • *Department of Biological Sciences, University of Calgary, Calgary, Alberta, Canada*
ERWIN HEBERLE-BORS • *Institute of Microbiology and Genetics, Vienna Biocenter, University of Vienna, Vienna, Austria*
CHRISTOPHER S. HUNTER • *Faculty of Applied Sciences, University of the West of England, Frenchay, Bristol, UK*
KAZUTOSHI ITO • *Plant Bioengineering Research Laboratories, Sapporo Breweries Ltd., Nitta-Machi, Nitta-Gun, Gunma, Japan*

EVERT JACOBSEN • *Department of Plant Breeding, Wageningen Agricultural University, Wageningen, The Netherlands*

ALWINE JÄHNE-GÄRTNER • *Angewandte Molekularbiologie der Pflanzen, Institut fuer Allgemeine Botanik, Hamburg, Germany*

DIAAN C. L. JAMAR • *Department of Plant Physiology, Wageningen Agricultural University, Wageningen, The Netherlands*

DANIEL JONES • *Fermentation Biochemistry, USDA, ARS, NCAUR, Peoria, IL*

JOAN P. JOTHAM • *Department of Life Sciences, University of Nottingham, University Park, Nottingham, UK*

MAKOTO KIHARA • *Plant Bioengineering Research Laboratories, Sapporo Breweries Ltd., Nitta-Machi, Nitta-Gun, Gunma, Japan*

NIGEL J. KILBY • *Faculty of Applied Sciences, University of West of England, Bristol, Bristol, UK*

GEERT-JAN M. DE KLERK • *Centre for Plant Tissue Culture Research, Lisse, The Netherlands*

ERHARD KRANZ • *Center for Applied Plant Molecular Biology, Institute for General Botany, University of Hamburg, Hamburg, Germany*

FRANS A. KRENS • *Department of Cell Biology, DLO-Centre for Plant Breeding and Reproduction Research, CPRO-DLO, Wageningen, The Netherlands*

MEREL M. LANGENS-GERRITS • *Centre for Plant Tissue Culture Research, Lisse, The Netherlands*

PAUL A. LAZZERI • *Biochemistry and Physiology Department, IACR-Rothamsted, Harpenden, Hertfordshire, UK*

ROSITA LE PAGE • *Plant Genetic Systems, Gent, Belgium*

HORST LÖRZ • *Angewandte Molekularbiologie der Pflanzen, Institut fuer Allgemeine Botanik, Hamburg, Germany*

KENNETH C. LOWE • *Department of Life Sciences, University of Nottingham, University Park, Nottingham, UK*

PAUL T. LYNCH • *Division of Biological Sciences, University of Derby, Derby, UK*

INGRID M. VAN DER MEER • *Department of Cell Biology, DLO-Centre for Plant Breeding and Reproduction Research, CPRO-DLO, Wageningen, The Netherlands*

ANNETTE OLESEN • *Department of Agricultural Sciences, Plant Breeding and Biotechnology, Royal Veterinary and Agricultural University, Copenhagen, Denmark*

WAGNER OTONI • *Department of Life Sciences, University of Nottingham, University Park, Nottingham, UK*

LINUS H. W. VAN DER PLAS • *Department of Plant Physiology, Wageningen Agricultural University, Wageningen, The Netherlands*

J. BRIAN POWER • *Department of Life Sciences, University of Nottingham, University Park, Nottingham, UK*

KRIT J. J. M. RAEMAKERS • *Department of Plant Breeding, Wageningen Agricultural University, Wageningen, The Netherlands*

KAMISETTI S. RAMULU • *Department of Developmental Biology, DLO-Centre for Plant Breeding and Reproduction Research, CPRO-DLO, Wageningen, The Netherlands*

SONRIZA RASCO-GAUNT • *Biochemistry and Physiology Department, IACR-Rothamsted, Harpenden, Hertfordshire, UK*

ALAN H. SCRAGG • *Department of Environmental Health and Science, University of the West of England, Frenchay, Bristol, UK*

HARI C. SHARMA • *Department of Agronomy, Purdue University, West Lafayette, IN*

GARY TALLMAN • *Department of Biology, Willamette University, Salem, OR*

TREVOR A. THORPE • *Department of Biological Sciences, University of Calgary, Calgary, Alberta, Canada*

BRENT TISSERAT • *Fermentation Biochemistry, USDA, ARS, NCAUR, Peoria, IL*

JUDITH TOLEDO • *International Potato Centre (CIP), Genetic Resources Department, Lima, Peru*

ALISHER TOURAEV • *Vienna Biocentre, Institute of Microbiology and Genetics, Vienna University, Vienna, Austria*

INDRA K. VASIL • *Laboratory of Plant Cell and Molecular Biology, Department of Horticultural Sciences, University of Florida, Gainesville, FL*

VIMLA VASIL • *Department of Horticultural Sciences, University of Florida, Gainesville, FL*

HARRIE A. VERHOEVEN • *Department of Cell Biology, DLO-Centre for Plant Breeding and Reproduction Research,CPRO-DLO, Wageningen, The Netherlands*

RICHARD G. F. VISSER • *Department of Plant Breeding, Wageningen Agricultural University, Wageningen, The Netherlands*

STEPHEN YARROW • *Biotechnology Strategies and Coordination Office, Agriculture and Agri-Food Canada, Nepean, Ontario, Canada*

I

Introduction

1

An Introduction to Plant-Cell Culture

Pointers to Success

Robert D. Hall

1. Introduction

With the continued expansion of in vitro technologies, plant-cell culture has become the general title for a very broad subject. Although in the beginning it was possible to culture plant cells either as established organs, such as roots, or as disorganized masses, it is now possible to culture plant cells in a variety of ways: individually (as single cells in microculture systems); collectively (as calluses or suspensions, on Petri dishes, in Erlenmeyer flasks, or in large-scale fermenters); or as organized units, whether this is shoots, roots, ovules, flowers, fruits, and so forth. In the case of *Arabidopsis,* it is even possible to culture complete plants for generations from seed germination to seed set without having to revert to an in vivo phase.

In its most general definition, plant-cell culture covers all aspects of the cultivation and maintenance of plant material in vitro. The cultures produced are being put to an ever-increasing variety of uses. Initially, cultures were used exclusively as experimental tools for fundamental studies on plant cell division, growth, differentiation, physiology, and biochemistry *(1)*. Such systems were seen as ways to reduce the degree of complexity associated with whole plants, providing additional exogenous control over endogenous processes, to enable more reliable conclusions to be made through simpler experimental designs. However, more recently, this technology has been increasingly exploited in a more applied context, and successes in a number of areas have resulted both in a major expansion in the number of people making use of these techniques and also in an enhancement of the degree of sophistication associated with in vitro technology. Techniques for micropropagation and the

From: *Methods in Molecular Biology, Vol. 111: Plant Cell Culture Protocols*
Edited by: R. D. Hall © Humana Press Inc., Totowa, NJ

production of disease-free plant stocks have been defined and refined to such an extent that they have become standard practice for a range of (usually vegetatively propagated) horticultural and ornamental crop plants, such as gerberas, lilies, strawberries, ferns, and so on, thus creating what is now a multimillion dollar industry.

Nevertheless, the discipline within this technology that will eventually have the greatest impact on both fundamental and applied plant science is that of genetic modification of plant cells. Although this methodology is effectively still only in its infancy, it is now already possible, using a range of different techniques, to modify genetically virtually every plant species that has been tested so far, albeit with widely divergent degrees of efficiency *(2)*. Without doubt, this technology provides us with the most powerful single tool with which to study all aspects of plant-cell physiology, metabolism, and development by allowing the molecular dissection of individual components of the (sub)cellular organization of plants. In addition, the application of genetic modification techniques has already enabled us to produce crop plants with altered phenotypes, concerning e.g., herbicide resistance, insect resistance, and yield parameters *(2)*. Many additional applications are at the experimental/precommercial stage.

In simple terms, plant-cell culture can be considered to involve three phases: first, the isolation of the plant (tissue) from its usual environment; second, the use of aseptic techniques to obtain clean material free of the usual bacterial, fungal, viral, and even algal contaminants, and third, the culture and maintenance of this material in vitro in a strictly controlled physical and chemical environment. The components of this environment are then in the hands of the researcher, who gains a considerable degree of external control over the subsequent fate of the plant material concerned. An extra, fourth phase may also be considered where recovery of whole plants for rooting and transfer to soil is the ultimate goal.

The success of this technology is to a great extent, dependent on abiding by a number of fundamental rules and following a number of basic protocols. For those who have no experience at all with in vitro technology, it is strongly recommended, prior to initiating a first research project, that some basic knowledge be gained by visiting a working lab, preferably one doing similar work to that which is planned. This will not only save time, but also will help to avoid many of the pitfalls that could arise. Researchers can then also make direct contact with an experienced scientist who may later act as mentor. To proceed, a straightforward, well-tested protocol can be used to become acquainted with the manipulations required to achieve a particular goal. Then, having gotten this protocol to work, the researcher can begin with the modifications needed to achieve the original goal. The aim of the rest of this chapter is to act as a reference giving some basic guidelines concerning how to initiate a research program based on in vitro technology for plant tissues. The remaining chapters in this book will then describe individual protocols for specific techniques in detail.

2. Materials

2.1. Plant Material

Probably the worst thing that any researcher can do when embarking on a new in vitro technique is to use material that is suboptimal. This not only means using the wrong species/variety/genotype, but also using the right material, but which has been grown under substandard conditions. Thus, choosing in vivo-grown material from plants that are diseased or too old, or have not been maintained in an active growth phase during their entire life should be avoided. With suboptimal material, problems can be encountered in obtaining sterile cultures, excessive variability in in vitro response can result, and at worst, a complete failure of the experiment may occur. For most applications, explants from very young plants will respond best. For this reason, in vitro germinated seedlings are a frequently favored choice. Seed is often also much more readily sterilized than softer plant tissues. This, therefore, maximizes the likelihood of obtaining explants that are not only healthy, but are also guaranteed to be free of undesirable contaminants. However, species producing small seed can give rise to problems in obtaining sufficient experimental material. Furthermore, seed from outbreeders can also be genetically heterogenous, entailing an undesired variation in in vitro response that otherwise might be avoided by using explants from a single, larger greenhouse-grown individual.

For specific applications, precise growth conditions may be essential, particularly with regard to the period directly before the plants are to be used. Similarly, even when plants are healthy and at the desired stage for use, it is often the case that only a specific part of these plants will give the best explants, e.g., a particular internode, the youngest fully expanded leaf, flower buds within a certain size range, and so forth. A good search of the literature and paying close attention to the recommendations of experienced researchers are always to be strongly recommended.

2.2. Equipment

A plant-cell culture laboratory does not differ greatly from most other botanical laboratories in terms of layout or equipment. However, the requirement for sterility dominates. Plant cell cultures require rich media, but are relatively slow-growing. This places them in great danger of being lost, within days, through the accidental introduction of contaminating microorganisms. Plant-cell cultures also quickly exhaust their nutrient source, and therefore, sterile transfer to fresh media is a weekly to monthly requirement.

A cell-culture laboratory should be kept tidy, and dust-free with clean working surfaces. Some type of sterile culture transfer facility is essential. A laminar flow cabinet is preferable but a UV-sterilized transfer room or glove box, both of which are used solely for this purpose, and which are UV irradiated at all times when not in use, can also be employed effectively. Such facilities, when used for plant material, should never be used by colleagues for work on other organisms, such as yeast or *Escherichia coli.* It should also be held as a general rule that everything going into the sterile working area should already be sterile or, in the case of instruments, should be sterilized immediately on entry. This means also that in vivo grown plant material should only enter the transfer area after it has been submerged in the sterilizing solution.

The other equally important piece of equipment is the autoclave which is needed to sterilize glassware, media, and so forth. This should be of a size sufficient to cope with daily requirements. However, very large autoclaves should be avoided unless they are specifically designed for rapid heating and cooling before and after the high-pressure period to avoid long delays and also to prevent media being severely "cooked" as well as being autoclaved.

Although specialized techniques have specific equipment requirements (noted in the relevant chapters), in addition to the sterile transfer and autoclaving facilities, the following are generally needed to perform basic cell-culture procedures:

1. Tissue-culture-grade chemicals with appropriate storage space at room temperature, 4°C and –20°C.
2. Weighing and media preparation facilities: Balances to measure accurately mg to kg quantities should be available.
3. A range of sterilization facilities: In addition to the autoclave, a hot-air sterilizing oven is useful. Sterile filters (0.22-µm) are required for sterilizing heat-labile compounds. If large volumes of sterile liquids are required, a peristaltic or vacuum pump is also to be strongly recommended.
4. A source of (double) distilled water.
5. Stirring facilities that allow a number of different media to be made simultaneously.
6. A reliable pH meter with solutions of HCl and KOH (0.01, 0.1, 1.0, and 10 *M*) to adjust the pH accurately.
7. Culture vessels either of (preferably borosilicate) glass or disposable plastic, tubes, Erlenmeyer flasks, jars, and so on.
8. Plastic disposables, e.g. Petri dishes (9, 6, 3 cm), filter units, syringes, and so forth, as well as plastic bottles of various sizes for freezing media and stock solutions for long-term storage.

9. Sealants, e.g., aluminium foil, Parafilm/Nescofilm, clingfilm/Saranwrap.
10. Basic glassware (measuring cylinders, volumetric flasks), dissection instruments, hot plate/stirrer, gas, water, and electricity supply, microscopes, and so forth.
11. Microwave: Although not essential, the ability to make solidified media in "bulk" and remelt it for pouring when required not only saves time, but also avoids the risk of undesired condensation building up in culture vessels (especially Petri dishes) on prolonged storage.

2.3. Washing Facilities

The importance of cleaning glassware in a tissue-culture laboratory should never be underestimated. Furthermore, incorrect rinsing is equally as bad as incorrect washing. Traces of detergent or old media can cause devastation the next time the glassware is used. If not to be washed immediately, all glassware should be rinsed directly after use and should not be allowed to dry out. Therefore, keep a small amount of water in each vessel until it is cleaned. Certain media components (e.g., phytohormones), which are only poorly soluble in water when dried onto the inside of a flask, may not be removed by the normal washing procedures, but can redissolve the next time the vessel is autoclaved and contaminate the medium. For this reason, flasks used to make or store concentrated stocks of medium components should not be used for any other purpose.

New automatic washing machines can be programmed to wash at temperatures approaching 100°C, rinse extensively with warm and then cold water, and finally demineralized water before even blow-drying! However, if such equipment is not available, washing by hand is equally as good, if a little time-consuming. In this case, glassware should be soaked overnight in a strong detergent before being thoroughly scrubbed with a suitable bottle brush and then rinsed two to three times under running tap water and finally at least once with demineralized water. All glassware should then be dried upside down before being stored in a dust-free cupboard until required. It is generally recommended that glassware be thoroughly washed in an acid bath on a regular basis.

2.4. Media

There is a small number of standard culture media that are widely used with or without additional organic and inorganic supplements (*see* Appendix; *3–7*). However, next to these, there is an almost unending list of media that have been reported to be appropriate for specific purposes ***(8)***. Protoplast culture media, for example, can have a wide variation in composition, reflective of the often critical conditions required by these highly sensitive and fully exposed cells. However, even these are to a large extent derived from one of the standard recipes. Plant-culture media generally consist of several inorganic salts, a (small) number of organic supplements (e.g., vitamins, phytohormones), and a carbon source. In addition to these standard components, the specific needs of particular species or tissues, or the precise conditions required to initiate a desired in vitro response dictate which additional supplements are required. Today, with the wealth of knowledge concerning a very divergent list of plant species that has been built up over the last 20 years and that is readily available in the literature, the choice of medium with which to begin for a particular plant should be made only after referring to previous publications on the same or related species.

It can be seen, from the standard media recipes listed in the Appendix, that the micro- and macroelements and organic supplements can vary considerably. The species to be used will generally determine which medium to choose and, of course, the aim of the experiment (e.g., callus production, plant regeneration, somatic embryogenesis, anther culture, and so on) will determine which additional supplements are required. This is especially so for the phytohormones, which can play an extremely important role in determining the response of plant cells/tissues in vitro. Indeed, in many cases, it is only the number, concentration, type, and balance of the phytohormones used that distinguishes one experimental design from another. Of the macrocomponents, the source of nitrogen (N) is often considered to be of particular influence. Most media have N in the form of both nitrate and ammonia, but the ratio of one to the other can vary enormously to the extremes that one of the two sources is absent. Alternatively, both sources can be omitted and replaced by organic N sources in the form of amino acids, as in the case of AA medium ***(9)***. Although many media are composed as a fine balance to promote and maintain cell growth in vitro, temporary divergence from using the usual media components is often employed to direct growth and morphogenesis in particular directions. For example, by limiting or removing the N or phosphate source, secondary metabolite production can be stimulated, and through the qualitative and quantitative manipulation of the sugar supplement, organogenesis or embryogenesis may be induced.

Briefly, the importance of the different media components can be given as follows:

1. Inorganics
   a. Macronutrients: Ca, K, Mg, N, P, and S are included in anion and/or cation form and are generally present at m*M* concentrations. All are essential for sustained growth in vitro.
   b. Micronutrients: B, Co, Cu, Fe, I, Mo, Mn and Zn are generally included at μ*M* concentrations. Ni and Al may also be included, but the miniscule amounts required are possibly already present as contaminants in, e.g., agar.

2. Organics
   a. Vitamins: Generally, thiamine (vitamin B_1), pyridoxine (vitamin B_6), nicotinic acid (vitamin B_3) and myoinositol are included, but only thiamine is considered to be essential. The others have growth-enhancing properties. The concentrations of each can vary significantly between the different media compositions (*see* Appendix).
   b. Amino acids: Some cultured plant cells can synthesize all amino acids, none are considered essential. However, some media do contain certain amino acids for their growth-enhancing properties, e.g., glycine in MS media ***(3)***. However, high concentrations of certain amino acids can prove toxic. Crude amino acid preparations (e.g., casamino acids; ***10***) can also be used (e.g., for protoplast culture), but their undefined nature makes them less popular.
   c. Carbon source: Generally, most plant-cell cultures are nonautotrophic and are therefore entirely dependent on an external source of carbon. In most cases, this is sucrose, but occasionally glucose (e.g., for cotton cultures) or maltose (e.g., for anther culture) is preferred.
   d. Phytohormones: The most commonly used phytohormones for plant-cell culture are the auxins and cytokinins. However, for specific applications with certain species, abscisic acid or gibberellic acid may be also used. Auxins induce/stimulate cell division in explants and can also stimulate root formation. Both natural (indole-3-acetic acid, IAA) and synthetic (e.g., indole-3-butyric acid, IBA; 1-naphthalene acetic acid, NAA; 2,4-dichlorophenoxyacetic acid, 2,4-D; *p*-chlorophenoxyacetic acid, pCPA) forms are used.

      Although the synthetic forms are relatively stable, IAA is considered to be rapidly inactivated by certain environmental factors (e.g., light). In addition, auxin-like compounds, such as Dicamba and Picloram, can be used to the same effect. Cytokinins play an influential role in cell division, regeneration, and phytomorphogenesis, and are believed to be involved in tRNA and protein synthesis. Although the natural form, Zeatin (or Zeatin riboside) is available commercially and is widely used for certain applications, the synthetic cytokinins (benzyladenine, BA, or 6-benzylaminopurine, BAP; kinetin, K; and isopentyl adenine, 2-iP) are more generally used. Other compounds, such as Thidiazuron and phenylurea derivatives, also have cytokinin activity with the former, for example, gaining increasing popularity for woody species. Gibberellin (usually GA_3) is occasionally used to stimulate shoot elongation in cultures that contain meristems or stunted plantlets. Abscisic acid (ABA) is sporadically used, but its mode of action is unclear. In some cases, it is used for its inhibitory and, in some cases, for its stimulatory effect on cell-culture growth and development.

      Altering the qualitative and quantitative balance of the phytohormones included in a culture medium, and especially in relation to the auxin/cytokinin balance is one of the most powerful tools available to the researcher to direct in vitro response. In many cases, making the correct choice, right from culture initiation, is all-determining.

e. Others: In the past, a wide range of relatively indefinable supplements have been used for plant-cell culture ranging from protein hydrolysates to yeast extracts, fruit (e.g., banana) extracts, potato extracts, and coconut milk. However, the use of such components, through their unknown composition combined with our improved knowledge of cellular requirements in vitro, together with the increasing availability of components, such as zeatin, is now greatly reduced. Coconut milk, however, is still widely used for protoplast culture and is now commercially available.

3. Antibiotics: Both synthetic and naturally occurring antibiotics can be used for plant-cell culture. These play an essential role, for example, in eliminating *Agrobacterium* species after cocultivation in transformation experiments or in providing selection pressure for stably transformed cells. However, for standard practices, the use of antibiotics is usually avoided, since these can have unknown physiological effects on cell development. Low levels are nevertheless often used in the more risky/expensive large-scale operations, e.g., in fermenters and in micropropagation programs.
4. Gelling agent: It is becoming increasingly evident that not only the concentration, but also the type of agent used to make solid media influences the in vitro response of cultured plant tissue. Both natural products extracted from seaweeds (e.g., agar, agarose, and alginate) and their more recently emerged substitutes (e.g., Gelrite, Phytagel), obtained from microbial fermentation, can be used. Each has its advantages and disadvantages, and the choice is usually determined by the species and the application. Agars and agaroses generally produce gels that are stable for prolonged periods and are considered not to bind media components excessively. Products with various degrees of purity are available, and low-gelling temperature types can even enable the embedding of sensitive cells, such as protoplasts. On the other hand, Gelrite/phytagel produces a rigid gel at much lower concentrations than agar or agarose. They are also almost transparent, which makes it easier, e.g., to identify contamination at an early stage. These gels do, however, tend to liquify in long-term cultures owing to pH changes or the depletion of salts necessary for crosslinking. Higher concentrations of antibiotics (e.g., kanamycin) may also be required in Phytagel/Gelrite solidified media in comparison to those solidified with agar/agarose.

In most countries, the most commonly used media are now commercially available (e.g., from Sigma, Duchefa) at competitive prices, saving a lot of time and effort. Furthermore, when the exploratory work is completed and a specific modified medium has been designed for use, some companies (e.g., Duchefa, Haarlem, The Netherlands) will even make this medium to order.

2.5. Culture Facilities

It is to be strongly recommended that plant-cell cultures be incubated under strictly controlled and defined environmental conditions. Although certain cultures (e.g., shoot cultures) will have a set of optimum conditions for growth, they may continue to survive and grow under other, suboptimal conditions.

Other cultures, however, e.g., protoplast or microspore cultures require very precise treatments. Deviations from this, by 1–2° in temperature can mean complete experimental failure. Facilities are therefore required that allow good and reliable regulation of light quality and intensity, photoperiod, temperature (accuracy to ±1°C), air circulation, and in certain countries, humidity. The space available should also be sufficient to allow the execution of experiments under uniform conditions. The choice of facility is often difficult. Several small incubators give flexibility, but generally increase variability in culture conditions and can also prove expensive. A large walk-in growth room in which can be placed not only shelves, but also rotary shakers, bioreactors, and so on, reduces flexibility, but is generally more economical. The extra equipment then no longer needs expensive stand-alone, controlled environment units. However, the failure (through an electrical fault, power cut, and so forth) of such a large growth room could be disastrous, and therefore, safety features should always be included, so that technical personnel can immediately be warned, 24 h/d, when the environmental conditions seriously deviate from the chosen settings.

In incubators without lighting, obtaining uniform conditions is realtively easy. However, when light is introduced into a culture room, variation almost inevitably arises. Not every culture vessel can be placed at an equal distance from the light source. Limited space also often necessitates piling Petri dishes two or three deep. Furthermore, even with the best air circulation, local temperature differences at culture/shelf level can be significant. Although little can be done about this, it is certainly important to be aware of these inequalities. Consequently, it is recommended to carry out related experiments in the same place in the culture room if at all possible. The most uniform provision of light in a culture room is through fluorescent tubes placed above the shelves. However, since space usually has to be used efficiently, shelves are usually stacked above each other. This often results in significant localized increases in temperature on the upper shelves. This is not only undesirable, but also can result in the frequently occurring problem of Petri dish condensation. This can be so extreme that the explants end up sitting in a pool of liquid, which can prove highly detrimental to culture development/survival. Insulating materials placed above the lights or channeled air flows along shelves can help to some extent, but the latter may increase the risk of contamination. The problem is immediately solved if the lights are placed vertically on the walls behind the shelving, but this entails the disadvantage of a significant variation in light intensity across each shelf. The importance of these different factors to the plant material to be used and the nature of the work to be done determines which type of facility should be chosen and how it should be organized.

3. Methods

3.1. Sterilization of Equipment

1. Transfer facilities: On installation, transfer areas (laminar flow cabinets, inoculation rooms, glove boxes) should be thoroughly decontaminated using a suitable disinfectant and, then, if the material allows, 70% ethanol (**Note:** any object made of perspex should never be exposed, however brief, to alcohol, since it will become brittle and crazed). New flow cabinets should be left running overnight to clean the filters thoroughly before being brought into circulation. Once in use, it should become standard practice for every user to spray down the transfer area with 70% alcohol both before and after use. Furthermore, for transfer rooms and glove boxes, which are sterilized by UV light, an exposure of at least 15 min between each user is required to ensure complete decontamination.
2. Glassware: Before sterilizing open glassware (e.g., beakers, Erlenmeyer flasks, and so on), these should be capped with a double layer of aluminum foil to ensure that sterility is maintained after treatment. Glassware with screw caps should always have these loosened half a turn before treatment to prevent high pressures building up, which can lead to the vessel exploding. Glassware can routinely be autoclaved at 121°C at a pressure of 15 psi for 15 min. Alternatively, dry heat can be used at 160°C for 3 h. The latter should, however, be avoided when plastic caps are used (e.g., for closing culture tubes), since these cannot withstand the prolonged high temperatures. Dry heat sterilization is also to be recommended for glassware destined for use with protoplast media. The osmolality of these media is often very critical, and even small amounts of condensation, which can result from autoclaving, can prove detrimental.
3. Instruments: We routinely flame the lower parts of instruments (e.g., scalpels, forceps, and so on) in the laminar flow cabinet directly before use. These are then always allowed to cool before bringing into contact with plant tissue. Between manipulations, the instruments are stored with their working surfaces submerged in 70% ethanol in a glass vessel (e.g., a 100-mL measuring cylinder or beaker) kept in the transfer area for this purpose. The alcohol is replaced at least once a day. Instruments and other metal objects can also be sterilized using dry heat after first wrapping them in aluminum foil or heavy brown paper. Autoclaving is to be avoided, since the combination of elevated temperatures and steam quickly leads to corrosion.
4. Heat-labile components: Certain plastics (e.g., PVC, polystyrene) and other materials may not tolerate the high temperatures generally required for sterilization. If it is not known what material a component is made of or if it is unclear whether a known material is autoclavable, it is always unwise to gamble. Check with a single item first if possible. Otherwise, use the alternative of a chemical method (e.g., immersion for several minutes in 70% ethanol or in one of the solutions listed below for plant material) or UV radiation. However, the latter is only suitable if the UV rays can penetrate to all surfaces of the object concerned.

3.2. Sterilization of Complete Media and Media Components

1. Autoclaving: The easiest and most widely used method to sterilize culture media is to autoclave for 15–20 min at 121°C and a pressure of 15 psi. However, this is only possible if all the components in the medium are heat-stable. Longer times are to be avoided to prevent the risk of chemical modification/decomposition. For certain components, e.g., when glucose is used instead of sucrose, a lower temperature (110°C) is often recommended to avoid caramelization of the carbon source. The autoclaving time should be measured from the moment that the desired pressure is reached and not from the moment that the autoclave is switched on. To avoid excessively long periods before maximum pressure is reached, it is advisable never to overload the autoclave nor to autoclave large volumes in single flasks. Dividing the medium over a number of smaller flasks (preferably 500-mL flasks and only if absolutely necessary, 1000-mL flasks) increases the surface area/volume ratio and, therefore, allows the medium to heat through quicker. This reduces the time needed to reach the desired steam pressure. For this reason also, larger volumes need longer autoclaving times (e.g., 30 min are recommended for volumes of 1000 mL and 40 min for 2000-mL vol). After autoclaving, the pressure should be allowed to fall relatively slowly to avoid the media from boiling over in the flasks. In this regard, flasks should never be filled to more than 90% of their total volume.
2. Filter sterilization: Media containing heat-labile components should either be filter-sterilized in their entirety, or the heat-labile components should be dissolved separately and added after autoclaving the other components. In the latter case, care must be taken to ensure that:

   The pH of the solution to be filter-sterilized is the same as that of the desired final pH of the medium.

   All components are fully dissolved before filtering.

   The temperature of the autoclaved fraction is as low as possible before adding the filter-sterilized components, i.e., room temperature for liquid media, 50°C for agar-based and 40°C for agarose-based media.

   If one or more of the components is poorly soluble, thus requiring a significant volume to fully dissolve, the volume of the autoclaved components should be reduced accordingly in order to end up with the desired final volume and concentration of all components. For example, it is standard practice when requiring solidified versions of heat-labile media to make a double-concentrated medium stock for mixing with an equal volume of double-concentrated agar/agarose stock in water, after the latter has been autoclaved and allowed to cool to the required temperature.

   Solutions for protoplast isolation and culture media should routinely be filter-sterilized. Autoclaving can result in a reduced pH, an altered osmolality, and undesirable chemical modifications, all of which can prove detrimental to these very sensitive cells. For filtration, various filters are now commercially available for filtering different types and volumes of media.

The usual pore size is 0.22 μm, which is appropriate for excluding microbial contaminants while allowing the medium to flow through relatively easily. Care should be taken to chose the right membrane type to use as certain solutions, e.g., those containing proteins, DNA, or alcohol have special requirements. To this end, the manufacturer's recommendations should always be followed. Disposable filter units available from a wide range of sources are the ideal choice, but a cheaper alternative is to buy autoclavable filter units, which can be used with a wide range of membrane inserts. The solution to be sterilized is placed in a syringe that is connected to the sterile filter unit in the sterile transfer area. The solution is then forced through by hand and collected in an appropriate presterilized vessel. Since the pressures needed can sometimes be quite high, a syringe/filter combination with a Luer-Lok system or a spring clamp is to be recommended. For large volumes (>500 mL) peristaltic or vacuum pump-based systems will save a lot of energy and frustration.

3.3. Sterilization of Plant Material

1. Choice of explant: The first critical step in initiating a new cell culture is the obtainment of plant tissue free of all contaminating microorganisms. For some species and explant types, this is exteremly easy, whereas for others, it can be a desperately frustrating experience. The ease with which plant material can be sterilized is directly related to the way the plants providing the explants have been grown. Diseased plants or plants that have been attacked by biting or sucking insects are likely to be contaminated both externally and internally, and it may prove impossible to obtain microbe-free material without the continued/prolonged use of antibiotics. Seed is a favored choice, but even here, weather conditions before and during seed harvest can also prove to be greatly influential to the subsequent success of sterilization.
2. Choice of treatment: For material that is endogenously "clean," the usual procedure is to apply an exogenous chemical treatment for a specified period. Different sterilants are available for use at a range of different concentrations, and the choice of protocol is usually determined by the nature of the explant and the extent of external contamination. A wetting agent is usually included (e.g., a drop of household detergent or Tween) to improve contact with the sterilant, since the ease of wetting can be critical. Very hairy or heavily waxed tissues can prove difficult to sterilize successfully. Ideally, a treatment should be chosen that is as mild as possible while still guaranteeing decontamination. For this reason, seed is often the material of choice, since these generally survive stronger sterilization procedures better than softer tissues, such as leaves or stems.

   A standard procedure is, for example, treatment with 70% ethanol for 30–60 s followed by a 15-min immersion in sodium hypochlorite (1% available chlorine). This is then followed by three washes in sterile water to remove the sterilant, which can prove to be toxic to the explants if it is carried over into culture. Variations to this involve the use of different times of exposure to the sterilant, different concentrations of hypochlorite, or the use of one of the alter-

Table 1
A Summary of the Most Commonly Used Sterilants for Decontaminating Plant Material

Sterilant	Typical concentrations	Comments
Ethanol	70%	70% has better wetting properties than 96%
Sodium hypochlorite[a]	0.5–5% free chlorine	Most widely used
Calcium hypochlorite	1–10% saturated solution	Frequently used, prepare fresh
Hydrogen peroxide	1.0–10%	Prepare fresh
Mercuric chloride	0.1–1.0%	Very toxic, requires special handling and waste disposal
Silver nitrate	1%	Requires special waste disposal
Bromine water	0.5–2%	Shorter exposure times (<10 min) recommended
Combinations		Ethanol is usually used as a short pretreatment in combination with one of the others; hydrogen peroxide or mercuric chloride is sometimes used sequentially with sodium hypochlorite or calcium hypochlorite for particularly difficult material

[a]Household bleach (e.g., Domestos, Clorox) is often used. Chose one that does not include a thickening agent, and observe the stock concentration closely, because this varies among batches/brands/countries.

native sterilants. The most widely used compounds and concentrations are listed in **Table 1**.

When working with material for the first time, the best approach is to choose seed or young, healthy plants and perform a small sterilization experiment where the hypochlorite concentration is varied (e.g., 0.5–5% available chlorine) along with the exposure time (e.g., 5–30 min). The mildest conditions that give 100% sterile explants should then be used for all subsequent work.

3.4. Preparation of Media

Plant-culture media are made either from commercially available powders or self-made concentrated stock solutions. All should be prepared using purified water that is de-ionized and distilled or the equivalent thereof, so that all pyrogens, organics, salts, and microorganisms have been removed. Commercial media only need to be dissolved in an appropriate volume of water, and after the additional supplements have been added the pH is adjusted to the required value using HCl or KOH. For solid media, a gelling agent is required, and this can be added either before or after the pH is adjusted. No clear guideline can be given concerning which is better, but it is important to be consistent. However, the presence of gelling agents will slowly decrease the efficiency of the pH electrode, which then requires more frequent cleaning in HCl.

If a particular medium is not commercially available or if the relative concentrations of individual medium components are to be changed, the most economical option is to make stock solutions for repeated use. Generally, separate stocks of the different groups of components are made with the macronutrients being prepared as a ×10 or ×20 stock, the micronutrients and organics as a 100× stock, and the phytohormones as ×1000 stocks. These stocks are stored separately in the freezer in suitable aliquots. As a general rule, individual stock components should first be fully dissolved separately before adding to the main solution. This reduces the likelihood of precipitation. It is also best to dissolve the inorganic N components first. If precipation does occur, check that the pH is close to 5.5, and try changing the order in which the components are added. Owing to their very low concentrations, micronutrient stocks are usually prepared from individual stock solutions, which are diluted and mixed to give the desired concentrations. These individual stocks may be stored frozen for many months without detrimental effects. However, as with all stored stock solutions, when a precipitate appears, the solution must be discarded and a new one prepared.

Phytohormones are, in general, poorly soluble in water, and 1000× stocks should be made by first dissolving in 1–5 mL of an appropriate solvent before making up to the final volume with distilled water and adjusting the pH to approx 5.5. HCl (0.1 *N*), NaOH (0.1 *M*), or ethanol can be used as solvent. Although most synthetic phytohormones can usually be autoclaved together with the rest of the medium, the naturally occurring forms (e.g., IAA, GA, ABA, Zeatin) are generally filter-sterilized.

Prepared media stocks can usually be stored frozen for up to 6 mon without precipitation or reduced growth responses. Phytohormone stocks can similarly be stored frozen, but stocks for daily use can also be stored in the refrigerator. However, these should be replaced monthly. Prepared media in Petri dishes and jars should be used within 2–3 wk of storage at room temperature. Storage at 4°C is to be avoided, since this leads to excessive condensation inside the vessels.

3.5. Preparation of Explants and Subculture Techniques

Explants vary considerably in terms of size and shape in relation to the type of material and the aim of the experiment. Very small explants may not survive except on rich media or on medium that has been "conditioned" (i.e., preexposed in some way to growing cultures). Initially grouping small explants can help. For culture initiation, new explants are prepared under aseptic conditions from presterilized material. The outside edges are usually removed, since these are likely to have been damaged by the sterilant. The tissue is cut into pieces using a sharp (new) scalpel blade to give relatively uniform explants of an appropriate size (dimensions usually <1 cm). The cut surfaces are important for nutrient uptake and for stimulating callus growth. Petri dishes or sterilized ceramic tiles make suitable cutting surfaces. On preparation, explants should be transferred to culture medium as quickly as possible to avoid desiccation. The orientation on the medium, in relation to orientation in the original plant, may be important (in relation to polar phytohormone transport, for example). In this regard, literature guidance should be sought, or otherwise all possibilities should be tested in an initial experiment to determine the best choice for a particular application. Once a culture has been established, regular transfer of all or part of the plant material to fresh medium is necessary. Before doing so, all cultures should be checked for sterility, and all contaminated cultures discarded. This is especially important for recently initiated cultures. A sharp scalpel blade should be used to remove that part that is to be transferred, since a clean cut results in the least damage and gives the best growth response. The amount of material to be transferred is determined entirely by the growth rate of the culture concerned with the general rule being that the faster the growth, the less material needs to be transferred. Growth rate usually also determines the subculture interval. Rapidly growing suspensions, for example, may need subculture every 4–7 d, whereas callus cultures may survive quite happily with transfers only every 1–2 mon.

For established cultures, the risk of loss through contamination is of course greatest during subculture. Consequently, it is recommended always to keep a number of old cultures after each transfer, until it can be confirmed that the new cultures are sterile or, alternatively, transfers should be staggered so that not all cultures of a particular type are transferred at the same time. In this way, disastrous losses can be avoided.

3.6. Precautions and Hints to Success

The following list represents a series of hints, precautions, and recommendations that should help in avoiding some of the common pitfalls.

1. Make culture medium several days before it is required. If the autoclave has been faulty, contamination will become evident before subculturing has taken place.
2. When preparing solid media, allow the temperature to fall to at least 50°C before pouring in order to prevent excessive condensation forming in the dishes and jars.

3. When initiating new cultures, always begin with healthy plant material.
4. When preparing culture media, stocks, and so forth, follow a strict routine regarding the way components are added, the order of addition, and so on, to avoid the risk of precipitation.
5. When making media and stock solutions, work with checklists, and tick off each component as it is added. This is particularly important when making stocks that are to be stored and used for a long time. A fault when making the stock can mean months of problems.
6. To maintain maximum sterility in a culture lab, always autoclave contaminated cultures before discarding. Never try to "rescue" cultures infected with a fungus that is already sporulating.
7. The waste bin is a frequent source of contamination in a lab. Clean it out, and sterilize it with ethanol regularly.
8. When autoclaving media in bottles with screw caps always loosen these beforehand, since this will avoid both the risk of the bottles exploding and also the creation of a vacuum after cooling, which can be so strong that it becomes impossible for the cap to be removed.
9. When remelting solidified media in a microwave, always loosen the cap and ensure that there is sufficient air space to avoid the medium boiling out of the flask as it melts.
10. Before using a new type of pen or marker, make sure that the ink is not light-sensitive (if to be used for labeling cultures to be grown in the light for long periods) and that it does not disappear on autoclaving.
11. When adding filter-sterilized (i.e., heat-labile) supplements to autoclaved medium, always allow the latter to cool to approx 10°C above gelling temperature, and then, after thorough mixing, pour the medium immediately into plates/jars to ensure that it cools as rapidly as possible.
12. When adding filter-sterilized supplements to autoclaved medium, adjust the pH of the solution to that of the medium before filtration. This is particularly important with, for example, hormone stocks that are dissolved in HCl or KOH/NaOH solutions.
13. If forced to use suboptimal culture conditions that result in considerable condensation within culture vessels, allow the medium to solidify as a slope by placing the dishes/jars at a slight (10°) angle. This allows the condensation to collect at the lowest point, thus keeping the explants high and dry.
14. When wishing to cut up tiny explants, e.g., meristems or embryos, even a standard scalpel blade is often not fine enough. Use the sharpened point of a syringe needle, or alternatively, cut up pieces of a double-edged razor blade with an old pair of scissors and fix them in an appropriate holder. Replace these regularly.
15. When using alcohol to store sterilized instruments between manipulations in a laminar flow cabinet, make sure that the vessel used has an even, flat top with no chips or pouring lips (as with beakers). Accidentally setting the alcohol alight (a not infrequent occurrence) can then immediately be remedied by placing any flat surface on top of the vessel, immediately starving the flame of oxygen.

16. Since work in many tissue-culture labs is often now a combination of cell biology and molecular biology techniques, it is recommended to use separate glassware for each type of work. This avoids the risk of suboptimal washing practices resulting in the contamination of culture media with trace amounts of toxic compounds, such as SDS or ethidium bromide.
17. Even the richest labs often use cheap glassware (jam pots, honey pots, baby food jars) for mass culture purposes. These vessels are usually made of soda glass rather than the recommended borosilicate (Pyrex-type) glass. In such cases, it has been recommended ***(11)*** that these vessels be discarded after 1 yr of use or, alternatively, be coated with dichloro-silane for longer usage.
18. New glassware should always be thoroughly washed before using for the first time. Street ***(1)*** even recommends autoclaving new flasks filled with distilled water twice before bringing them into circulation.
19. Since good suspension cultures often take a long time to produce, keep a number of flasks on a second shaker, if space/expense allows, to avoid complete loss through motor failure. Backup cultures on solid medium are also to be recommended.
20. When transferring cultures to a new medium to test, e.g., growth rate/regeneration response, bear in mind that nutrient carryover can be significant. This is especially true for synthetic hormones, such as 2,4-D. Consequently, when making assessments of a new medium, subculture the tissues at least twice beforehand.
21. Medium pH significantly influences nutrient availability and culture response. Consequently, for certain applications, it should be borne in mind that the pH generally decreases 0.2–0.5 U on autoclaving. Furthermore, the stiffness of solid media is also pH-dependent. If a low pH is desired, higher agar concentrations may be required to give the same gel strength and vice versa.

References

1. Street, H. E. (ed.) (1974) *Tissue Culture and Plant Science.* Academic, New York.
2. Christou, P. (1995) Strategies for variety-independent genetic transformation of important cereals, legumes and woody species utilizing particle bombardment. *Euphytica* **85,** 13–27.
3. Murashige, T. and Skoog, F. (1962) A revised medium for rapid growth and bioassays with tobacco tissue cultures. *Physiol. Plant.* **15,** 473–479.
4. Linsmaier, E. M. and Skoog, F. (1965) Organic growth factor requirements of tobacco tissue cultures. *Physiol. Plant.* **18,** 100–127.
5. Gamborg, O. L., Miller, R. A., and Ojima, K. (1968) Nutrient requirements of suspension cultures of soybean root cells. *Exp. Cell. Res.* **50,** 151–157.
6. Schenk, R. U. and Hildebrandt, A. C. (1972) Medium and techniques for induction and growth of monocotyledonous and dicotyledonous plant cell cultures. *Can. J. Botany* **50,** 199–204.
7. Nitsch, J. P. and Nitsch, C. (1969) Haploid plants from pollen. *Science* **163,** 85–87.
8. George, E. F., Puttock, D. J. M., and George, H. J. (eds.) (1987) *Plant Culture Media I.* Exegetics, Westbury, UK.

9. Toriyama, K., Hinata, K., and Sasaki, T. (1986) Haploid and diploid plant regeneration from protoplasts of anther callus in rice. *Theor. Appl. Genet.* **73,** 16–19.
10. Kao, K. N. and Michayluk, M. R. (1975) Nutritional requirements for growth of *Vicia hajastana* cells and protoplasts at a very low population density in liquid media. *Planta* **126,** 105–110.
11. de Fossard, R. A. (ed.) (1976) *Tissue Culture for Plant Propagators*. University of New England Press, Armidale, Australia.

II

Cell Culture and Plant Regeneration

2

Callus Initiation, Maintenance, and Shoot Induction in Rice

Nigel W. Blackhall, Joan P. Jotham, Kasimalai Azhakanandam, J. Brian Power, Kenneth C. Lowe, Edward C. Cocking, and Michael R. Davey

1. Introduction

Embryogenic suspension cultures provide the most widely employed source of totipotent cells for protoplast isolation in rice (*Oryza sativa* L.), since mesophyll-derived protoplasts of this important cereal rarely undergo sustained mitotic division leading to the production of tissues capable of plant regeneration. Cells from embryogenic suspensions provide an alternative to immature zygotic embryos for biolistic-mediated production of fertile transgenic rice plants *(1)* and are also amenable to transformation procedures employing agrobacteria *(2)*. Currently, protocols are available for regenerating fertile plants from cell suspension-derived protoplasts of the three major subgroups of rice varieties, namely japonica *(3)*, javanica *(4)* and indica *(5)* rices.

Previous reports have stated that genotype and explant source are important parameters in determining the success of plant regeneration from cultured tissues of rice *(6–10)*. In japonica rices, callus cultures can be produced relatively easily from almost any part of the plant, including roots, shoots, leaves, leaf-base meristems, mature and immature embryos, young inflorescences, pollen grains, ovaries, scutella, and endosperm. Such tissues can be induced to regenerate plants *(11)*. Conversely, indica rices are more recalcitrant in culture *(12,13)*. In the procedures described in this chapter, scutella from mature seeds are used as the source of callus for both indica and japonica rices.

The establishment and maintenance of embryogenic cell suspensions is generally difficult, with morphogenic competence of suspensions usually declining with successive subculture over prolonged periods *(14)*. However,

From: *Methods in Molecular Biology, Vol. 111: Plant Cell Culture Protocols*
Edited by: R. D. Hall © Humana Press Inc., Totowa, NJ

following the development of reproducible protocols for cryopreservation of rice cell suspensions ***(15,16)***, it has been possible to devise strategies to overcome difficulties associated with the loss of totipotency and the requirement to reinitiate periodically new cultures capable of plant regeneration. Samples of the cell suspensions should be cryopreserved as soon as possible after initiation. The cultures should be resurrected from frozen stocks immediately if there are indications of loss of embryogenic potential of the suspensions. Loss of totipotency can occur at any time in the development and maintenance of cell suspensions. However, it arises most frequently 9–12 mon after initiation of the suspension cultures.

Frequently, it has been noted that the initiation and growth of embryogenic callus require media containing auxin (specifically 2,4-dichlorophenoxyacetic acid 2,4-D), whereas for the development of embryos into plants, auxins should be omitted from the culture medium. The regeneration of rice plants has been found to be enhanced by media lacking auxin, but supplemented with reduced concentrations of cytokinins, such as 6-benzylaminopurine (BAP) or kinetin ***(17,18)***.

Micropropagation provides a means of rapidly multiplying material of both cultivated and wild rices, as well as genetically modified plants (e.g., transgenics, somatic hybrids, and cybrids). The ability to multiply plants in vitro is especially important for wild rices (*Oryza* species other than *O. sativa*), for which only limited supplies of seed may be available. Indeed, wild rices are an important genetic resource, since they possess resistances to biotic and abiotic stresses. These *Oryzae* can also be used to generate alloplasmic lines for the development of novel cytoplasmic male sterility systems.

2. Materials

2.1. Initiation of Embryogenic Callus from Mature Seed Scutella

1. Seeds of *O. sativa* cvs. Taipei 309 and Pusa Basmati 1 (obtained from the International Rice Research Institute [IRRI], Los Baños, Philippines).
2. "Domestos" bleach (Lever Industrial Ltd., Runcorn, UK) or any commercially available bleach solution containing approx 5% available chlorine.
3. Sterile purified water: water purified by distillation, reverse osmosis, or ion-exchange chromatography, which has been autoclaved (121°C, 20 min, saturated steam pressure).
4. MS basal medium: based on the formulation of Murashige and Skoog ***(19)***. This medium can be purchased in powdered form lacking growth regulators (Sigma, Poole, UK), to which 30 g/L sucrose is added. The medium is semisolidified by the addition of 8 g/L SeaKem LE agarose (FMC BioProducts, Vallensbaek Strand, Denmark), pH 5.8 (*see* **Table 1**). Autoclave.

Table 1
Formulation of Media—Macronutrients, Micronutrients, Vitamins, and Other Supplements

	Concentration, mg/L			
Component	MS	LS2.5	AA2	R2
Macronutrients				
$CaCl_2$	332.2	332.2	332.2	147.0
KH_2PO_4	170.0	170.0	170.0	170.0
$MgSO_4$	180.7	180.7	180.7	120.4
KCl			2940.0	
KNO_3	1900.0	1900.0		4040.0
$NaH_2PO_4 \cdot 2H_2O$				312.0
$(NH_4)_2SO_4$				330.0
Micronutrients				
KI	0.83	0.83	0.83	
H_3BO_3	6.20	6.20	6.20	500.00
$MnSO_4$	16.90	16.90	19.92	447.00
$NaMoO_4 \cdot 2H_2O$	0.25	0.25	0.25	50.00
$ZnSO_4 \cdot 7H_2O$	8.60	8.60	8.60	500.00
$CuSO_4 \cdot 5H_2O$	0.025	0.025	0.025	50.00
$CoCl_2 \cdot 6H_2O$	0.025	0.025	0.025	
$FeSO_4 \cdot 7H_2O$	27.85	27.85	27.85	
Na_2EDTA	37.25	37.25	37.25	
NaFeEDTA				2.5
Vitamins				
Myo-inositol	100.0	100.0	100.0	
Nicotinic acid	0.5		0.5	
Pyridoxine HCl	0.5		0.1	
Thiamine HCl	0.1	1.0	0.5	1.0
Glycine	2.0		75.0	
L-Glutamine			877.0	
L-Aspartic acid			266.0	
L-Arginine			228.0	
L-Proline				560.0
Other supplements				
2,4-Dichlorophenoxyacetic acid		2.5	2.0	2.0
Gibberellic acid			0.1	
Kinetin			0.2	
Sucrose	30000	30000	20000	
Maltose				30000
pH	5.8	5.8	5.8	5.8
Sterilization	Autoclave	Autoclave	Filter	Autoclave

5. LS2.5 medium: based on the formulation of Linsmaier and Skoog *(20)* supplemented with 2.5 mg/L 2,4-D and semisolidified by the addition of 8 g/L SeaKem LE agarose, pH 5.8 (*see* **Table 1**). Autoclave.
6. Nescofilm: Bando Chemical Ind. Ltd., Kobe, Japan.

2.2. Micropropagation of Cultivated and Wild Rices

1. Seeds of *O. sativa* cvs. Taipei 309 and Pusa Basmati 1, *Oryza australiensis* and *Oryza granulata* (obtained from IRRI).
2. "Domestos" bleach: as in **Subheading 2.1., item 2.**
3. Sterile purified water: as in **Subheading 2.1., item 3.**
4. MS basal medium: as in **Subheading 2.1., item 4.**
5. Autoclavable culture vessels, e.g., 175-mL capacity screw-capped "Powder-Round" glass jars (Beatson Clark and Co. Ltd., Rotherham, UK).
6. Micropropagation medium: MS medium (as in **Subheading 2.1., item 4**). with 2.0 mg/L BAP and 50 g/L sucrose, semisolidified by the addition of 8 g/L SeaKem LE agarose, pH 5.8. Autoclave.
7. Glazed white ceramic tiles (15 × 15 cm), wrapped in aluminum foil and autoclaved.

2.3. Initiation of Embryogenic Callus from Leaf Bases of Micropropagated Plants

1. LS2.5 medium: as in **Subheading 2.1., item 5.**
2. Nescofilm: as in **Subheading 2.1., item 6.**

2.4. Shoot Regeneration from Callus

1. Differentiation medium: MS medium (as in **Subheading 2.1., item 4.**) supplemented with 2.0 mg/L BAP and 30 g/L sucrose, semisolidified by the addition of 8 g/L SeaKem LE agarose, pH 5.8. Autoclave.
2. Nescofilm: as in **Subheading 2.1., item 6.**
3. Rooting medium: MS medium (as in **Subheading 2.1., item 4.**), supplemented with 1.5 mg/L α-naphthaleneacetic acid (NAA) and 30 g/L sucrose, semisolidified by the addition of 8 g/L SeaKem LE agarose, pH 5.8. Autoclave.
4. Maxicrop liquid fertilizer solution: Maxicrop Plus Sequestered Iron, Maxicrop Garden Products, Gr. Shelford, Cambridge, UK.
5. Initiation compost: a 12:1 (v:v) mixture of M3 soil-less compost (Fisons plc., Ipswich, UK) and Perlite (Silvaperl Ltd., Gainsborough, UK).
6. Maintenance compost: a 6:1:1 (v:v) mixture of M3 soil-less compost, John Innes No. 3 compost (J. Bentley Ltd., Barton-on-Humber, UK) and Perlite.

2.5. Initiation of Embryogenic Suspensions

1. AA2 medium: modified AA medium *(21)* supplemented with 2 mg/L 2,4-D, pH 5.8 (*see* **Table 1**). Filter-sterilize.
2. R2 medium: modified R2 medium *(22)* supplemented with 1 mg/L 2,4-D and 30 g/L maltose, pH 5.8 (*see* **Table 1**). Autoclave (*see* **Note 1**).

3. Disposable sterile plastic 10-mL pipets with a wide orifice (e.g., Sterilin 47110; Bibby Sterilin, Stone, UK), or 10-mL glass serological pipets (e.g., Sterilin 7079-10N) with the ends removed to produce a wider orifice.

3. Methods

3.1. Initiation of Embryogenic Callus from Mature Seed Scutella of Cultivated Rices

1. Dehusk 100 seeds each of *O. sativa* cvs. Taipei 309 and Pusa Basmati 1.
2. Surface-sterilize the seeds by immersion in a 30% (v/v) solution of "Domestos" bleach for 1 h at room temperature.
3. Remove the "Domestos" solution using five rinses with sterile purified water.
4. Germinate the seeds by laying on the surface of 20-mL aliquots of MS basal medium in 9-cm diameter Petri dishes (9 seeds/dish). Seal the dishes with Nescofilm and incubate in the dark at 28 ± 1°C.
5. After 14 d, remove the coleoptiles and radicles, and transfer the explants to 20-mL aliquots of LS2.5 medium in 9-cm diameter Petri dishes (9 explants/dish). Seal the dishes with Nescofilm, and incubate as in **Subheading 3.1., step 4.**
6. Subculture every 14–28 d (*see* **Note 2**) by selecting the most embryogenic callus, i.e., tissue with a dry, friable appearance, and transferring 1–5 calli (each approx 5 mm in diameter) to 20-mL vol of LS2.5 medium in 9-cm diameter Petri dishes.

3.2. Micropropagation of Wild Rices

1. Dehusk 10 seeds each of *O. australiensis* and *O. granulata* (the number of seeds can be varied depending on supply).
2. Surface-sterilize the seeds and germinate as in **Subheading 3.1., steps 2–4.**
3. Subculture the seedlings after 14 d. Aseptically remove the seedlings and place onto the surface of a sterile white tile. Trim the roots to their base using a scalpel, and trim the leaves of the shoots to a length of 2 cm.
4. Transfer the shoot bases to screw-capped glass jars each containing 50-mL aliquots of micropropagation medium (*see* **Subheading 2.2., step 6.**). Immerse the bases of the shoots 5 mm below the surface of the medium.
5. Subculture the shoots every 28 d. At each subculture, use forceps and a scalpel to separate the multiple shoots (tillers), which develop from each explant. Select healthy micropropagules (tillers), and trim the roots and stems as in **Step 3.** Transfer to micropropagation medium as in **Step 4.**

3.3. Initiation of Embryogenic Callus from Leaf Bases of Micropropagated Shoots

1. Use separate micropropagules of *O. australiensis* and *O. granulata* plants obtained as in **Subheading 3.2., steps 3–5.** Each micropropagule has hard white tissue at its base. This tissue is embryogenic and is used for callus initiation. Excise the tissue, and cut into 4-mm^2 sections; culture the latter on 20-mL aliquots of LS2.5 medium in 9-cm diameter Petri dishes (eight tissue sections/dish).
2. Seal the Petri dishes with Nescofilm, and incubate in the dark at 28 ± 1°C.

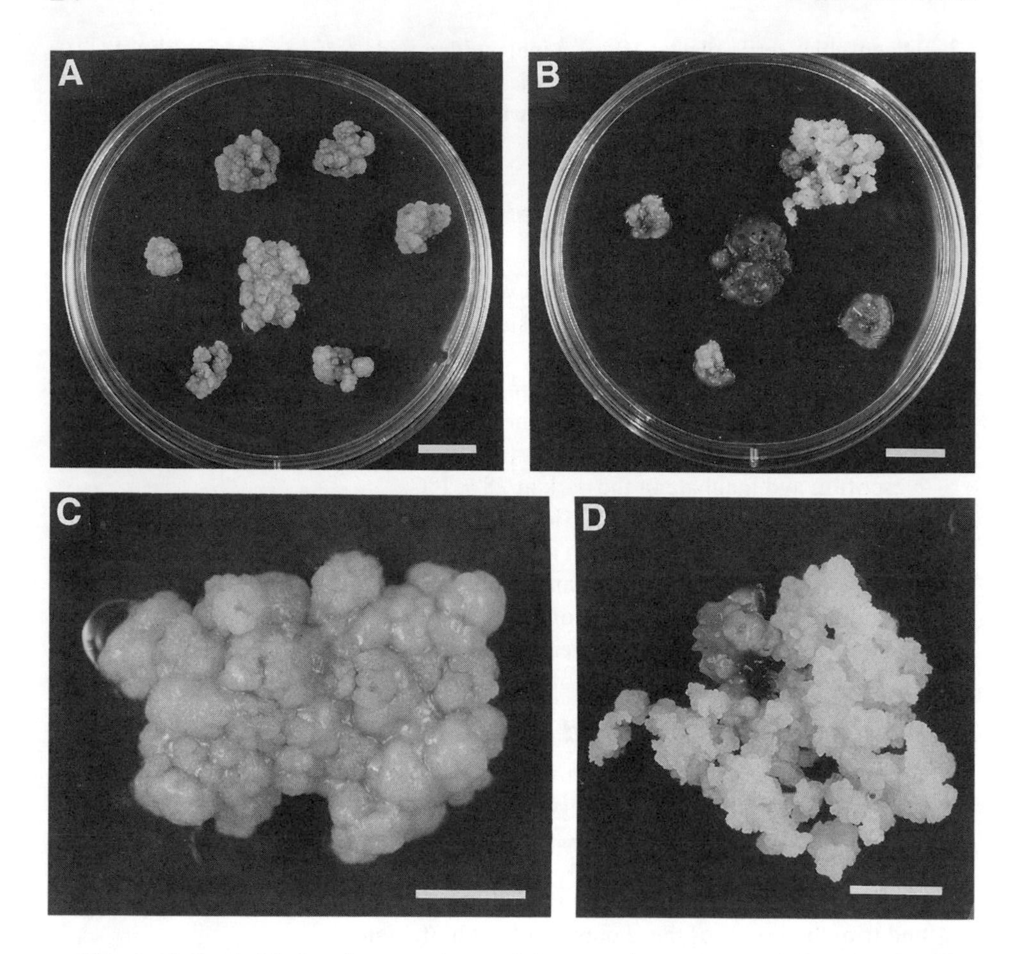

Fig. 1. Callus initiation from explants of *O. australiensis* **(A)** and *O. granulata* **(B)** after culture of the explants for 62 and 84 d, respectively (bars = 1 cm). Embryogenic callus of *O. australiensis* **(C)** and friable callus of *O. granulata* **(D)**, both suitable for the initiation of cell suspensions (bars = 0.5 cm).

3. After 28 d, inspect the dishes for callus production by the explants. Select yellow-colored, rapidly dividing calli composed of small cell clusters, excise the tissue from the parent explants, and transfer the tissues to LS2.5 medium every 28 d (8 tissues/dish; *see* **Note 3**, **Fig. 1A–D**).

3.4. Shoot Regeneration from Callus

1. Transfer individual pieces of callus, obtained either from mature seed scutella or from leaf bases as in **Subheading 3.3., step 1.**, to 9-cm diameter Petri dishes each containing 20-mL aliquots of differentiation medium (12 calli/dish, each approx 3 mm in diameter, **Fig. 2A**). Seal the dishes with Nescofilm and incubate

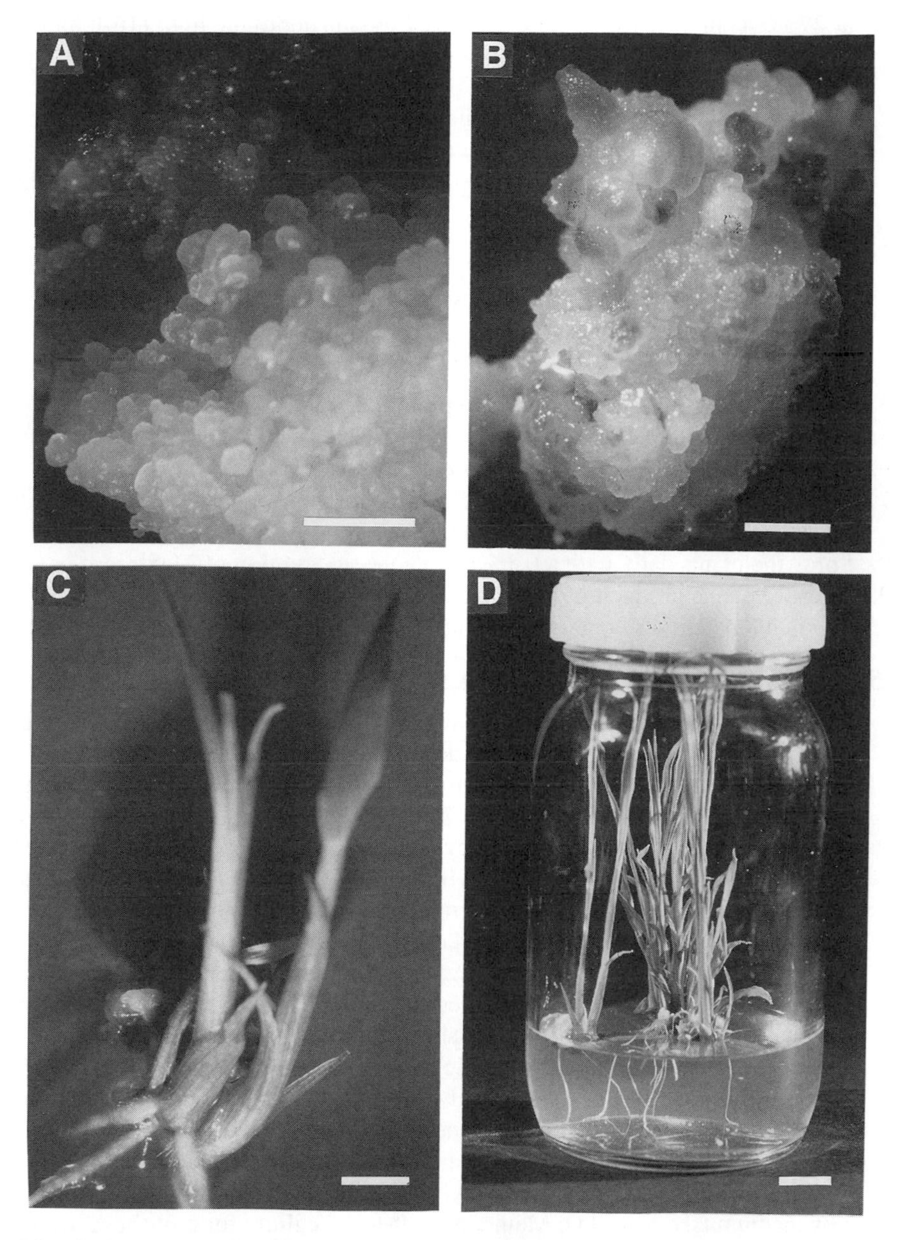

Fig. 2. Stages in plant regeneration from mature seed scutellum-derived callus of the indica rice cv. Pusa Basmati 1. **(A)** Embryogenic callus 28 d after sub-culture on LS2.5 medium (bar = 0.5 mm). **(B,C)** Stages in shoot regeneration, 15 d after transfer of callus to differentiation medium in the dark **(B)**, and 30 d after transfer to the light **(C)** (bars = 0.5 cm). **(D)** Rooted shoots ready for transfer to compost (bar = 1 cm).

at 27 ± 2°C in the dark for 14 d, followed by transfer to the light (16-h photoperiod, 55 µmol/s/m^2, Daylight fluorescent tubes). Examine every 7 d for the regeneration of shoots and roots.

2. When shoots appear, transfer each shoot, together with a 3-mm diameter piece of the adjacent parental callus, to rooting medium (1–8 shoots/dish; **Fig. 2B,C**; *see* **Note 4**). Seal the dishes with Nescofilm, and incubate at 27 ± 2°C in the light as in **Subheading 3.4., step 1.**
3. Transfer rooted shoots to initiation compost (*see* **Subheading 2.4.5.**) in 7.5-cm diameter plastic plant pots (**Fig. 2D**). Cover the regenerated plants with 20 × 20 cm clear polythene bags (*see* **Note 5**). Maintain the potted plants in a glasshouse under natural daylight with maximum day and night temperatures of 28 ± 2°C and 24 ± 2°C, respectively.
4. After 3 d, make five incisions with a pin into the top of the bags.
5. Four d later, remove, with scissors, one corner of each bag.
6. After a further 4 d, remove the other corner of each bag.
7. Every 2 d, cut off the top 1 cm of each bag, until the topmost leaves of the potted plants are exposed. Remove each bag.
8. Spray the plants daily with a 0.1% (v/v) aqueous solution of Maxicrop.
9. Transfer plants producing tillers and roots showing healthy, vigorous growth to 15-cm diameter pots containing maintenance compost.

3.5. Initiation of Embryogenic Suspensions

1. When sufficient dry, friable, callus has been obtained (approx 5 wk for cultivated rices and up to 24 wk for wild rices, depending on the growth rate of the callus), initiate cell suspension cultures by transferring 1.5 g fresh weight of embryogenic callus (*see* **Notes 6** and **7**) to a 75-mL Erlenmeyer flask containing 18 mL of either AA2 medium (for *O. sativa* cv. Taipei 309 and *O. granulata* [*see* **Note 8**]) or R2 medium (for *O. sativa* cv. Pusa Basmati 1 and *O. australiensis*). Incubate at 28 ± 2°C in the dark on a horizontal rotary shaker (120 rpm, 4-cm throw).
2. Subculture the suspensions every 3–4 d. Allow the cells to settle, remove 50% of the supernatant, and replace with an equal or slightly greater volume of medium. Gradually increase the volume of medium and the size of flasks in accordance with the rate of growth of the suspension cultures.
3. After 12–15 wk, there should be sufficient quantity of small clusters of cells (*see* **Note 9**) to transfer 1 mL packed cell volume (PCV; *see* **Note 10**) of the small clusters to produce a "pipetable" suspension (*see* **Note 9**). This "pipetable" culture is subcultured at 7-d intervals by transferring 1 mL PCV, together with 9 mL of spent medium, to a 250-mL capacity Erlenmeyer flask containing 42 mL of new medium (*see* **Note 11**). Maintain the "stock" culture for a further 2 wk until the new "pipetable" suspension has become established. Repeat this procedure if the "pipetable" culture fails to become established.
4. The growth characteristics of established cultures are determined by daily measuring the settled cell volume (SCV). Decant the cell suspension into a graduated centrifuge tube, allow to sediment under gravity (10 min), and note the SCV from the graduations (*see* **Note 12**).

4. Notes

1. R2 medium should be autoclaved at 116°C for 30 min, since sterilization at 121°C for 20 min causes excessive caramelization.
2. Regular subculture of scutellum-derived callus will produce fast-growing, globular tissues. Fourteen days are the optimal time between transfers, but the tissues can be left for 28 d before subculture.
3. Transfer of leaf base-derived calli to fresh LS2.5 medium every 28 d is important, since compact embryogenic callus becomes more globular with time in culture and, therefore, more suitable for the initiation of finely divided cell suspensions.
4. More than 30% of the shoots regenerated from some callus cultures may be albino. In such cases, the callus should be discarded, and new cultures initiated from seed. The regeneration of albinos appears to be cultivar specific. For example, it occurs in the japonica rice cv. Taipei 309, but not in the indica cv. Pusa Basmati 1.
5. Place polythene bags over the potted plants, and secure the bags by placing the pots inside other 7.5-cm pots. The bags are required to maintain high humidity around the plants and to reduce transpiration. The humidity is reduced gradually by opening the bags as the plants become acclimatized.
6. In the case of *O. granulata*, approx 2% of the stem bases produce fast-growing, fine yellow callus suitable for the initiation of cell suspension cultures. Transfer of this callus to new semisolidified medium results in rapid tissue proliferation, enabling initiation of cell suspensions after 8–12 wk of culture of the callus on semisolid medium.
7. For *O. australiensis*, approx 90% of the stem bases produce embryogenic callus. Initially, the cell suspension cultures of *O. australiensis* consist of large clusters of callus (each in excess of 2 mm diameter), but after 8–12 wk of subculture, the majority of the cell clusters are <1 mm in diameter.
8. Maintenance of cell suspensions of *O. granulata* is difficult because of the extreme viscosity of the cultures. It is necessary to shake the Erlenmeyer flasks vigorously immediately before subculture and to pay particular attention to the volume of cells (*see* **Note 9**) used for inoculation into new medium.
9. When a cell suspension is first initiated, the initial or "stock" suspension consists of large cell clusters. As these clusters grow, they release smaller clumps of cells into the liquid medium. For experimental use, a "pipetable" culture is initiated from the small cell clusters. The "stock" culture is discarded once a "pipetable" culture has been established.
10. PCV is measured by drawing a suspension of cells into a sterile 10-mL pipet (*see* **Subheading 2.5., item 3.**). The medium is expelled while holding the pipet tip against the bottom of the flask. The cells will be retained in the tip of the pipet, allowing the medium to escape. The PCV may be adjusted to 1 mL by drawing more suspension into the pipet or by releasing packed cells from the pipet tip.
11. The maintenance of cell suspension cultures requires careful attention to the subculture procedure. Regular subculture, usually at 7-d intervals and use of the

correct inoculation volume (1 mL PCV) are vital in maintaining totipotency. An excess of inoculum causes exhaustion of the nutrients early in the subculture cycle, resulting in necrosis within the centers of the cell clusters. Conversely, if too small an inoculum is transferred, the growth rate of the cells is reduced considerably, since the cells fail to reach the minimum inoculation density.

12. The doubling time of cultures varies, but is generally 2–4 d.

References

1. Sivamani, E., Shen, P., Opalka, N., Beachy R. N., and Fauquet, C. M. (1996) Selection of large quantities of embryogenic calli from indica rice seeds for production of fertile transgenic plants using the biolistic method. *Plant Cell Rep.* **15,** 322–327.
2. Hiei, Y., Ohta, S., Komari, T., and Kumashiro,T. (1994) Efficient transformation of rice (*Oryza sativa* L.) mediated by *Agrobacterium* and sequence analysis of the boundaries of the T-DNA. *Plant J.* **6,** 271–282.
3. Thompson, J. A., Abdullah, R., and Cocking, E. C. (1986) Protoplast culture of rice (*Oryza sativa* L.) using media solidified with agarose. *Plant Sci.* **47,** 179–183.
4. Suh, S. C., Kim, H. I., and Park, W. (1992) Plant regeneration from rice protoplasts culture in javanica cv. Texmont. *Res. Rep. Rural Dev. Adm. (Suweon)* **34,** 1–9.
5. Jain, R. K., Khehra, G. S., Lee, S-H., Blackhall, N. W., Marchant, R., and Davey, M. R. (1995) An improved procedure for plant regeneration from indica and japonica rice protoplasts. *Plant Cell Rep.* **14,** 515–519.
6. Lai, K. L. and Liu, L. F. (1982) Induction and plant regeneration of callus from immature embryos of rice plants (*Oryza sativa* L.). *Jpn. J. Crop Sci.* **51,** 70–74.
7. Lai, K. L. and Liu, L. F. (1986) Further studies on the variability of plant regeneration from young embryo callus cultures in rice plants (*Oryza sativa* L.). *Jpn. J. Crop Sci.* **55,** 41–46.
8. Lai, K. L. and Liu, L. F. (1988) Increased plant regeneration frequency in water-stressed rice tissue cultures. *Jpn. J. Crop Sci.* **57,** 553–557.
9. Abe, T. and Futsuhara, Y., (1984) Varietal difference of plant regeneration from root callus tissues in rice. *Jpn. J. Breed.* **34,** 147–155.
10. Abe, T., and Futsuhara, Y. (1986) Genotypic variability for callus formation and plant regeneration in rice (*Oryza sativa* L.). *Theor. Appl. Genet.* **72,** 3–10.
11. Bajaj, Y. P. S. (1991) Biotechnology in rice improvement, in *Biotechnology in Agriculture and Forestry, vol. 14, Rice* (Bajaj, Y. P.S., ed.), Springer-Verlag, Heidelberg, pp. 3–18.
12. Kyozuka, J., Otoo, E., and Shimamoto, K. (1988) Plant regeneration of protoplasts of Indica rice: genotypic differences in culture response. *Theor. Appl. Genet.* **76,** 887–890.
13. Lee, L., Schroll, R. E., Grimes, H. D., and Hodges, T. K. (1989) Plant regeneration from indica rice (*Oryza sativa* L.) protoplasts. *Planta* **178,** 325–333.
14. Abe, T. and Futsuhara Y. (1991) Regeneration of rice plants from suspension cultures, in *Biotechnology in Forestry and Agriculture, vol. 14, Rice* (Bajaj, Y. P.S., ed.), Springer-Verlag, Heidelberg, pp. 38–46.

15. Anthony, P., Jelodar, N. B., Lowe, K. C., Power J. B., and Davey, M. R. (1996) Pluronic F-68 increases the post-thaw growth of cryopreserved plant cells. *Cryobiology* **33,** 508–514.
16. Lynch, P. T., Benson, E. E., Jones, J., Cocking, E. C., Power, J. B., and Davey, M. R. (1994) Rice cell cryopreservation: The influence of culture methods and the embryogenic potential of cell suspensions on post-thaw recovery. *Plant Sci.* **98,** 185–192.
17. Raghava-Ram, N. V. and Nabors, M. W. (1984) Cytokinin mediated long-term, high-frequency plant regeneration in rice tissue culture. *Z. für Pflanzenphysiol.* **113,** 315–323.
18. Raghava-Ram, N. V. and Nabors, M. W. (1985) Plant regeneration from tissue cultures of Pokkali rice (*Oryza sativa*) is promoted by optimizing callus to medium volume ratio and by a medium-conditioning factor produced by embryogenic callus. *Plant Cell, Tiss. Org. Cult.* **4,** 241–248.
19. Murashige, T. and Skoog, F. (1962) A revised medium for rapid growth and bioassays with tobacco tissue cultures. *Physiol. Plant* **15,** 473–497.
20. Linsmaier, E. M. and Skoog F. (1965) Organic growth factor requirements of tobacco tissue cultures. *Physiol. Plant* **18,** 100–127.
21. Müller, A. J., and Grafe R. (1978) Isolation and characterisation of cell lines of *Nicotiana tabacum* lacking nitrate reductase. *Mol. Gen. Genet.* **161,** 67–76.
22. Ohira, K., Ojima K., and Fujiwara A. (1973) Studies on the nutrition of rice cell culture. I. A simple, defined medium for rapid growth in suspension culture. *Plant Cell Physiol.* **14,** 1113–1121.

3

Callus Initiation, Maintenance, and Shoot Induction in Potato

Monitoring of Spontaneous Genetic Variability In Vitro and In Vivo

Rosario F. Curry and Alan C. Cassells

1. Introduction

Potato tissue culture has been widely researched both from the perspectives of mass clonal propagation, e.g., in potato seed certification schemes, and of genetic manipulation *(1,2)*. Potato micropropagation is central to most potato certification schemes *(3)*, and genetic engineering has resulted in the release and sale to consumers in North America of transformed "Russet Burbank" potatoes (NatureMark Potatoes, Boise, ID).

Early potato tissue-culture studies showed that the genome was unstable in callus formation and subculture, and gave rise to high-frequency spontaneous mutations ("somaclonal variation") in adventitiously regenerated plants *(4)*. Various workers have sought to exploit somaclonal variation in potato improvement, but with limited success owing to the high level of useless variability produced, typical of mutation breeding, including polyploids, aneuploids, and unstable maturation mutants *(5)*. This problem of unwanted background variability is of concern to those involved in potato genetic transformation *(6)*.

Here, direct and indirect adventitious pathways for potato regeneration are described *(7)*. The use of flow cytometry is described for the detection of variability *(8)* in explants used to initiate cultures, and in the callus and adventitious shoots derived. This application of flow cytometry to monitor genetic drift is based on the high-frequency occurrence of polyploidy in potato tissues in vivo *(9)* and in callus and adventitious regenerants in vitro *(10)*. Flow

From: *Methods in Molecular Biology, Vol. 111: Plant Cell Culture Protocols*
Edited by: R. D. Hall © Humana Press Inc., Totowa, NJ

Table 1
Constituents of Potato Culture Media[a]

Compound	Callus initiation M1	Greening M2	Shoot induction M3	Potato nodal medium
MS medium	Full strength	Full strength	Full strength	Half strength
Myo-inositol	100	_	100	
Casein hydrolysate	—	—	1000	
Sucrose	30 g	10 g	3 g	15 g
Glucose	—	10 g	_	
Agar	7 g	10 g	10 g	6.0 g
Mannitol	–	40 g	36 g	
NAA	5	0.2	_	
IAA	-	0.1	0.1	
t - Zeatin	—	—	0.5	
Kinetin	—	—	0.5	0.1
6 - BAP	—	0.5	—	
GA_3	-	0.2	—	0.2
MES	—	—	1.0 g	
pH	5.8	5.7	5.7	5.8

[a]The medium (MS) used is that of Murashige and Skoog *(27)* (*see* Appendix). Concentrations expressed as mg/L, unless otherwise stated.

cytometry also has potential application in selection against polyploids and aneuploids in genetic transformation and mutation breeding studies. Image analysis can be used to measure variability in the adventitious regenerant populations *(11)* using the "standard deviation assay" of De Klerk *(12)*, which is based on the principle that random mutation in a population will increase character variability, e.g., variability in leaf shape, that is, in characters that are controlled by minor genes. Image analysis can also be used to differentiate some nonmorphological mutants from the control population, but in this case, owing to large environment–genotype interaction in potato, large clonal populations are required for sampling *(13)*.

2. Materials

1. 140-mL polyvinylchloride food-grade tubs (*see* **Note 1**) sealed with "Parafilm M" (*see* **Note 2**) containing 45 mL of potato nodal medium (*see* **Table 1**); 9-cm Petri dishes (*see* **Note 3**) containing 25 mL of callus initiation medium, callus greening medium, or shoot induction medium, as appropriate (media compositions are given in **Table 1**).
2. Potato plants for explants: these should be grown from certified virus-free seed tubers in the glasshouse in insect-proof cages to exclude aphid virus vectors.

These cages can be made from wooden frames covered with muslin. Alternatively, the plants can be grown in plastic bags with insect-proof vents (Sunbags: Sigma Chemical Co., Poole, Dorset, UK). To introduce these plants from the glasshouse into culture in the growth room, nodes (1–1.5 cm in length) are excised, surface-sterilized and used as explants.
3. 80% (v/v) aqueous ethanol, 10% (v/v) commercial bleach (Domestos—active ingredient 0.5% sodium hypochlorite; Lever Industrial Ltd., Ireland), sterile distilled water.
4. Potato microplants in culture are used as source explants for callus (*see* **Note 4**).
5. A suitably equipped laboratory for plant tissue culture to include autoclaving facilities; instruments and instrument-sterilizing equipment, and a transfer (laminar flow) cabinet (*see* **Note 5**).
6. Suitable plant tissue-culture growth conditions include: a temperature maintained at 22 ± 3°C; a 16-h photoperiod of 35–50 μmol/m^2/s, provided by "white" 65/80 W liteguard flourescent bulbs (Osram Ltd., UK).
7. Flow cytometer (*see* **Note 6**).
8. Mungbean and parsley seeds as internal standards for flow cytometry.
9. Buffer for extraction of nuclei. It consists of 15 m*M* Tris (hydroxymethyl aminomethane), 2 m*M* Na_2EDTA, 0.5 m*M* spermine, 80 m*M* KCl, 20 m*M* NaCl, 15 m*M* mercaptoethanol, and 0.1% (v/v) Triton X-100, pH 7.5 *(17)*.
10. 4', 6-diamidino-2-phenylindole (DAPI) stain: The solution of DAPI consists of 20 mg DAPI/1 mL of water. Ten milliliters of DAPI solution are added to 10 mL citrate buffer in a small flask with closure. The citrate buffer is prepared by adding Na_2HPO_4 to citric acid until a pH of 4.0 is reached (ca. 6:5 ratio). The DAPI mixture is covered with foil and stored in the refrigerator when not in use.
11. Glass Petri dishes.
12. Razor blades.
13. Image analysis system (*see* **Notes 8** and **9**).

3. Methods

3.1. Callus Initiation from In Vivo Material

1. Internode and petiole explants are taken from the mid region of plants, 18–25 cm in height, growing in the glasshouse.
2. Surface-sterilize explants in 80% (v/v) aqueous ethanol for 30 s followed by immersion for 15 min in 10% (v/v) commercial bleach. Rinse in sterile distilled water five times.
3. Trim back 3–5 mm from cut ends of sterilized material to remove damaged material.
4. Cut explant longitudinally before placing both halves, and cut surface down on callus initiation medium (M1) (*see* **Table 1**).

3.2. Callus Initiation from In Vitro Material

1. Choose healthy vigorous internode explants (*see* **Note 10**) from 6-wk-old in vitro plantlets. Internodes should be 1.0–2.0 cm in length *(7)*.
2. Place internodes on M1 (*see* **Table 1; Note 11**).

3.3. Maintenance of Callus

1. Maintain callus under growth room conditions (*see* **Subheading 2., item 6**).
2. Subculture every 4–5 wk onto fresh M1, by transferring single pieces of callus approx 1 cm in diameter.
3. Only subculture viable, healthy calli (*see* **Note 12**).

3.4. Induction of Shoots

1. Transfer callus from M1 onto greening medium (M2) for 4 wk (*see* **Table 1**; **Note 13**).
2. Once callus has greened, transfer to shoot induction medium (M3) (*see* **Table 1**).
3. When shoots begin to grow, excise and transfer to potato nodal medium (*see* **Table 1**; **Note 14**) ***(14,15)***.
4. Seal tubs with parafilm (*see* **Note 2**), and maintain under growth room conditions.
5. When shoots have developed 5–6 nodes (approx 4 wk in the growth room) remove from container. Wash off agar attached to roots with warm unsterile water (*see* **Note 15**).
6. Fill an unsterile plug tray with a mixture of vermiculite and potting compost (1:1 ratio) (*see* **Note 16**). Place one plantlet in each plug, and firm the compost at the crown. Mist gently with tepid water. Place a translucent plastic cover over the tray to keep the relative humidity high. Weaning is carried out for 3 wk in the glasshouse.
7. Mist the plantlets twice daily, and open air vents in the covers one-third of the way each week. Remove the covers after 3 wk.

3.5. DNA Quantitation by Flow Cytometry

3.5.1. Preparation of Nuclei Suspension Samples for DNA Measurement

1. Mungbean and parsley seeds are surface-sterilized, immersed in sterile distilled water for 24 h, and germinated on 2.2 g/L Murashige and Skoog basal medium containing 15 g/L sucrose and 8 g/L agar, pH 5.8, in the growth room.
2. Remove young leaf tissue (*see* **Notes 17** and **18**) from the internal standard and 10 mg of fresh tissue from test sample to be analyzed, and place on a clean Petri dish (*see* **Note 19**).
3. Chop standard and sample with approx 200 clean strikes of a sharp blade in 0.4 mL of ice-cold nuclei extraction buffer (*see* **Notes 20** and **21**) ***(22)***. Leave for 10–20 min.
4. Add 2.0 mL of DAPI stain mixture (*see* **Note 7**).
5. Filter through a nylon mesh of 40 μm into a sample tube.

3.5.2. Use of Flow Cytometer

1. Turn on machine, and allow lamp to stabilize for 30 min before beginning DNA measurements.

2. Check that parameter settings are correct (*see* **Notes 22** and **23**).
3. Place the sample tube in the sample holder.
4. Adjust the gain and lower limits until all peaks are apparent on the *x*-axis (*see* **Note 23**).
5. Clear the sample measurement thus far once the flow rate stabilizes (*see* **Note 24**). All measurements taken after this point will be taken as the result for that sample.
6. Do not allow the volume of sample to decrease to the point where the electrode taking measurements emerges from the sample. If this occurs, the sample data are cleared automatically.
7. Ensure Data Pool Application for Cytometry (DPAC) software is set to receive information from the flow cytometer (*see* **Note 25**).
8. Cease measurement, and transfer data to computer software.
9. The next sample can now be analyzed.
10. After all measurements are taken, clean the flow cuvet (*see* **Note 26**).
11. Switch off the flow cytometer, computer, printer, and UV lamp.

3.5.3. Use of DPAC Software

1. Analyze data on the software once the information has been transferred and stored on disk (*see* **Subheading 3.5.4.**).
2. Name the sample.
3. Analyze peaks. Try to align the reference line with the internal standard peak. This helps in the comparison of sample nuclei with internal standard nuclei.
4. If software does not identify all peaks, there is a facility to select peaks.
5. Print out graphs.

3.5.4. Interpretation of Results

1. Examine DNA histograms. Ideally graphs should be clear with sharp peaks and low coefficients of variation (*see* **Note 27**).
2. Decide which peaks represent the fluorescence of internal standard nuclei. This can be done by referring to the peak number and index values printed underneath the graph (*see* **Fig. 1.**, **Notes 28** and **29**). It may be necessary to calculate the DNA content of the internal standard peak to verify this (*see* **Note 30**).
3. Check if a second peak of the standard, the 4C amount, is also represented in the histogram.
4. The remaining peak(s) represents the fluorescence of potato sample nuclei (*see* **Fig. 1.**, **Notes 31** and **32**).
5. To estimate the nuclear DNA content, calculate the ratio of the mode of the target nuclei to internal standard nuclei. Multiply this by the nuclear DNA content in picograms of the standard. The result is the amount of DNA (pg) in that peak (*see* **Note 30**).
6. Analyze DNA content in picograms of each peak to determine if polyploid nuclei are present (*see* **Notes 33** and **34**).

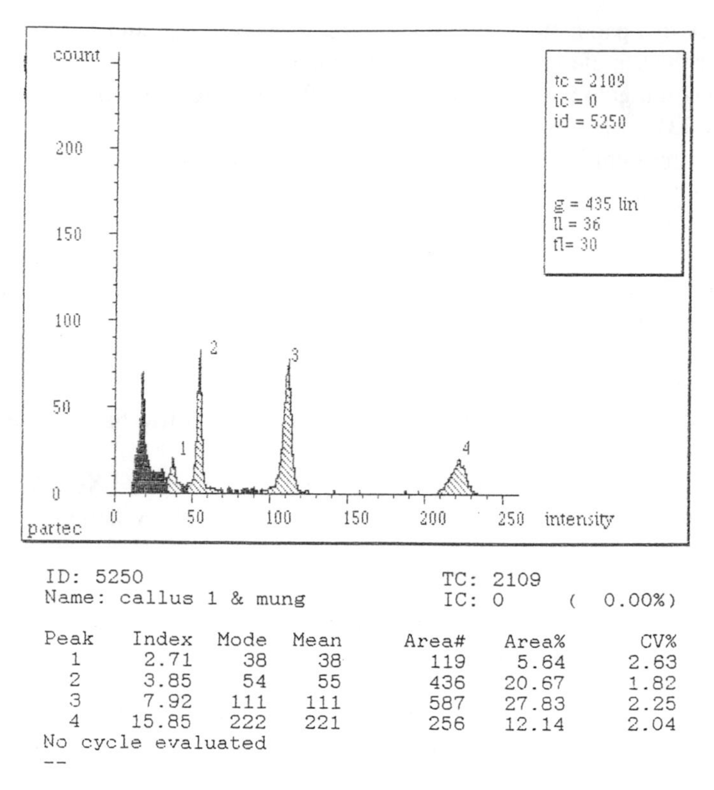

ID: 5250 TC: 2109
Name: callus 1 & mung IC: 0 (0.00%)

Peak	Index	Mode	Mean	Area#	Area%	CV%
1	2.71	38	38	119	5.64	2.63
2	3.85	54	55	436	20.67	1.82
3	7.92	111	111	587	27.83	2.25
4	15.85	222	221	256	12.14	2.04

No cycle evaluated
--

Fig. 1. DNA histogram of DAPI-stained nuclear preparation of internal standard (*Vigna radiata*) and callus tissue of *S. tuberosum* (*see* **Note 32**).

3.6. Image Analysis of Leaf Shape

3.6.1. Sampling of Plant Population

1. Plants should be grown in the glasshouse in randomized blocks with at least 50 replicates for each treatment, under uniform environmental conditions. Plants should be clearly coded/labeled.
2. Terminal leaflets are taken from each plant in the batch and coded with respect to the batch (*see* **Note 35**).
3. Harvested leaflets should be placed in labeled plastic bags to avoid desiccation during processing.

3.6.2. Image Analysis

1. Turn on the lightbox, focus the camera, and set a midrange aperture of approximately F-8 .
2. Open the system menu and click on “set defaults.” Set the parameter you want to measure, e.g., leaf shape factor, and press “enter.”

3. Calibrate the equipment (*see* **Note 36**) by selecting the "analysis menu" and "calibrate;" place a ruler in the measuring frame, and click the mouse cursor on the scale at the two ends. Key in the measured length, and enter the minimum object size. Enter units of measurement, e.g., cm.
4. Place a sample leaflet on a lightbox, and the raise or lower the video camera, such that the leaflet fits within the camera frame.
5. Click on the "image menu" and select "acquire image." Place a sample leaflet on the lightbox, and set the threshold. Adjust the controls until the image is fully thresholded with no extra light or dark highlighting outside the object. Press "enter" to acquire the image. Check the image has been acquired by placing your hand underneath the camera. If the image has been grabbed, the leaflet image will remain on the monitor when the camera is blocked.
6. Select "object" from the "analysis menu." Press "enter," and "save;" label the image.
7. Press appropriate key to show shape factor values (*see* **Note 37**).

3.6.3. Analysis and Interpretation of Image Data

1. Enter the individual data points from above in a statistical software spread sheet.
2. Calculate the standard deviations for the control and test populations. The smaller the SD, the less the variability in the character chosen, i.e., in the respective population (*see* **Note 38**) *(12)*.

4. Notes

1. All plantlets are grown and maintained in 140 mL clear sterile polyvinylchloride tubs (cat. no. 150c, Wilsanco Plastics Ltd., Killyman Road, Industrial Estate, Dungannon, Co. Tyrone DT7 16LN, Northern Ireland).
2. Each tub is sealed with "Parafilm M" (American Can Co. ,CT). Parafilm has high gas and vapor barrier properties, and can prevent contamination from entering the culture vessel. There is enough gaseous exchange through the walls of the container to prevent ethylene buildup or oxygen deficiency inside the vessel.
3. Each dish is wrapped in clear PVC "clingfilm wrap." This acts as a barrier to contaminants, yet allows gaseous exchange *(16)*.
4. Potato microplants are produced and maintained in vitro via nodal subculture. In sterile conditions, nodes are excised from microplants and placed on potato nodal medium. Subculturing takes place every 5–7 wk.
5. All media, glassware, and instruments are sterilized at 110 kPa for 15 min. It is necessary to filter-sterilize certain hormones into sterilized media, since autoclaving them with other media components may result in loss of activity (*see* **Table 2**). Subculturing is performed under sterile conditions in a laminar flow cabinet (cat. no. KD2S, Brassaire, John Bass Ltd., Hampshire, UK). Instruments are sterilized for 15–20 s at 250°C in a bead sterilizer placed in the flow cabinet.
6. In this study, the flow cytometer used is a Partec CA-III (Partec GmbH, Otto-Hahn-StraBe 32, D-48161, Münster, Germany).

Table 2
Plant Growth Regulators: Preparation and Storage of Stock Solutions

Growth regulators	NAA	IAA	Zeatin	Kinetin	BAP	GA_3
Mol. wt.	186.2	175.2	219.2	215.2	225.3	346.4
solvent	1 *N* NaOH	ETOH/ 1 *N* NaOH	1 *N* NaOH	1 *N* NaOH	1 *N* NaOH	ETOH
Diluent	Water	Water	Water	Water	Water	Water
Liquid storage	2–8°C	< 0°C	< °C	< 0°C	2–8°C	2–8°C
Sterilization	A	A/F	A/F	A/F	A/F	A/F

NAA, 1-naphthalene acetic acid; IAA, indole-3-acetic acid; BAP, benzyladenine; GA_3, gibberellic acid.
A, Autoclavable with other media components.
F, Filter-sterilize.
A/F, Autoclavable, but may suffer some loss of activity.

7. The fluorescent dye DAPI has been demonstrated in the past to give reproducible results in staining released nuclei, although some argue that the use of base preference fluorochromes, such as DAPI, can lead to errors in calculations. It is slow-fading and has greater specificity for double-stranded DNA, making it unnecessary to pretreat the sample with RNase. Other stains used frequently include: propidium iodide (intercalating dye); Mithramycin (fluorescent antibiotic); Hoechst dyes (A-T specific stains) ***(18)***.
8. The image analysis system described here is the DIAS system (Delta-T Instruments, Cambridge, UK).
9. Any basic statistical software pack will be able to provide values for the standard deviations of population means, e.g., Prism (GraphPad Software, San Diego, CA).
10. In vitro plantlets should be bacterial-indexed routinely to ensure the health status of the material ***(19)***.
11. When internode explants are first placed on M1 (callus initiation medium), it may take up to 8 wk before significant callus forms at cut ends. After the first subculture, callus will grow rapidly and can be subcultured every 4 wk.
12. If callus turns brown or a translucent-gray color, this means it has senesced. Do not subculture senesced callus. It is important to wrap plates in clingfilm wrap and not parafilm, so ethylene emitted from the cultures will not build up inside the Petri plate. Thus callus is maintained in a less stressful environment, thereby decreasing the incidence of senescence.
13. The "greening" stage is a necessary step in shoot induction. It will be observed that callus continuously subcultured on M1 is mostly friable. However, after 4 wk on greening medium, callus changes its texture to a hard, crusty mass.

14. It can take up to 6 wk for shoots to be induced. Once one shoot is removed from a callus, many more shoots will be produced within days. Therefore, it is important to excise the first shoot as soon as it appears.
15. It is important to wash agar off roots to prevent fungal contamination in the glasshouse.
16. Vermiculite has water-retention properties that prevents the compost from drying out in the glasshouse.
17. The amounts of standard used vary with each type of sample. It is advisable to run a preliminary sample to decide on the correct amount of test tissue and standard to use. At least 4000 nuclei are required.
18. Flow cytometry provides only relative values. Simultaneous measurement of the fluorescence of internal standard stained nuclei of known DNA content is necessary to determine DNA content in picograms of the sample nuclei. Mungbean (*Vigna radiata* cv. Berken, 2C:1.0pg) or parsley (*Petroselinum crispum* 2C:3.6 pg) nuclei are used as an internal standard. (DNA amounts in plants are expressed in picograms, 1 pg = 10^{-12} g). These were selected as standards because their peak positions do not coincide with either the G0 + G1 or the G2 peak of the potato sample ***(21)***.
18. Flow cytometric analysis has been performed on fixed tissues ***(20)***. Samples "fixed" with formaldehyde can be stored for weeks, without loss of resolution, before analysis. The number of released nuclei can be higher with fixed than with fresh tissues, so smaller samples can be used.
20. Nuclei will be damaged if the blade is blunt, and this debris will lead to increased "background noise" on graphs ***(23)***. It may be difficult to chop callus in nuclei extraction buffer, because it tends to clump in the liquid. Incubation in the buffer and mechanical disruption of cells by syringing may prove a more reliable method for isolation of nuclei.
21. The sample must be kept at ice-cold temperatures after isolation of nuclei in order to decrease nuclease activity ***(24)***.
22. The speed of the flow of nuclei analyzed is best adjusted to as low a setting as possible, i.e., 20–40 nuclei/s. For greater accuracy, i.e., sharper peaks and with low coefficients of variation, speed can be reduced even further, but analysis is more time-consuming.
23. The gain is reduced to shift the readings to the left, until all peaks on the *x*-axis are visible. (Most machines have a screen that allows you to see the peaks form as each nucleus is being analyzed.) Adjusting the lower limits removes the background reading that arises from debris in the sample.
24. It is necessary to clear all previous measurements when the flow rate stabilizes to ensure accurate results.
25. Follow the manufacturer's guidelines to ensure the computer attached is ready to receive data from the flow cytometer. This may involve setting the software to a transfer command.
26. The flow cuvet can be cleaned after each sample is analyzed and when all the measurements have been taken. It allows water to flow through the injection tubes to remove nuclei and debris.

27. For accurate interpretation of results, coefficients of variation (CV) should be below 5%. Sample preparation is very important in this respect, since nuclei damaged during mechanical isolation will result in higher CV values ***(25)***. One could centrifuge each sample and discard debris before analysis, but this is time-consuming. Over- or understaining nuclei with DAPI can also result in larger coefficients of variation and, therefore, more variable results. In **Fig. 1**., CV values range from 2.63–1.82. Peak 1 representing mungbean nuclei have the largest CV values.
28. The index value of the internal standard does not always approximate to its 2C DNA content, e.g., 1.06 pg in the case of mungbean ***(21)***, since it depends on where the reference line is positioned during the analysis. This line should be aligned with the peak of the internal standard. It is found sometimes that the nuclei of the internal standard tissue are dividing, giving a second peak, representing the 4C amount of DNA.
29. The C value refers to the DNA amount of the unreplicated haploid genome. 2C is the DNA amount of an unreplicated diploid genome. 4C is the DNA amount in nuclei with two replicated copies of a genome, i.e., nuclei after division and DNA synthesis, in the G_2 and *M* (mitotic) phase ***(21)***.
30. Calculation of nuclear DNA content:

$$\frac{\text{Mode of sample}}{\text{mode of reference standard}} \times \text{DNA (pg) of reference standard} \qquad (1)$$

The "Mode" represents the most frequently occurring number in a sample.

Peak 1 $38 \div 38 \times 1.06 = 1.06$ pg
Peak 2 $54 \div 38 \times 1.06 = 1.50$ pg
Peak 3 $111 \div 38 \times 1.06 = 3.09$ pg
Peak 4 $222 \div 38 \times 1.06 = 6.19$ pg

31. Exercise caution when determining the 2C DNA content of the sample. A previous knowledge of ploidy level of material is important. *Solanum tuberosum* is a tetraploid ($2n = 48$). 4C is the DNA amount of the unreplicated tetraploid genome (4C:3.6 pg) ***(17)***.
32. This graph shows a minor peak of 2C nuclei (peak 1) of mungbean cells. Peak 2 (1.50 pg) represents the unreplicated genome (4C DNA content) of potato nuclei in $G_{0/1}$ phase. Peak 3 (3.09 pg) represents the 8C DNA content of potato nuclei in G_2 phase. The nuclei have doubled the normal DNA content after division and DNA synthesis. Peak 4 represents 6.19 pg of DNA. This is approximately twice the value of peak 3, indicating that some of these nuclei have undergone division.
33. If the DNA amount is greater than twice the 2C DNA amount, polyploid nuclei may be present. For example, the DNA content of peak 3 is slightly higher (3.09 pg) than twice the DNA content of peak 2 (3.00 pg). One could hypothesize that the extra DNA is from aneuploid nuclei, or it could be the result of poor sample preparation. It is extremely difficult to determine aneuploidy (± chromosome differences) using flow cytometry.

34. Accuracy of any result from flow cytometry is questionable because of the large portion of nuclear DNA in native chromatin that is unstained. The amount of DNA remaining unstained depends on chromatin structure ***(24)***.
35. It is important that the leaves are fully expanded and are taken from equivalent leaf positions on the plants to reduce variability in the data.
36. The instructions are for the MS-DOS version of the DIAS image analysis system (Delta-T Instruments, Cambridge, UK). The principles apply to other systems.
37. Depending on the apparatus used, there will be a choice of system-generated parameters. It is recommended that a quantitative character is used, although for highly variable populations, this may not be essential.
38. The "standard deviation assay" ***(12)*** may be used to detect variability in a population at the percentage level. To discriminate between clones, variability in the data owing to the genotype environment interaction requires that large sample sizes may be analyzed. (For analysis of Gaussian data *see* **ref. *26***).

References

1. Bajaj, Y. P. S., (ed.) (1987) Potato, in *Biotechnology in Agriculture and Forestry 3.* Springer Verlag, Berlin.
2. Salamini, F. and Motto, M. (1993) The role of gene technology in plant breeding, in *Plant Breeding—Principles and Prospects* (Hayward, M. D., Bosemark, N. O., and Romagosa, I. eds.) Chapman and Hall, London, pp. 138–159.
3. Cassells, A. C. (1997) Pathogen and microbial contamination management, in micropropagation: an overview, in *Pathogen and Microbial Contamination Management in Micropropagation,* Kluwer, Dordrecht, pp. 1–14.
4. Kumar, A. (1994) Somaclonal variation, in *Potato Genetics* (Bradshaw, J. E. and Mackay, G. R., eds.), CAB International, Wallingford, pp. 197–212.
5. Cassells, A. C. and Sen, P. T. (1996) *In vitro* production of late blight (*Phytophthora infestans*)-resistant potato plants, in *Biotechnology in Agriculture and Forestry* 36 (Bajaj, Y. P. S., ed.), Springer Verlag, Berlin, pp. 107–118.
6. Dale, P. J. and Hampson, K. K. (1995) An assessment of morphological and transformation efficiency in a range of varieties of potato (*Solanum tuberosum L.*), in *Plant Genetic Manipulation in Plant Breeding: Criteria for Decision Making* (Cassells, A. C. and Jones, P. W., eds.), Kluwer, Dordrecht, pp. 101–108.
7. Cassells, A. C., Austin, S., and Goetz, E. M. (1987) Variation in tubers in single cell-derived clones of potato in Ireland, in *Biotechnology in Agriculture and Forestry* 3 (Bajaj, Y. P. S., ed.), Springer-Verlag, Berlin, pp. 375–391.
8. Kubalakova, M., Dolezel, J., and Lebeda, A. (1966) Ploidy instability of embryogenic cucumber callus culture. *Biol. Plant* **38,** 475–480.
9. D'Amato, F. (1952) Polyploidy in the differentiation and function of tissues and cells in plants. *Caryologia* **4,** 311–358.
10. Karp, A. (1990) Somaclonal variation in potato, in *Biotechnology in Agriculture and Forestry* 11 (Bajaj, Y. P. S., ed.,), Springer Verlag, Berlin, pp. 379–399.
11. Cassells, A. C., Walsh, C., and Periappuram, C. (1993) Diplontic selection as a positive factor in determining the fitness of mutants of *Dianthus* "Mystere" derived from x-irradiation of nodes in *in vitro* culture. *Euphytica* **70,** 167–174.

12. De Klerk, G.-J. (1990) How to measure somaclonal variation. *Acta Bot. Neerl.* **39,** 129–144.
13. Cassells, A. C., Croke, J.T., and Doyle, B. M. (1997) Evaluation of image analysis, flow cytometry and RAPD analysis for the assessment of somaclonal variation and induced mutation in tissue culture-derived pelargonium plants. *Angewandte Bot.* **71,** 125–130.
14. Ahloowalia, B. S. (1982) Plant regeneration from callus culture in potato *Euphytica* **31,** 755–759.
15. Skirvin, R. M., Lam, S. L., and Janick, J. (1975) Plantlet formation from potato callus *in vitro. Hort. Science* **10,** 413.
16. Cassells, A. C. and Roche, T. (1993) The influence of the gas permeability of the vessel lid and growthroom light intensity on the characteristic of *Dianthus* microplants *in vitro* and *ex vitrum,* in *Physiology, Growth and Development of Plants and Cells in Culture* (Lumsden, P. J., Nicholas, J.R., and Davies, W. J., eds.), Kluwer, Dordrecht, pp.204–214.
17. Dolezel, J., Binarová, P., and Lucretti, S. (1989) Analysis of nuclear DNA content in plant cells by flow cytometry. *Biol. Plant.* **31,** 113–120.
18. Dolezel, J., Sgorbati, S., and Lucretti, S. (1992) Comparison of three DNA fluorochromes for flow cytometric estimation of nuclear DNA content in plants. *Physiol. Plant.* **85,** 625–631.
19. Cassells, A. C. (1998) Production of pathogen-free plants, in *Biotechnology in Agriculture* (Altman, A., ed.), Marcel Dekker, New York, pp. 43–56.
20. Sgorbati, S., Levi, M., Sparvoli, E., Trezzi, F., and Lucchini, G. (1986) Cytometry and flow cytometry of 4', 6-diamidino-2-phenylindole (DAPI)-stained suspensions of nuclei released from fresh and fixed tissues of plants. *Physiol. Plant.* **68,** 471–476.
21. Bennett, M. D. and Leitch, I. J. (1995) Nuclear DNA amounts in Angiosperms. *Ann. Bot.* **76,** 113–176.
22. Galbraith, D. W., Harkins, K. R., Maddox, J. M., Ayres, N. M., Sharma, D. P., and Firoozabady, E. (1983) Rapid flow cytometric analysis of the cell cycle in intact plant tissues. *Science* 220, 1049–1051.
23. Ulrich, I. and Ulrich, W. (1991) High-resolution flow cytometry of nuclear DNA in higher plants. *Protoplasma* 165, 212–215.
24. Dolezel, J. (1991) Flow cytometric analysis of nuclear DNA content in higher plants. *Phytochem. Anal.* **2,** 143–154.
25. De Laat, A. M. M., Göhde, W., and Vogelzang, M. J. (1987) Determination of ploidy of single plants and plant populations by flow cytometry. *Plant Breeding* **99,** 303–307.
26. Motulsky, H. (1995) *Intuitive Biostatistics,* Oxford University Press.
27. Murashige, T. and Skoog, F. (1962) A revised medium for rapid growth and bioassays with tobacco tissue cultures. *Physiol. Plant.* **15,** 473–497

4

Somatic Embryogenesis in Barley Suspension Cultures

Makoto Kihara, Hideyuki Funatsuki, Kazutoshi Ito, and Paul A. Lazzeri

1. Introduction

Barley is an important crop both for brewing and for animal feed. In addition to conventional breeding, in vitro culture is a useful technology for the improvement of barley quality. For example, a number of methods for barley anther and microspore culture have been published for the production of doubled haploid lines for use in plant breeding and in genetic analysis *(1)*.

Recently, novel technologies, such as protoplast fusion and DNA uptake by protoplasts, are being integrated into plant breeding programs *(2)*. For the utilization of these novel technologies in the molecular breeding of cereals, embryogenic suspension cultures provide a valuable tool, because they are good sources of protoplasts. These allow manipulations at the single cell level, such as direct DNA uptake or somatic hybridization *(3)*. In addition, such suspensions are a suitable target for gene transfer by particle bombardment *(4)*.

The establishment of embryogenic suspension cultures has been reported for several graminaceous species. In rice, reliable and efficient protocols for the establishment of suspension cultures have been developed, and transgenic, somatic hybrid, and cybrid plants have been obtained by using protoplasts isolated from embryogenic suspension cells *(5,6)*. In contrast to the significant progress in rice, barley has been classified as a difficult species for the establishment of embryogenic suspension cultures *(2)*. In 1991, the first successful plant regeneration from barley protoplasts was reported *(7)*. In this work, embryogenic suspension cells, as a source of regenerable protoplasts, were initiated from anther-derived calluses. However, these cells lost their regeneration capacity quickly, and the efficiency of plant regeneration was not high

From: *Methods in Molecular Biology, Vol. 111: Plant Cell Culture Protocols*
Edited by: R. D. Hall © Humana Press Inc., Totowa, NJ

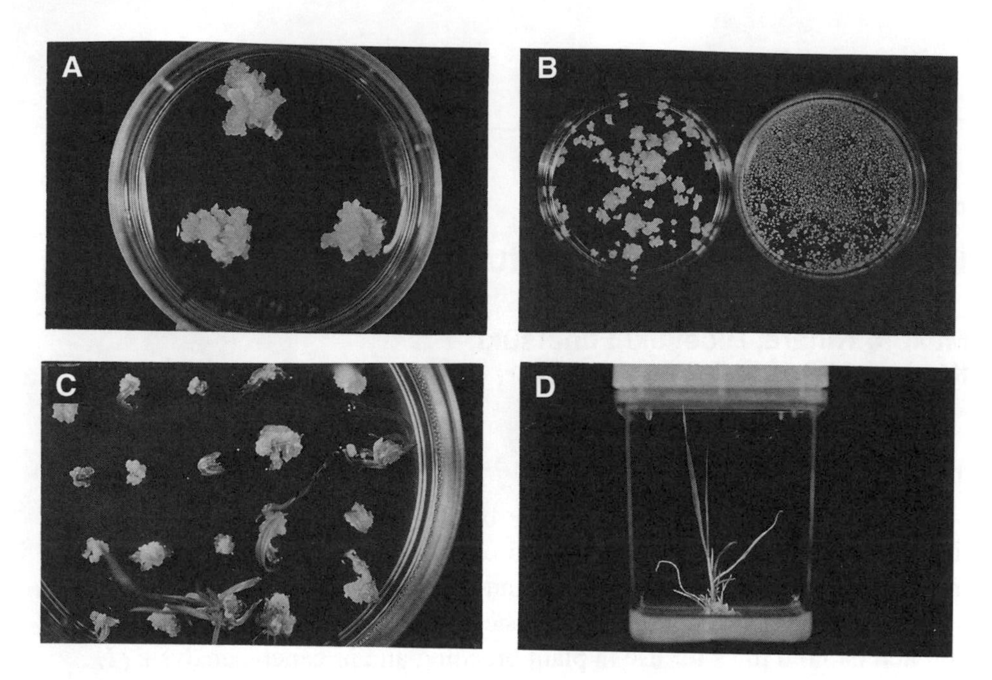

Fig. 1. Different stages in the initiation and establishment of embryogenic cell suspension cultures of barley. **(A)** Primary calluses on callus induction medium (approx 4 wk after callus initiation). **(B)** Suspension cells subcultured in modified AA (left) and L1 (right) medium. **(C)** Shoot regeneration from embryogenic structures on regeneration medium. **(D)** Regenerated plantlet in Magenta box.

enough for genetic manipulation ***(8)***. In the last few years, however, with the combination of established transformation techniques ***(9)*** and improved suspension culture protocols using immature embryo-derived calluses (***10,11***; *see* Fig. 1), transgenic barley plants have been produced by direct gene transfer to suspension cell-derived protoplasts ***(12)***.

Improvements in the efficiency of production of embryogenic barley suspensions have been achieved by analysis and optimization of limiting factors at different steps in the overall process, as follows:

1. Callus induction.
   a. Genotype and growth condition of donor plants.
   b. Developmental stage of explants.
   c. Callus induction medium.
2. Suspension establishment.
   a. Subculture technique.
   b. Culture vessel and subculture medium.

3. Plant regeneration.
   a. Plant regeneration medium.
   b. Analysis of regeneration ability of suspension cells and colonies derived from protoplasts isolated from suspension cells.

The following protocol has proven very successful in our lab for the establishment of embryogenic barley suspension cultures.

2. Materials

2.1. Callus Induction

1. Immature embryos, cv. Igri (*see* **Note 1**).
2. 70% Ethanol in water.
3. Sodium hypochlorite solution (1.0% active chlorine) in distilled water.
4. Sterile distilled water.
5. Sterile working surface, scalpel, forceps.
6. Stereo microscope.
7. Modified L2 medium (*see* **Table 1**) in 60-mm Petri dishes.
8. Culture room (25°C, 16 h light, 50–100 lx).

2.2. Establishment of Suspension Cultures

1. Sterile 100-mL conical flasks (Iwaki Glass, Japan) capped with foil.
2. Sterile macroplates with six wells (Falcon, NJ).
3. Sterile 180-mL plastic vessels (Greiner, Germany).
4. Modified AA or L1 medium (*see* **Table 1**).
5. Rotary shaker.
6. Culture room (25°C, 16 h light, 50–100 lx).

2.3. Plant Regeneration

1. Modified L3 medium (*see* **Table 1**) in 90-mm Petri dishes.
2. Stereo microscope.
3. Hormone-free modified L3 medium in 90-mm Petri dishes and Magenta boxes (Magenta Corp., IL).
4. Culture room (25°C, 16 h light, 50–100 lx and 2500–3000 lx).

3. Methods

3.1. Callus Induction

1. Sterilize immature seeds of donor plants (*see* **Note 1**) with 70% ethanol for 30 s and then with sodium hypochlorite containing 1.0% active chlorine for 1–3 min. Rinse these seeds three times with sterilized distilled water.
2. Excise immature embryos of length 0.5–1.0 mm (*see* **Note 2**) from the seeds under a stereo microscope, and place on modified L2 medium (*see* **Table 1** and **Notes 3** and **4**) in 60-mm Petri dishes. Incubate these cultures at 25°C, under 16 h of dim light (50–100 lx).

Table 1
Composition of Culture Media[a]

	Modified L1	Modified L2	Modified L3	Modified AA
Macro salts (mg/L)				
NH_4NO_3	700	1500	200	
KNO_3	1750	1750	1750	
KH_2PO_4	200	200	200	
$MgSO_4 \cdot 7H_2O$	350	350	350	252
$CaCl_2 \cdot 2H_2O$	450	450	450	150
$NaH_2PO_4 \cdot 2H_2O$				150
KCl				2960
Micro Salts (mg/L)				
$MnSO_4 \cdot 4H_2O$	15	15	15	10
H_3BO_3	5	5	5	3
$ZnSO_4 \cdot 7H_2O$	13.4	7.5	7.5	2
Kl	0.75	0.75	0.75	0.8
$Na_2MoO_4 \cdot 2H_2O$	0.25	0.25	0.25	0.25
$CuSO_4 \cdot 5H_2O$	0.025	0.025	0.025	0.025
$CoCl_2 \cdot 6H_2O$	0.025	0.025	0.025	0.025
FeNaEDTA (mg/L)				
Na_2EDTA	37	37	37	37
$FeSO_4 \cdot 7H_2O$	28	28	28	28
Vitamins (mg/L)				
Myo-Inositol	100	100	100	100
Thiamine HCl	10	10	10	10
Pyridoxine HCl	1	1	1	1
Nicotinic acid	1	1	1	1
Ascorbic acid	2		2	2
Ca pantothenate	1		1	1
Choline chloride	1		1	1
Folic acid	0.4		0.4	0.4
Riboflavin	0.2		0.2	0.2
p-Aminobenzoic acid	0.02		0.02	0.02
Biotin	0.01		0.01	0.01
Amino acids (mg/L)	Gln 750 Pro 150 Asn 100	Gln 750 Pro 150 Asn 100	Gln 750 Pro 150 Asn 100	Gln 876 Asp 266 Arg HCl216 Gly 7.5
Sugars (g/L)	Maltose 50	Maltose 30	Maltose 30	Sucrose 30
Hormone (mg/L)	2,4-D 2.0	2,4-D 2.5	BAP 1.0	2,4-D 2.0
Agarose (g/L)		4	4	
pH	5.6	5.6	5.6	5.8

[a]All media are sterilized by ultrafiltration. For solid cultures (modified L2 and modified L3 media), double-concentrated media are mixed with double-concentrated type I-A (Sigma) agarose solution.

3.2. Establishment of Suspension Cultures

1. After 3–4 wk of callus induction, transfer the calluses to 100-mL conical flasks containing 5 mL of modified AA medium (*see* **Table 1**) or macroplates with six wells containing 0.5 mL of modified AA medium (*see* **Note 5**). Incubate the cultures in conical flasks and macroplates (25°C, under 16 h of dim light) on a rotary shaker at 100–120 and 80–100 rpm, respectively (*see* **Note 6**).
2. Gradually increase the volume (0.3–0.5 mL in conical flasks and 0.1–0.2 mL in macroplates) of the subculture medium each time. Maximum volume of medium is 12 and 2 mL in conical flasks and macroplates, respectively (*see* **Note 7**).
3. In the case of macroplates, after 2–6 mon of subculture, transfer the suspension lines, which have released small cell aggregates, to plastic vessels containing 5 mL of modified AA medium. Incubate the cultures (25°C, under 16 h of dim light) on a rotary shaker at 80–100 rpm, remove old medium, and replace with fresh medium at intervals of 7–10 d. Gradually increase the volume (0.3–0.5 mL) of the subculture medium each time. Maximum volume of medium is 12 mL.
4. At each subculture, remove elongated empty or brownish cells, and subculture only cytoplasmically rich and yellow or white cells (*see* **Note 8**).

3.3. Plant Regeneration

1. Directly plate suspension aggregates drained of suspension medium on 25 mL of modified L3 regeneration medium in 90-mm Petri dishes.
2. Transfer embryogenic structures, which are observed under a stereo microscope, to fresh regeneration medium 3–4 wk later, and culture at 25°C under dim light until distinct green shoots appear.
3. Incubate the green shoots (about 10 mm in length) at 25°C under lighting of 2500–3000 lx from cold white fluorescent tubes, 16 h day length. After 2–3 wk of exposure to light, transfer the green shoots (20–30 mm length) to hormone-free modified L3 medium in 90-mm Petri dishes for the induction of roots.
4. Transfer shoots more than ca. 40 mm length to 25 mL hormone-free modified L3 medium in Magenta boxes.
5. After visible root induction, pot the plantlets in soil (*see* **Note 9**).
6. For the detailed analysis of regeneration ability of suspension cultures, isolate protoplasts from suspension cells (*see* **Note 10**) and culture (*see* **Note 11**). After about 1 mon of protoplast culture, check the regeneration ability of protoplast-derived colonies as described above (*see* **Note 12**).

4. Notes

1. Growth conditions of donor plants have a major influence on the efficiency of cell suspension establishment and their regeneration ability. In our study, after 2–3 wk from germination, donor plants (cv. Igri) are vernalized for 6–8 wk at 5 ± 2°C under 2000 lx from cold white fluorescent tubes (16 h day length), and then grown in a growth chamber (12–16°C, 20,000–30,000 lx from high-intensity discharge lamps, 16 h day length). These plants are ready for removal of the immature seeds after 2–3 mon after vernalization. These conditions provide much better

results than using greenhouse-grown plants ***(10,11)***. However, growth conditions must be optimized for each genotype used in the experiments, because the same tendency was not found for another cultivar (cv. Dissa) ***(11)***.

2. The developmental stage of immature embryos affects the efficiency of callus induction and cell suspension establishment ***(11)***. Optimal embryo size depends on the genotype, and in Igri, optimal length is 0.5–1.0 mm (10–20 d after anthesis).
3. One to three immature embryos should be plated scutellum-side up on 10 mL callus induction medium in 60-mm Petri dishes (Falcon). All Petri dishes used in this protocol are sealed with Nesco film (Nippon Shoji, Japan). The embryos with a length of more than 1.0 mm tend to germinate. The removal of root and shoot meristems from embryos prior to culture prevents germination and promotes callus induction.
4. Higher concentrations (12.5 and 5.0 mg/L) of 2,4-D in callus induction medium suppress the proliferation of callus, resulting in a lower efficiency of cell suspension establishment in cv. Igri ***(11)***.
5. Modified L1 medium (*see* **Table 1**) is also suitable for suspension establishment. In the experiments performed in the years 1994–1995, 22, 43, and 31 cell lines showed consistent proliferation by using modified AA medium in conical flasks, modified AA medium in macroplates and modified L1 medium in macroplates, respectively. Among them, 2, 8, and 4 cell lines showed green shoot regeneration ability from protoplasts.
6. In the case of conical flasks, intervals of first and second subculture are approx 14 d. After the second subculture, intervals are 9–11 d. In the case of macroplates, intervals of subculture are 7–10 d.
7. After 1 mon of transfer to liquid medium, suspension cells consist of elongated empty cells and cytoplasmically rich cells. After a minimum time of 2 mon and maximum time of 6 mon, cytoplasmically rich cells release small cell aggregates. Until the release of small cell aggregates, two-thirds to three-quarters of the old medium is removed and replaced with fresh subculture medium, and suspension cultures are maintained at a high callus:medium ratio. After the release of small cell aggregates, cultures are gradually brought to a lower density. The cell density is finally 4–6 g cells/10–12 mL liquid medium.
8. Suspension cells subcultured in modified L1 medium tend to grow as homogenous, yellowish small cell aggregates. In contrast, suspension cells subcultured in modified AA medium tend to be heterogeneous and whitish with larger aggregates.
9. Immediately after transfer to pots, regenerated plants are often damaged by conditions of low humidity, so these plants are grown in a growth chamber controlled at 18°C (20,000 lx from cold white fluorescent tubes, 16 h day length) and 80% humidity for 2–3 wk. Flowering plants are obtained by the same vernalization and growth conditions as described for donor plants.
10. For protoplast isolation, incubate approx 2 g of suspension cells (at day 3–6 after subculture) in 20 mL filter-sterilized enzyme solution for 3–4 h at 25°C. The enzyme solution comprises 1% Cellulase Onozuka RS (Yakult, Japan) and 0.1%

Pectolyase Y-23 (Seishin, Japan) dissolved in LW solution ***(9)***. LW solution contains the macro- and microsalts and amino acids of modified L1 medium and 0.6 *M* mannitol.

11. Wash the isolated protoplasts three times with LW solution by centrifugation. After the last centrifugation, suspend the protoplast pellets in modified L1 medium containing 0.4 *M* maltose and 1.8% sea plaque agarose (FMC, ME). Culture these protoplasts with nurse cells ***(10)***. Fast-growing suspension cells should be used as the nurse cells.
12. The methods for the analysis of regeneration ability of protoplast-derived colonies are the same as described for those of suspension cells. These colonies show a rapid response on regeneration medium, and after about 2 wk of transfer, somatic embryogenesis or shoot regeneration can be observed.

References

1. Pickering, A. R. and Devaux, P. (1992) Haploid production: Approaches and use in plant breeding, in *Barley: Genetics, Biochemistry, Molecular Biology and Biotechnology* (Shewry, P. R., ed.), CAB International, Wallingfold, UK, pp. 519–547.
2. Vasil, I. K. (1994) Molecular improvement of cereals. *Plant Mol. Biol.* **25,** 925–937.
3. Krautwig, B. and Lörz, H. (1995) Cereal protoplasts. *Plant Sci.* **111,** 1–10.
4. Fromm, M. E., Morrish, F., Armstrong, C., Williams, R., Thomas, J., and Klein, T. M. (1990) Inheritance and expression of chimeric genes in the progeny of transgenic maize plants. *Bio/Technology* **8,** 833–839.
5. Shimamoto, K., Terada, R., Izawa, T., and Fujimoto, H. (1989) Fertile transgenic rice plants regenerated from transformed protoplasts. *Nature* **338,** 274–276.
6. Yang, Z. Q., Shikanai, T., Mori, K., and Yamada, Y. (1989) Plant regeneration from cytoplasmic hybrids of rice (*Oryza sativa* L.). *Theor. Appl. Genet.* **77,** 305–310.
7. Jähne, A., Lazzeri, P. A., and Lörz H. (1991) Regeneration of fertile plants from protoplasts derived from embryogenic cell suspensions of barley (*Hordeum vulgare* L.). *Plant Cell Rep.* **10,** 1–6.
8. Lörz, H. and Lazzeri, P. A. (1992) In vitro regeneration and genetic transformation of barley. *Barley Genet. VI* **II,** 807–815.
9. Lazzeri, P. A., Brettschneider, R., Lührs, R., and Lörz, H. (1991) Stable transformation of barley via PEG-induced direct DNA uptake into protoplasts. *Theor. Appl. Genet.* **81,** 437–444.
10. Kihara, M. and Funatsuki, H. (1994) Fertile plant regeneration from barley (*Hordeum vulgare* L.) protoplasts isolated from long-term suspension culture. *Breeding Sci.* **44,** 157–160.
11. Funatsuki, H. and Kihara, M. (1994) Influence of primary callus induction conditions on the establishment of barley cell suspensions yielding regenerable protoplasts. *Plant Cell Rep.* **13,** 551–555.
12. Funatsuki, H., Kuroda, H., Kihara, M., Lazzeri, P. A., Müller, E., and Lörz, H. (1995) Fertile transgenic barley generated by direct gene transfer to protoplasts. *Theor. Appl. Genet.* **91,** 707–712.

5

Somatic Embryogenesis in *Picea* Suspension Cultures

Ulrika Egertsdotter

1. Introduction

Somatic embryogenesis is a nonsexual propagation process where somatic cells differentiate into somatic embryos *(1)*. In gymnosperms, somatic embryogenesis was first reported in *Picea abies* *(2,3)*, but has now been induced in a variety of cycads and conifers *(4)*. The development of somatic embryos is similar in different coniferous species, and in principle, it resembles their zygotic counterparts during development.

The technique of multiplying conifers via somatic embryos offers many advantages, both in the mass propagation of selected genotypes and as a part of breeding programs. Some of the advantages of somatic embryogenesis are: the somatic embryos can be propagated on a large scale in bioreactors, a high yield of plants can be obtained in a short time, the embryos already have a tap root, the embryos can be encapsulated and treated like seeds, true rejuvenation can be obtained even if the somatic embryos are regenerated from mature trees, and the somatic embryos can be cryopreserved *(5)*. By using somatic embryos in breeding programs, it is possible to keep simultaneously all genotypes cryopreserved until valuable clones have been identified in field tests. The cryopreserved material can then be mass propagated.

Somatic embryos from conifers have been stably transformed using microprojectile bombardment *(6–8)*. Also, several *Agrobacterium* strains are capable of infecting conifers and forming tumors and roots, but the regeneration of plantlets from transformed embryos has been less successful. The first transgenic conifer (*Picea glauca*) was obtained after the bombardment of mature somatic embryos. Embryogenic cultures were subsequently established and plants regenerated *(9)*.

From: *Methods in Molecular Biology, Vol. 111: Plant Cell Culture Protocols*
Edited by: R. D. Hall © Humana Press Inc., Totowa, NJ

The regeneration of plants from somatic embryos in *Picea* can be divided into four main steps: initiation of somatic embryos from the initial explant, proliferation of the embryogenic cultures, maturation of the somatic embryos, and regeneration of plants from the somatic embryos. Embryogenic cultures have been successfully initiated and proliferated in several different *Picea* species ***(4)***. However, the maturation and regeneration steps still contain many difficulties, and are at present areas of intensive research.

Initiation: Embryogenic cultures can be induced from seedlings, and both immature and mature zygotic embryos. It has also been implied that embryogenic cultures can be initiated from more mature tissue ***(10)***. The different explants have been successfully used within different species ***(4)***. The initiation frequency is genotype-dependent, and it declines with the increasing age of the primary explant.

The cells in the primary explant are stimulated to form somatic embryos on a medium containing auxin and cytokinin. The LP medium (***11***; **Table 1**) has been successfully used for the initiation and subsequent proliferation of several *Picea* species (**Table 2**).

Proliferation: The somatic embryos formed on the initiation medium continue to proliferate when transferred to fresh medium. The embryos can then, after a few subculture intervals, be transferred to growth in liquid medium. At the proliferation stage, embryos within different cell lines reach different developmental stages, which is reflected in the number of cells, both in the embryonic and suspensor regions, and by the degree of organization of the cells into a polarized structure. The embryogenic cell lines appear to be of two main types, A and B. Embryos in type A cell lines are composed of densely packed, large embryonic regions and many suspensor cells, whereas in type B, they are composed of small, loosely packed embryonic regions and few suspensor cells (***12***; **Fig. 1**). In general, the embryos in suspension cultures have a less polarized morphology than embryos grown on solid medium.

The embryo morphology is partially dependent on the hormonal balance present in the medium ***(13)***. Also, different proteins are secreted into the medium in suspension cultures of types A and B, which influence the developmental stage of the embryos under proliferation ***(14)***. Changing the hormonal balance, or adding conditioned medium does not change the original type of embryo in the cell line. However, embryo morphology can be strongly improved by adding extracts of mature seed ***(15)***. The seed extracts stimulate the formation of larger embryonic regions, more suspensor cells, and a more polarized embryo. The size of the embryonic region and how polarized the embryo is are both important characters for the subsequent maturation process.

Maturation: In the maturation process, the somatic embryos stop proliferating, increase in size, and start to accumulate storage materials, including starch,

Table 1
Components of 1/2 LP Medium[a] for Initiation and Proliferation of *Picea*

Component	Concentration in stock, g/L	Volume stock, mL, for 1 L of final medium	Concentration in final medium, mg/L
1. KNO_3	95	10	950
2. NH_4NO_3	60	5	300
3. $MgSO_4 \cdot 7\ H_2O$	37	5	185
4. KH_2PO_4	34	5	170
5. $CaCl_2 \cdot 2\ H_2O$	44	5	220
6. Micronutrients[b]	—	0.5	
7. Amino acids[c]	—	10	
8. Fe-EDTA	0.07342	1.25	6.95
9. Pyridoxine-HCl	0.010	0.5	0.5
10. Nicotinic acid	0.020	0.5	1.0
11. Glycine	0.020	0.5	1.0
12. 2, 4-D[d]	0.10	10	1
13. BA[e]	0.10	5	0.5
14. Thiamine-HCl	—		25
15. Myo-inositol	—		50
16. D-Glucose	—		90
17. D-Xylose	—		75
18. L-Arabinose	—		75
19. Glutamine	—		500
20. Sucrose	—		10×10^3
21. Gelrite[f]			4×10^3

[a]Culture medium 1/2 LP: Mix all components in water, adjust to pH 5.8, and autoclave. Stock solutions nos. 1–11 can be stored at +4°C for several months; hormonal stocks are stored at –20°C and kept at +4°C for a maximum of 4 wk. Prepared medium is stored at +4°C for a maximum of 4 wk.

[b]1000X Micronutrient stock: Zn-EDTA (4.716 g/L), $MnSO_4.2\ H_2O$ (2.23 g/L), H_3BO_3 (0.63 g/L), $Na_2MoO_4.2\ H_2O$ (0.025 g/L), $CuSO_4.5\ H_2O$ (0.0025 g/L), $CoCl_2.6\ H_2O$ (0.0025 g/L), KI (0.75 g/L).

[c]50X Amino acid stock: L-glutamine (0.02 g/L), L-alanine (0.0025 g/L), L-cysteine. HCl (0.001 g/L), L-arginine (0.0005 g/L), L-leucine (0.0005 g/L), L-phenylalanine (0.0005 g/L), L-tyrosine (0.0005 g/L).

[d]2, 4-Dichlorophenoxyacetic acid.

[e]N^6-benzyladenine.

[f]Gelrite is a trademark of Merck and Co., Inc. Rahway, NJ, Kelco Division USA. Gelrite is used in initiation medium or in solid proliferation medium. For standard purposes, 25 mL of gelrite medium is poured/10-cm Petri plate.

proteins, and lipids ***(16)***. The ability of the somatic embryo to mature varies significantly among different cell lines, and this appears to be closely correlated to embryo morphology during the proliferation stage. In general, type A

Table 2
Reported Variations in Culture Requirements Within *Picea*

Species	Initial explant[a]	Medium, initiation and proliferation
P. abies	IE	1/2 LP
	ME	1/2 LP
	S	1/2 LP
P. engelmanni	IE	LP
P. glauca	IE	LP
	ME[e]	LP
	S	LP
P. glauca-engelmanni	IE	LP
	ME[e]	LP
P. glehni	ME	Lepoivre[b]
P. jezoensis	ME	Lepoivre
P. mariana	IE	LP
	ME	1/2 LM[c]
	S	LP
P. omorika	S	
P. pungens	ME	LP
P. rubens	ME	LP
P. sitchensis	IE	1/2 MS[d]
	ME	1/2 LP

[a]IE, immature embryos; ME, mature embryos; S, seedling material.
[b]Lepoivre medium ***(24)***.
[c]Litvays medium ***(25)***.
[d]Murashige-Skoog medium ***(26)***.
[e]Initiation in light.

cell lines produce many mature embryos, whereas type B cell lines seldom produce mature embryos. Aberrant mature embryos are common among most cell lines, and these embryos often fail to form a tap root.

The embryos are stimulated to mature by abscisic acid (ABA) ***(17)*** and a raised osmotic potential ***(18)***. The medium used for maturation is similar to that used for proliferation, but with different growth regulators, e.g., ABA, a raised osmotic potential and, in a few cases, auxin and/or cytokinin. Embryogenic cultures of conifers have until now been most commonly grown on a solidified medium, with the maturation treatments, as reported for different species, generally being developed for solid grown cultures. However, meth-

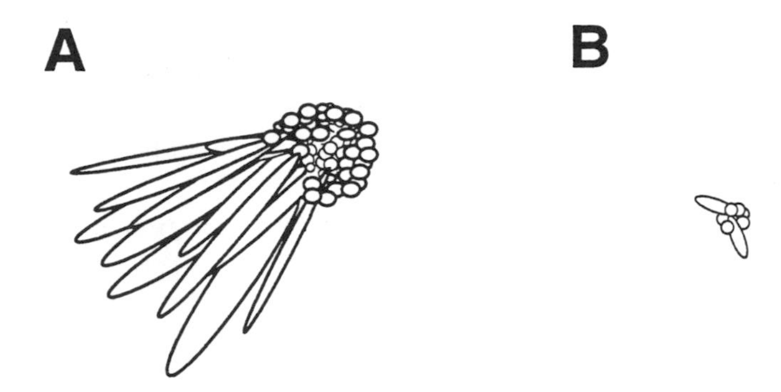

Fig. 1. Schematic drawing of the main types of somatic embryos that appear in suspension cultures of *Picea*. **(A)** Type A somatic embryo, **(B)** Type B somatic embryo.

ods for successful maturation and regeneration from suspension cultures have been reported for *P. glauca* ***(18,19)***, *P. abies* and *Picea sitchensis* ***(20,21)***, *Picea mariana* ***(22)***, and *Picea glauca-engelmanni* ***(23)***. Basically, the suspension is successively transferred to a prematuration medium lacking auxin and cytokinin. The embryos are then collected and transferred on a filter paper, to a medium containing ABA and a raised osmotic potential. The concentration of ABA required for maturation to start differs depending on the species and cell lines being used. Sucrose and/or polyethylene glycol (PEG) are commonly used to raise the osmotic potential. To improve further development, the mature embryos can then be partially desiccated. In *P. abies*, between 50 and 400 mature embryos can be produced/g of embryogenic tissue.

Regeneration: Plantlet regeneration from mature embryos is routine in many laboratories, although for a limited number of cell lines. A common problem is that the mature somatic embryos can only form cotyledons and a hypocotyl, but not a root.

The mature embryos are initially imbibed on a solidified medium under low light intensity. As the root develops, the plantlets can be transferred in an upright position to a vermiculite medium, then to nonsterile conditions in growth chambers, then to greenhouses, and finally to the field. Regenerated plants have been extensively compared to seed-derived plants. In field tests, *P. abies* plants derived from somatic embryos and from seeds flowered at the same time after 10 growth seasons.

Suspension cultures offer many advantages over solid-grown cultures, both in pure and applied research. Studies on the developmental regulation of proliferating somatic embryos provide information on embryo development in general and are also necessary to increase the number of genotypes that can be used for regeneration. The response of the somatic embryos to different environmental stimuli are best studied in suspension cultures as all embryos are exposed to the same conditions. Studies on their response to different compounds are easily undertaken, and samples removed from the suspension should represent the whole culture. For the application of somatic embryogenesis to forestry, suspension cultures using bioreactors is a prerequisite for future, large-scale plantlet production.

The following descriptions on initiation and proliferation are based on protocols optimized for *P. abies*. There are at present no general methods for maturation and regeneration that are likely to be successful in all *Picea* species and genotypes. Therefore, it is advisable to find current publications on the species of interest, which will then allow the development of suitable methods for maturation and regeneration of whichever cell lines are being used.

The time schedule for different steps involved may vary depending on the species and cell line. It is therefore important to follow carefully the growth and development of the embryos at the different stages.

1. Initiation and proliferation of embryogenic tissue:
   a. Sterilize seeds.
   b. Imbibe seeds.
   c. Isolate zygotic embryos, and initiate embryogenic tissue: 4–8 wk.
   d. Isolate embryogenic tissue from initial explant and subculture onto fresh medium.
   e. Proliferate on solid medium: 4–8 wk.
2. Start of suspension culture:
   a. Subculture into small aliquots and proliferate: 2–4 wk.
   b. Transfer to, and establish growth in, liquid medium: 4–12 wk.

2. Materials

1. Commercial bleach, Tween, and sterile water.
2. Flow cabinet, forceps, and scalpel.
3. Seeds (*see* **Note 1**).
4. Dissecting microscope.
5. Petri plates containing proliferation medium (**Table 1**).
6. Erlenmeyer flasks (*see* **Note 2**) containing proliferation medium (**Table 1**).
7. Wide-mouthed pipets (*see* **Note 3**).

3. Methods

3.1. Initiation of Embryogenic Tissue

1. Sterilize the seeds by soaking them in 15% commercial bleach (6% [v/v] sodium hypochlorite) with a few drops of Tween for 15 min with agitation. Remove the bleach solution, and rinse the seeds with plenty of sterile water (*see* **Note 4**).
2. Imbibe seeds overnight in distilled water at +4°C in darkness.
3. Isolate the zygotic embryos with a scalpel and forceps under a dissecting microscope. Put the embryos gently onto the proliferation medium, making sure that they are in good contact with the medium. Seal the plates with strips of cling film.

4. Leave the plates at room temperature under darkness. Subculture embryos onto fresh medium at least once a month. Continuously check for the formation of embryogenic tissue (*see* **Note 5**). When the somatic embryos have reached a size of 5 mm, the embryogenic tissue can be isolated and transferred onto fresh medium with sterile forceps. Avoid transferring the initial explant.
5. Subculture onto fresh medium at 2–4 wk intervals.
6. Take samples of the somatic embryos, and examine under a microscope to determine the morphology of the proliferating embryos (**Fig. 1**).

3.2. Establishing Suspension Cultures

1. Two to 4 wk after the last subculturing, when new embryogenic tissue has been formed, carefully remove small (3-mm) pieces of embryogenic tissue from the outer part of the embryogenic mass (*see* **Note 6**). Place 10 pieces on single Petri plates containing 25 mL of fresh, solid proliferation medium. Subculture onto fresh medium every 2 wk until new embryogenic tissue has formed.
2. Remove the embryogenic tissue masses from the Petri plate, and place them in 250-mL Erlenmeyer flasks containing 20 mL of liquid proliferation medium. Incubate on a rotary shaker at 50 rpm in darkness.
3. After two weeks, add 5 mL of fresh medium. (*see* **Note 7**).
4. After another two weeks, let the embryos settle out, use a wide-mouthed pipet to move the loosened embryos to a 250-mL Erlenmeyer flask containing 25 mL of proliferation medium, and incubate on a rotary shaker at 100 rpm. Avoid transferring the initial embryogenic tissue mass.
5. Subculture the embryos to fresh medium every week (*see* **Note 8**). Transfer 10 mL of the embryos to 50 mL of freshly made medium as described above. If growth and embryo morphology (*see* **Note 9**) are unsatisfactory, add mature seed extract (*see* **Note 10**).

4. Notes

1. Mature seeds are generally the most accessible and therefore the most commonly used as initial explant. However, the initiation frequency from immature seed is generally higher.
2. Glass or plastic Erlenmeyer flasks can be used for suspension cultures. When establishing suspension cultures, baffleflasks can be used to facilitate loosening of the embryos from the initial masses. All flasks should be capped with cotton plugs (optional) and aluminum foil.
3. Measuring pipets with wide mouths allow the transferring of the embryos with less damage, and they can be necessary to enable the transfer of larger embryos.
4. The seeds can be sterilized and rinsed in a Petri plate. If sterility is a problem, the seeds can be sterilized in a presterile Erlenmeyer flask. The seeds can then be poured onto a sterile nylon mesh, and rinsed while on the mesh.
5. The embryogenic tissue is white to translucent in appearance. Under the dissecting microscope, the embryos can be seen to branch out from the initial explant.

6. The embryogenic tissue is growing most actively on the periphery. Only transfer pieces of tissue from this area; rapidly proliferating embryos are more easy to establish in liquid medium.
7. The formation of new embryos in the suspension culture can stop if the suspension is diluted too much. However, a too high density will cause browning of the culture and eventually death of embryos. The optimal cell density varies between cell lines, and depends on the proliferation rate and the type of embryo. To become familiar with cell line requirements, it is advisable to view samples of the suspension under the microscope to determine the viability of the embryos present.
8. Depending on the type of embryos in the cell line, different subculturing intervals should be employed. Generally, type A requires subculturing weekly, whereas type B is insensitive to prolonged subculturing intervals and can be left for months in the same medium without loss of viability.
9. Cell lines composed of large, well-developed embryos (type A) can be difficult to establish as suspension cultures, whereas cell lines composed of small, less developed embryos (type B) are easy to establish. Embryos cultured in liquid medium generally develop fewer suspensor cells and tend to loose their polarized appearance.
10. Seed extracts can be prepared from mature seed by grinding 10 g of seed in a food mixer together with 100 mL of proliferation medium. Larger debris is removed by filtering through cheesecloth or Miracloth. The extract is filter-sterilized into the suspension cultures to a concentration equivalent to 50–100 mg of seeds/mL of medium. The treatment can be repeated over a few subculturing intervals, but this seems to be detrimental if applied for longer periods. The seed extract stimulates the growth and development of somatic embryos of both type A and B. Additionally, the frequency of the formation of mature embryos is increased dramatically in all cell lines tested.

References

1. Bajaj, Y. P. S. (1995) *Biotechnology in Agriculture and Forestry 30: Somatic Embryogenesis and Synthetic Seed I.* Springer Verlag, Berlin.
2. Hakman, I. and von Arnold, S. (1985) Plantlet regeneration through somatic embryogenesis in *Picea abies* (Norway spruce). *J. Plant Physiol.* **121,** 149–158.
3. Chalupa, V. (1985) Somatic embryogenesis and plantlet regeneration from cultured immature and mature embryos of *Picea abies* (L.) Karst. *Commun. Inst. For.* **14,** 57–63.
4. Jain, S., Gupta, P., and Newton, R. (1995) *Somatic Embryogenesis in Woody Plants.* Kluwer Academic Publishers, Dordrecht, The Netherlands.
5. Nörgaard, J., Duran, V., Johnsen, Ö., Krogstrup, P., Baldurson, S., and von Arnold, S. (1993) Variations in cryotolerance of embryogenic *Picea abies* cell lines and the associations to genetic, morphological and physiological factors. *Can. J. For. Res.* **23,** 2560–2567.

6. Robertson, D., Weissinger, A. K., Ackley, R., Glover, S., and Sederoff, R. R. (1992) Genetic transformation of Norway spruce (*Picea abies* [L.] Karst) using somatic embryo explants by microprojectile bombardment. *Plant Mol. Biol.* **19,** 925–935.
7. Bercetche, J., Dinant, M., Matagne, R. F., and Påques, M. (1994) Selection of transformed embryogenic tissues of *Picea abies* (L.) Karst after microprojectile bombardment, in *Abstracts VIIIth International Congress of Plant Tissue and Cell Culture*, Firenze, June 12–17, 1994.
8. Charest, P. J., Lachance, D., Devantier, Y., and Klimaszewska, K. (1993) Transient gene expression and stable genetic transformation in *Picea mariana* (black spruce) and *Larix laricines* (tamarack), in: *Abstracts Fifth International Workshop, IUFRO Working Party* S2.04–07, Spain.
9. Ellis, D. D., McCabe, D. E., McInnes, S., Ramachandran, R., Russell, D. R., and Wallace, K. M. (1993) Stable transformation of *Picea glauca* by particle acceleration. *Bio/Technology* **11,** 84–89.
10. Westcott, R. J. (1994) Production of embryogenic callus from nonembryogenic explants of Norway spruce (*Picea abies* [L.] Karst). *Plant Cell Rep.* **14,** 47–49.
11. von Arnold, S. and Eriksson, T. (1981) *In vitro* studies of adventitious shoot formation in *Pinus contorta*. *Can. J. Bot.* **59,** 870–874.
12. Egertsdotter, U. (1996) Regulation of somatic embryo development in Norway spruce (*Picea abies*). Doctoral thesis, Swedish University of Agricultural Sciences, Uppsala.
13. Bellarosa, R., Mo, L. H., and von Arnold, S. (1992) The influence of auxin and cytokinin on proliferation and morphology of somatic embryos of *Picea abies* (L.) Karst. *Ann. Bot.* **70,** 199–206.
14. Egertsdotter, U., Mo, L. H., and von Arnold, S. (1993) Extracellular proteins in embryogenic suspension cultures of Norway spruce (*Picea abies*). *Physiol. Plant.* **88,** 315–321.
15. Egertsdotter, U. and von Arnold, S. (1995) Importance of arabinogalactan proteins for the development of somatic embryos of Norway spruce (*Picea abies*). *Physiol. Plant.* **93,** 334–345.
16. Hakman, I. (1993) Embryology in Norway spruce (*Picea abies*). An analysis of the composition of seed storage proteins and deposition of storage reserves during seed development and somatic embryogenesis. *Physiol. Plant.* **87,** 148–159.
17. von Arnold, S. and Hakman, I. (1988) Regulation of somatic embryo development in *Picea abies* by abscisic acid (ABA). *J. Plant Physiol.* **132,** 164–169.
18. Attree, S. M., Pomeroy, M. K., and Fowke, L. C. (1995) Development of white spruce (*Picea glauca* [Moench.] Voss) somatic embryos during culture with abscisic acid and osmoticum, and their tolerance to drying and frozen storage. *J. Exp. Bot.* **285,** 433–439.
19. Attree, S. M., Pomeroy, M. K., and Fowke, L. C. (1992) Manipulations of conditions for the culture of somatic embryos of white spruce for improved triacylglycerol biosynthesis and desiccation tolerance. *Planta* **187,** 395–404.

20. Krogstrup, P. (1990) Effect of cellular densities on cell proliferation and regeneration from embryogenic cell suspensions of *Picea abies*. *Plant Sci.* **72,** 115–123.
21. Krogstrup, P., Eriksen, E. N., Møller, J. D., and Roulund, H. (1988) Somatic embryogenesis in sitka spruce (*Picea sitchensis* [Bong.] Carr.). *Plant Cell Rep.* **7,** 594–597.
22. Tautorus, T. E., Lulsdorf, M. M., Kikcio, S. I., and Dunstan, D. I. (1992) Bioreactor culture of *Picea mariana* Mill. (black spruce) and the species complex *Picea glauca-engelmannii* (interior spruce) somatic embryos. Growth parameters. *Appl. Microbiol. Biotechnol.* **38,** 46–51.
23. Tautorus, T. E., Lulsdorf, M. M., Kikcio, S. I., and Dunstan, D. I. (1994) Nutrient utilization during bioreactor culture, and maturation of somatic embryo culture of *Picea mariana* and *Picea glauca-engelmannii. In Vitro Cell Dev. Biol.* **30,** 58–65.
24. Aitken-Christie, J. and Thorpe, T. A. (1984) Clonal propagation: gymnosperms. in *Cell Culture and Somatic Cell Genetics of Plants* (Vasil, I. K., ed.), Academic, London, pp. 82–95.
25. Litvay, J. D., Johnson, M. A., Verma, D., Einspahr, D., and Weyrauch, K. (1981) Conifer suspension culture medium development using analytical data from developing seeds. *IPC Technical Paper Series* **115,** 1–17.
26. Murashige, T. and Skoog, F. (1962) A revised medium for rapid growth and bioassays with tobacco tissue cultures. *Physiol. Plant.* **15,** 473–497.

6

Direct Cyclic Somatic Embryogenesis of Cassava for Mass Production Purposes

Krit J. J. M. Raemakers, Evert Jacobsen, and Richard G. F. Visser

1. Introduction

Cassava (*Manihot esculenta* Crantz) is a perennial shrub of the same family, the Euphorbiaceae, as castor bean (*Ricinus communis*) and rubber tree (*Hevea brasiliensis*). It is grown for its starch-containing tuberized roots, which are used for human consumption, as animal feed, and as raw material for the starch industry. The centers of diversity are in central and North eastern Brazil, southwestern Mexico, and eastern Bolivia *(1)*. Cassava is also grown on the African and Asian continents *(2)*. The average yield in the world is 8.8 tonnes/hectare *(3)*, which is only a fraction of the potential yield of 90 tonnes/hectare *(4)*. One of the reasons for this low yield is the use of cuttings infected with diseases and pests as the starting material.

Meristem culture *(5,6)* is used to produce healthy planting material (*see* **Note 1**), and multiple shoot culture *(7)* is used to produce large numbers of healthy plants. However, this multiple-shoot culture technique requires several tissue-culture steps (mechanical isolation of cuttings, rooting of cuttings, and hardening of plants), which makes the procedure labor-intensive. Somatic embryos have the developmental program to grow into complete plants without mechanical isolation and separate shooting and rooting steps *(8)*. Therefore, the multiplication of planting material by somatic embryogenesis would reduce labor input significantly.

In cassava, both a direct and indirect form of somatic embryogenesis exists. In the direct form, the somatic embryos develop directly into mature somatic embryos with large green cotyledons (**Fig. 1A–D**). In the indirect form of somatic embryogenesis, also called friable embryogenic callus, the embryogenic propa-

From: *Methods in Molecular Biology, Vol. 111: Plant Cell Culture Protocols*
Edited by: R. D. Hall © Humana Press Inc., Totowa, NJ

gules do not develop beyond the preglobular stage, but instead break up into new embryogenic propagules (*see* **Note 2**).

Both types of somatic embryogenesis start with the culture of leaf explants on Murashige and Skoog *(9)* medium supplemented with auxins *(10–17)*. Clumps of globular embryos (**Fig. 1A**) are formed after 10–14 d of culture. Transfer of these clumps with globular embryos to a Greshoff and Doy medium *(18)* supplemented with auxins initiates friable embryogenic callus *(16)*, and transfer of these clumps to a Murashige and Skoog *(9)* medium without auxins allows the globular embryos to develop into mature somatic embryos (**Fig. 1B;** *10–15,17*).

Somatic embryogenesis can only be used for plant multiplication if there is an efficient system of plant regeneration. In the direct form of somatic embryogenesis, mature embryos germinate efficiently into plants (**Fig. 1E;** *12,19,20*). In the indirect form, the friable embryogenic callus should first produce mature somatic embryos (maturation) before they can be germinated. However, because maturation occurs at very low frequencies *(16,20)*, friable embryogenic callus formation cannot be used for efficient plant multiplication.

Another advantage of direct somatic embryogenesis in cassava is that mature somatic embryos are ideal explants for direct secondary/cyclic somatic embryogenesis, thus allowing the production of millions of propagules *(11,19,21–25)*. For this, mature somatic embryos are chopped into small pieces and are subcultured in a medium supplemented with auxins. Both liquid and solid medium can be used. Depending on the auxin used, after 2–4 wk, new mature embryos are formed, which can then be used for a new cycle of secondary somatic embryogenesis (*see* **Note 3**).

This chapter describes methods to initiate somatic embryogenesis in elite genotypes of cassava, the subsequent multiplication of the embryogenic tissue by secondary/cyclic somatic embryogenesis, the germination of somatic embryos into plants, and finally, the transfer to the greenhouse. An overview of the whole procedure is given in **Table 1**.

2. Materials

2.1. General Requirements

1. Rotary shaker (120 rpm).
2. Sterile 300-mL Erlenmeyer flasks, capped with aluminum foil.
3. Sterile tubes and Petri plates.
4. Bottles (0.5, 1 L) that can be autoclaved.
5. Microwave oven to dissolve solid medium.
6. pH meter.
7. 1 *M* KOH and 0.1 *M* HCl for setting pH of the medium.
8. Pipet (100–1000 μL) with sterile tips to add stock solutions of growth regulators to the medium.

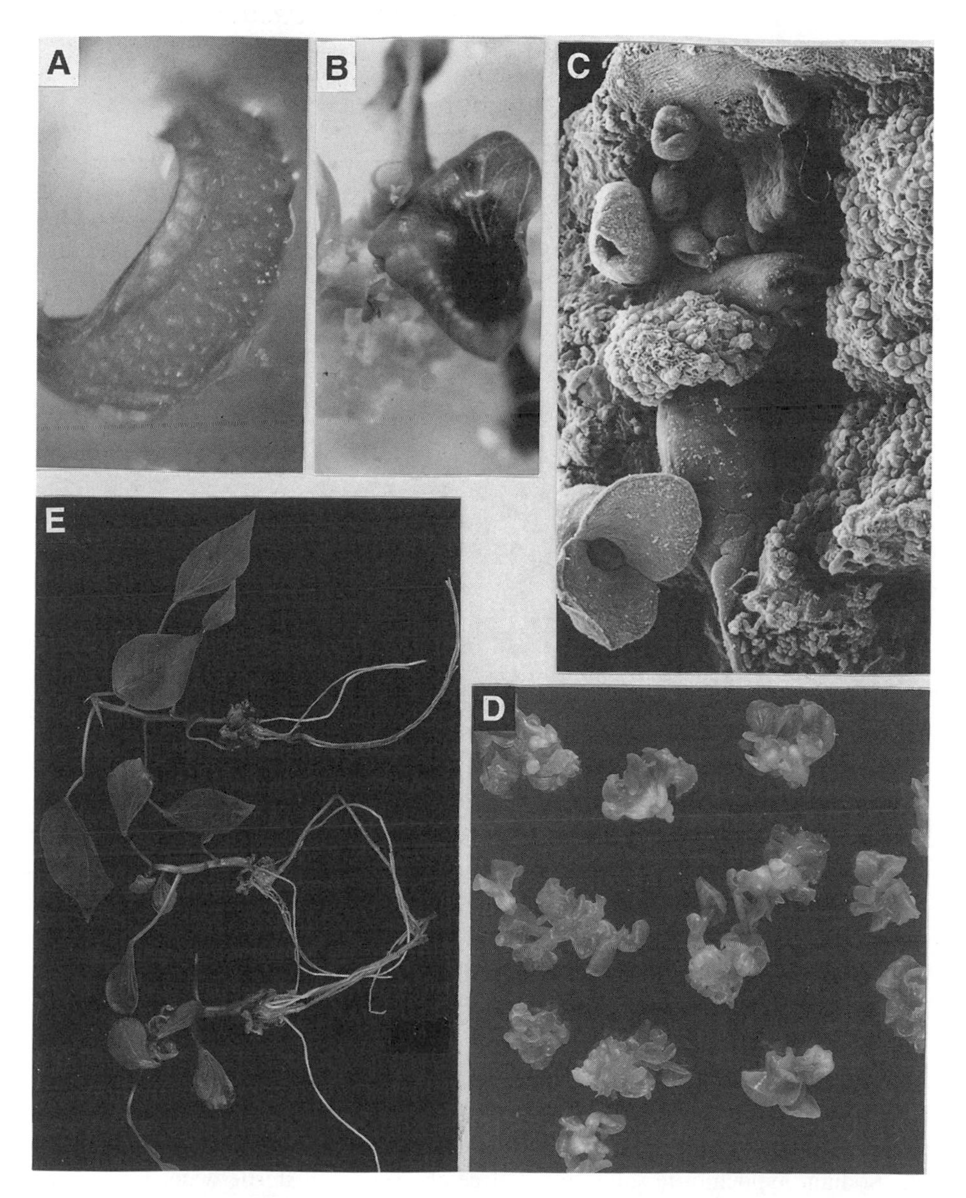

Fig. 1. Somatic embryogenesis in cassava (*M. esculenta* Crantz): **(A)** formation of globular somatic embryos on leaf explant cultured for 10 d on solid MS2 + 8 mg/L 2,4-D medium, **(B)** formation of mature somatic embryos on leaf explant cultured for 21 d on solid MS2 + 8 mg/L 2,4-D medium followed by 10 days on solid MS2 + 0.1 mg/L BAP medium, **(C)** scanning electron microscopy of the formation of secondary somatic embryos formed on primary somatic embryo cultured for 15 d on solid MS2 + 8 mg/L 2,4-D medium, **(D)** mature cyclic somatic embryos maintained in liquid MS2 + 10 mg/L NAA medium, **(E)** germinated cyclic somatic embryos (induced in liquid NAA supplemented medium and cultured after desiccation on solid MS2 + 0,1 mg/L BAP medium).

Table 1
Characteristics of the Culture Procedures Used in Somatic Embryogenesis in Cassava

Step	Medium	Transfer period, wk	Genotype dependency[a]	Refs.
Transfer of plants to in vitro	—	—	28/28	—
Multiplication of plants	MS2	3–25	28/28	—
Isolation of leaf explants	—	—	—	***14,27***
Primary somatic embryogenesis	MS2 + 8 mg/L 2,4-D[b]	3	—	—
	MS2+0.1 mg/L BA[b]	2–3	24/28	***15,25***
Secondary/cyclic somatic embryogenesis				
establishment	1st: MS2 + 8 mg/L 2,4-D[b]	—	—	
	2nd: MS2 + 0.1 mg/L BA[b]	1–2	19/20	***15,25***
maintenance	1st: MS2 + 8 mg/L 2,4-D[c]	3	—	—
	2nd: MS2 + 0.1 mg/L BA[c]	1–2	19/20	***15,25***
	or			
	MS2 + 10 mg/L NAA[c]	2	13/20	***19***
Germination	MS2 + 0.1–1.0 mg/L BA[b]	1	7/7	***19,20***

[a]Genotype dependency: number of successful genotypes/number of tested genotypes.
[b]Only solid medium can be used.
[c]Both solid and liquid media can be used.

9. 100-mL volumetric flask, for preparation of stock solutions of the growth regulators.
10. 0.2-µ*M* filter units for sterilization of stock solutions.
11. Dimethyl sulfoxide for dissolving the growth regulator picloram.
12. 1 *M* NaOH for dissolving the growth regulators NAA, 2,4-D, and BAP.
13. Sodium hypochlorite solution (1% free chlorine) and sterile water for sterilization of plant material.
14. Scalpel and forceps.
15. Binocular microscope for isolation of explants.
16. Sterile transfer facilities.
17. Refrigerator for storage of stock solutions.
18. Growth chamber: temperature 30°C, day length of 12 h, light intensity of 40 µmol/m^2/s.
19. Rock wool plugs.
20. Sterile soil and an incubator for transfer of plants from in vitro to the greenhouse.

Table 2
Preparation of Stock Solutions of Growth Regulators

	BAP	NAA	Picloram	2,4-D
Desired concentration	1 mg/mL	10 mg/mL	10 mg/mL	8 mg/mL
Powder for 100 mL	100 mg	1000 mg	1000 mg	800 mg
Solvent	1 *M* NaOH	1 *M* NaOH	DMSO	1 *M* NaOH
Storage of stocks	4°C	–20°C	–20°C	–20°C

2.2. Plant Materials and Culture Media

1. Cassava plants grown in the greenhouse.
2. Distilled water.
3. Preparation of stock solutions of the growth regulators (details for each growth regulator are given in **Table 2**): Add growth regulator to a 100-mL volumetric flask, dissolve the powder in a 0.2–25 mL of solvent, fill volumetric flask to 100 mL with distilled water, sterilize solution by filtering it through a double 0.2-μM filter unit, dispense sterile solution in 1-mL Eppendorf tubes and store Eppendorf tubes.
4. Preparation of MS2-medium: Add 800 mL to 1-L bottle, and dissolve Murashige and Skoog *(9)* basal salts and vitamins, and 20 g sucrose. Set pH to 5.7 by adding droplets of 1 *M* KOH or 0.1 *M* HCl, fill bottle to 1 L, autoclave it for 20 min at 121°C and 1.2 kg/cm^2. In the case of solid medium, 8 g/L agar is added after setting the pH. After sterilization, the bottles are stored in a dark cabinet. Before use in tissue culture, the solid medium is melted in a microwave oven, and growth regulators are added. The volume of the stock solution added to the medium can be calculated using the following formula:

   [Concentration of growth regulator in medium (mg/L) × volume of the medium (L)]/concentration of stock solution (mg/mL) = volume of stock solution required for medium (mL) (1)

   The concentration of stock solutions is chosen, in such a way that in most cases, 1 mL is needed for 1 L of MS2 medium. An exception is the MS2 + 6 mg/L NAA and 6 mg/L Picloram medium used for primary somatic embryogenesis by Taylor (personal communication): 0.6 mL of both 10 mg/mL NAA and 10 mg/mL Picloram are added to 1 L of MS2 medium.

3. Methods

3.1. Transfer of Plants to In Vitro and Multiplication

1. Isolate the upper single nodes (1–3 cm) from healthy and vigorously growing plants, preferably on a warm and sunny day.

2. Transfer each node to one tube with a 1% solution of sodium hypochlorite (20 min). Wash each node three times with sterilized, distilled water.
3. Remove discolored parts of the stem and transfer nodes to tubes with solid MS-2 medium.
4. Multiply plant material in-vitro every 3 wk by single-node cuttings until there are about 30 single plants.

3.2. Isolation of Leaf Explants and Culture for Primary Somatic Embryogenesis

1. Transfer single node cuttings to liquid MS2 medium. The plants are kept upright using rock wool plugs. After 12 d of growth, add 2,4-D to the liquid medium to a final concentration of 8 mg/L. The shoot tips are harvested 2 d later.
2. Under a binocular microscope, remove leaves larger than 4 mm from the shoot tips. Isolate single leaves smaller than 4 mm (if possible, divide them into single leaf lobes) and the remains of the shoot tip and culture these in the dark on solid MS2 + 8 mg/L 2,4-D medium (*see* **Notes 4** and **5**).
3. After 14–21 d transfer the explants to solid MS2 + 0.1 mg/L BA medium in the light, and subculture every 2–3 wk.
4. Cut off with a scalpel mature primary somatic embryos. Mature somatic embryos possess large green cotyledons (**Fig. 1B–D**).

3.3. Establishment and Maintenance of Secondary/Cyclic Somatic Embryogenic Cultures

1. Cut with a scalpel the mature primary somatic embryos into fragments of 1-6 mm^2, and culture on solid MS2 + 8 mg/L 2,4-D medium for 20 d. Then transfer to solid MS2 + 0.1 mg/L BA medium.
2. Harvest again the mature somatic embryos, cut into small fragments, and culture 0.2 g in 300-mL Erlenmeyer flasks containing 50 mL of liquid MS2 + 8 mg/L 2,4-D or 10 mg/L NAA medium. The Erlenmeyer flasks are cultured in the light on a shaker at 120 rpm.
3. After 21 d, transfer the material cultured in 2,4-D-supplemented medium to liquid MS2 + 0.1 mg/L BA medium (*see* **Notes 6** and **7**).
4. Every 14 d (NAA-supplemented medium) or 30 d (2,4-D-supplemented medium) mature somatic embryos can be isolated, and the procedure described in **steps 1**, **2**, and **3** can be repeated.

3.4. Germination of Cyclic Somatic Embryos

1. Mature somatic embryos maintained in NAA-supplemented medium can be used directly for germination (*see*, for example, **Fig. 1D**). Mature embryos maintained in 2,4-D-supplemented medium must be multiplied first for at least 1 mon in NAA-supplemented medium (*see* **Note 8**).
2. Dry the mature embryos first on sterile filter paper in the flow hood. After that, transfer them to an empty sterile Petri dish for desiccation in the growth cabinet. The lid of the Petri dish is changed every other day to remove condensed water.

3. After 4–8 d, well-desiccated mature embryos are cultured in the light on solid MS2 medium supplemented with 0.1–1 mg/L BA (*see* **Note 9**).
4. The mature somatic embryos will first form roots and then shoots. Depending on the quality of the roots, transfer the plants directly to the greenhouse (for example, plants shown in **Fig. 1E**) or after one round of multiplication on solid MS2 medium.

3.5. Transfer of Plants to the Greenhouse

1. Plantlets, 2–4 cm tall, can be transferred to the greenhouse. Remove the plantlets carefully from the tubes or containers, and wash away the agar from the roots thoroughly.
2. Grow plants in pots (1:3 sterilized mixture of soil with fine sand), and keep at high humidity in an incubator. After 1 wk, the plants can be acclimatized to the greenhouse conditions by gradual reduction of the humidity.

4. Notes

1. Healthy plants can also be obtained by growing them in vitro under a day/night temperature of 40/35°C and subculturing the shoot tips every 12 d ***(26)***. Another option is to combine meristem culture with primary somatic embryogenesis by using meristems as explants. Different methods are available to screen the plants for diseases (for details, *see 27*).
2. Friable embryogenic callus has been shown to be an excellent source of tissue to obtain transgenic plants ***(28–30)*** and to obtain regenerable protoplasts ***(31)***, which can also be used to obtain transgenic plants by electroporation (Raemakers et al., unpublished results).
3. In a well established culture, one mature somatic embryo multiplied in liquid medium via cyclic somatic embryogenesis will yield more than 20 new mature somatic embryos in 2–4 wk, resulting in 20^{13}–20^{26} mature somatic embryos a year.
4. Leaf explants from greenhouse-grown plants can also be used as the starting material for primary somatic embryogenesis. However, it has been shown that the embryogenic response of these leaf explants depends on the growth conditions in the greenhouse ***(25)***.
5. Other auxins, such as Dicamba and Picloram, can also be used to induce primary somatic embryogenesis ***(13)***. However, in our laboratory, these media were not as efficient as the procedure described in **Subheading 3.2.** Another option (Taylor, personal communication) is to culture leaf explants (harvested from donor plants grown on MS2 medium) on solid MS2 + 6 mg/L NAA + 6 mg/L Picloram medium to induce primary somatic embryogenesis. In our laboratory this method proved also to be very efficient.
6. The developmental range at which somatic embryos can be used to initiate new cyclic somatic embryogenesis varies from globular to young mature somatic embryos (light green cotyledons) for NAA-supplemented medium and from globular to mature somatic embryos (with dark green cotyledons) for 2,4-D-supplemented medium ***(19)***.

7. In our laboratory, it is found that in about 70% of the genotypes, NAA-supplemented medium can be used to maintain cyclic embryogenic cultures. In the other 30%, prolonged culture in NAA-supplemented medium resulted in root formation ***(19)***. In the first instance, cultures will contain somatic embryos with tap-roots, but later this will shift to the formation of adventitious roots. Such cultures can be induced to be embryogenic again by isolating the remaining embryogenic tissue and culturing this for cyclic/secondary somatic embryogenesis in MS2 medium + 8 mg/L 2,4-D medium.
8. Desiccation has no positive effect on germination of cyclic somatic embryos, which are initiated in liquid MS2 medium + 8 mg/L 2,4-D. Desiccation only stimulates root formation ***(19,20)***. However, Mathews et al. ***(12)*** showed that secondary somatic embryos that were maintained on solid MS2 medium + 4 mg/L 2,4D germinated properly after desiccation.
9. The desiccation procedure described in **Subheading 3.4.** is used in our laboratory. Each laboratory has to find the optimal conditions for their own material. Furthermore, the procedure described in **Subheading 3.4.**, can only be used for small amounts of somatic embryos. The method has to be adjusted for large-scale desiccation. In our hands, mature somatic embryos, which had lost 80% of their fresh weight, germinated at the highest frequencies ***(20)***. These optimally desiccated mature secondary somatic embryos require a lower level of BA (0.1 mg/L) for germination than suboptimally desiccated ones. Higher BA levels (>1 mg/L) stimulate the formation of thick roots and highly branched plantlets with short internodes. Because of these characteristics, such plantlets cannot be transferred directly to the greenhouse. It has also been found that germination in the dark decreases the level of BA needed for normal germination ***(19,20)***.

References

1. Nasser, N. M. A. (1978) Conservation of genetic resources of cassava (*Manihot esculenta*): determination of wild species localization with emphasis on possible origin. *Econ. Bot.* **32,** 311–320.
2. Byrne, D. (1984) Breeding cassava, in *Plant Breeding Reviews,* vol. 2 (Janick, J., ed.), AVI, Westport CT, pp. 73–134.
3. FAO (1993) *Yearbook 1992*. Rome, pp 101–102.
4. Cock, J. H. (ed.) (1985) *Cassava: New Potential for a Neglected Crop.* Wetview Press, Boulder and London.
5. Kartha, K. K. and Gamborg, O. L. (1975) Elimination of cassava mosaic virus disease by meristems. *Plant Sci. Lett.* **2,** 107–113.
6. Roca, W. M. (1984) Cassava, in *Handbook of Plant Cell Culture,* vol. 2 (Sharp, W. R., Evans, D. A., Ammirato, P. V., and Yamada, Y., eds.), Macmillan, New York, pp. 269–301.
7. Smith, M. K., Biggs, B. J., and Scott, K. J. (1986) In vitro propagation of cassava (*Manihot esculenta* Crantz). *Plant Cell Tiss. Org. Cult.* **6,** 221–229.
8. Parrot, W. A., Merkle, S. A., and Williams, E.G (1991) Somatic embryogenesis: potential for use in propagation and gene transfer systems, in *Advanced Methods*

in Plant Breeding and Biotechnology (Murray, D. R., ed.), CAB Int. Wallingford, Oxon UK, pp. 158–200.

9. Murashige, T. and Skoog, F. (1962) A revised medium for rapid growth and bioassay with tobacco cultures. *Physiol. Plant.* **15,** 473–497.
10. Stamp, J. A. and Henshaw, G. G. (1987) Somatic embryogenesis from clonal leaf tissue of cassava. *Ann. Bot.* 59, 445–450.
11. Szabados, L., Hoyos, R., and Roca, W. (1987) *In vitro* somatic embryogenesis and plant regeneration of cassava. *Plant Cell Rep.* **6,** 248–251.
12. Mathews, H., Schöpke, C., Carcamo, R., Chavarriaga, P., Fauquet, C., and Beachy, R. N. (1993) Improvement of somatic embryogenesis and plant regeneration in cassava. *Plant Cell Rep.* **12,** 328–333.
13. Sudarmonowati, E. and Henshaw, G. G. (1993) The induction of somatic embryogenesis of recalcitrant cassava cultivars using Picloram and Dicamba, in *Proceedings of the First International Scientific Meeting of the Cassava Biotechnology Network* (Roca, W. M. and Thro, A. M., eds.), Centro International de Agricultura Tropical, Cartagena de Indias, Columbia, pp. 128–134.
14. Raemakers, C. J. J. M., Bessembinder, J., Staritsky, G., Jacobsen, E., and Visser, R. G. F. (1993) Induction, germination and shoot development of somatic embryos in cassava. *Plant Cell Tiss. Org. Cult.* **33,** 151–156.
15. Sofiari, E. (1996) Regeneration and transformation of cassava. PhD thesis, Wageningen Agricultural University, Netherlands.
16. Taylor, N. J., Edwards, M. Kiernan, R. J., Davey, C. D. M., Blakesley, D., and Henshaw, G. G. (1996) Development of friable embryogenic callus and embryogenic suspension culture systems in cassava (*Manihot esculenta* Crantz). *Nature Biotechnol.* **14,** 726–730.
17. Taylor, N. J., and Henshaw, G. G., (1993) The induction of somatic embryogenesis in 15 African and one South American cassava cultivars, in *Proceedings of the First International Scientific Meeting of the Cassava Biotechnology Network* (Roca, W. M. and Thro, A. M., eds.), Centro International de Agricultura Tropical, Cartagena de Indias, Columbia, pp. 229–240.
18. Gresshoff, P. M. and Doy, C. H. (1974) Development and differentiation of haploid *Lycopersicon esculentum* (tomato). *Planta* **107,** 161–170.
19. Sofiari, E., Raemakers, C. J. J. M., Kanju, E., Danso, K., van Lammeren, A. M., and Jacobsen, E. (1997), Comparison of NAA and 2,4–D induced somatic embryogenesis in cassava. *Plant Cell Tiss. Org. Cult*, in press.
20. Raemakers, C. J. J. M., Rozenboom, M. G. M., Danso, K., Jacobsen, E., and Visser, R. G. F. (1997) Regeneration of plants from somatic embryos and friable embryogenic callus of cassava (*Manihot esculenta* Crantz). *Afr. Crop Sci. J.*, in press.
21. Stamp, J. A. and Henshaw, G. G. (1987) Secondary somatic embryogenesis and plant regeneration in cassava. *Plant Cell Tiss. Org. Cult.* **10,** 227–233.
22. Raemakers, C. J. J. M., Amati, M., Staritsky, G., Jacobsen, E., and Visser, R. G. F. (1993) Cyclic somatic embryogenesis in cassava. *Ann. Bot.* **71,** 289–294.
23. Raemakers, C. J. J. M., Schavemaker, C. M., Jacobsen, E., and Visser, R. G. F. (1993) Improvements of cyclic somatic embryogenesis of cassava (*Manihot esculenta* Crantz). *Plant Cell Rep.* **12,** 226–229.

24. Li, H. Q., Huang, Y. W., Liang, C. Y., and Guo, J. Y. (1995) Improvement of plant regeneration from secondary somatic embryos of cassava, in *Proceedings of Second International Meeting of Cassava Biotechnology Network, Bogor, Indonesia, August 22–26*, Centro Internacional de Agricultura Tropical, Cali, Columbia, pp. 289–302.
25. Raemakers, C. J. J. M. (1993) Primary and cyclic somatic embryogenesis in cassava. PhD thesis Wageningen Agricultural University, Netherlands.
26. Villegas, L. and Bravato, M. (1990) Conservation *in vitro* of cassava germplasm, in *In Vitro methods for Conservation of Plant Genetic Resources* (Dodds, J., ed.), Chapman and Hall, London, pp. 111–121.
27. Raemakers, C. J. J. M., Jacobsen, E., and Visser, R. G. F. (1997) Micropropagation of *Manihot esculenta* Crantz (cassava), in *Biotechnology in Agriculture and Forestry,* vol. 39 (Bajaj, Y. P. S., ed.), Springer Verlag, Berlin, pp. 77–103.
28. Schöpke, C., Taylor, N., Carcamo, R., Konan, N. K., Marmey, P., and Henshaw, G. (1996) Regeneration of transgenic cassava plants (*Manihot esculenta* Crantz) from microbombarded embryogenic suspension cultures. *Nature Biotechnol.* **14,** 731–735.
29. Raemakers, C. J. J. M., Sofiari, E., Taylor, N., Henshaw, G. G., Jacobsen, E., and Visser, R. G. F. (1996) Production of transgenic cassava (*Manihot esculenta* Crantz) plants by particle bombardment using luciferase activity as selection marker. *Mol. Breeding* **2,** 339–349.
30. Snepvangers, S. C. H. J., Raemakers, C. J. J. M., Jacobsen, E., and Visser, R. G. F. (1997) Optimization of chemical selection of transgenic friable embryogenic callus of cassava using the luciferase reporter gene system. *Afr. Crop Sci. J.*, in press.
31. Sofiari, E., Raemakers, C. J. J. M., Bergervoet, J. E. M., Jacobsen, E., and Visser, R. G. F. (1997) Plant regeneration from protoplasts isolated from friable embryogenic callus of cassava. *Plant Cell Rep.*, in press.

7

Immature Inflorescence Culture of Cereals

A Highly Responsive System for Regeneration and Transformation

Sonriza Rasco-Gaunt and Pilar Barcelo

1. Introduction

Success in obtaining transgenic cereals depends largely on efficient plant regeneration protocols. It is not trivial to establish regenerable tissue cultures in cereals, since only immature tissues of a limited age range can be induced to regenerate efficiently. Generally, the tissue-culture factors that determine the success of regeneration and transformation in cereals are: genotype, choice of explant, physiological state of the explants/donor plants, culture conditions, and medium composition.

Although scutellar tissues are now conventionally used for cereal transformation, immature inflorescences also provide a practicable source of young tissues for somatic embryogenesis and plant regeneration. Inflorescence cultures of rice ***(1,2)***, maize ***(3)***, wheat ***(4–8)***, sorghum ***(9,10)***, barley ***(11)***, millet ***(12,13)***, rye ***(6,14,15)***, and *Triticeae* sp. (tritordeum ***[16]***) are documented. However, cereal transformation using inflorescence tissues is novel and has, so far, only been demonstrated by Barcelo et al. ***(17)***.

The advantages of using inflorescence explants for tissue culture and transformation over the use of scutellar tissues are:

1. Their isolation is easier and quicker, particularly if the embryo axis must be excised to prevent precocious germination in scutellum cultures.
2. Their physiological state seems to be relatively stable in response to plant growth conditions. Therefore, variability in regeneration is generally less than is observed with scutellum cultures.
3. It is quicker to grow donor plants for inflorescence culture, resulting in a more efficient use of greenhouse/growth space, and there is less time for pest and pathogen problems to develop.

From: *Methods in Molecular Biology, Vol. 111: Plant Cell Culture Protocols*
Edited by: R. D. Hall © Humana Press Inc., Totowa, NJ

4. Plants from genotypes that normally require vernalization may not require vernalization when regenerated from inflorescences ***(18)***.
5. In certain genotypes, inflorescence tissues may be more amenable for culture and/or transformation than scutellar tissues.
6. Inflorescence tissues seem to withstand better the damage incurred by gene delivery via particle bombardment.

The limitation of inflorescence culture is that only a narrow range of cereal genotypes are responsive either to both tissue culture and transformation or to the latter alone.

This chapter presents procedures for the tissue culture (embryogenic callus induction and regeneration) and transformation via particle bombardment of tritordeum and wheat using immature inflorescence tissues. It includes recent findings and modifications to the authors' original protocol ***(17)***.

2. Materials

2.1. Donor Plant Materials and Growth Conditions

To produce inflorescences, grow plants of tritordeum (a fertile cereal amphiploid obtained from crosses between *Hordeum chilense* and durum wheat cultivars, containing the genome $H^{CH}H^{CH}AABB$ ***[19]***) and wheat (*Triticum aestivum*, e.g., variety Baldus [*see* **Note 1**]) in growth rooms under a 16-h photoperiod, air temperatures of 18–20°C during the day and 14°C at night, and at 50-70% relative humidity. If using winter wheat varieties, vernalize imbibed seeds (sown in vermiculite) for a period of 8 wk at 5–6.5°C prior to transferring to the above conditions (*see* **Note 2**).

2.2. Tissue-Culture Media Components and Other Solutions

2.2.1. Culture Media Components

1. L7 Macrosalts (10X stock): 2.5 g NH_4NO_3, 15.0 g KNO_3, 2.0 g KH_2PO_4, 3.5 g $MgSO_4 \cdot 7H_2O$, 4.5 g $CaCl_2 \cdot 2H_2O$. Dissolve salts separately in a small volume of distilled water, mix together, and make up to 1 L. Autoclave and store at 4°C (*see* **Note 3**).
2. L Microsalts (1000X stock): 15.0 g $MnSO_4$, 5.0 g H_3BO_3 , 7.5 g $ZnSO_4 \cdot 7H_2O$, 0.75 g KI, 0.25 g $Na_2MoO_4 \cdot 2H_2O$, 0.025 g $CuSO_4 \cdot 5H_2O$, 0.025 g $CoCl_2 \cdot 6H_2O$. Make up to 1 L with distilled water. Filter-sterilize through a 0.2-µm medium filter (e.g., MediaKap, Laguna Hills, CA) and store at 4°C.
3. Ferrous sulfate/chelate (100X stock, Sigma F-0518, UK).
4. L Vitamins/inositol (200X stock): 40.0 g myoinositol, 2.0 g thiamine HCl, 0.20 g pyridoxine HCl, 0.20 g nicotinic acid, 0.20 g Ca-pantothenate, 0.20 g ascorbic acid. Make up to 1 L with distilled water. Store solution at –20°C in 10 mL aliquots.

5. Amino acids, 3AA (25X stock): 18.75 g L-glutamine, 3.75 g L-proline, 2.50 g L-asparagine. Make up to 1 L with distilled water. Store solutions at –20°C in 40 mL aliquots.
6. Gelling agent: Agargel—make (10 g/L or 2X concentrated stock, Sigma A-3301, UK), agarose—make (8 g/L or 2X stock): Prepare stock solutions of Agargel and agarose. Autoclave and store at room temperature (*see* **Note 4**).
7. Silver Nitrate (20 mg/mL stock): Prepare $AgNO_3$ solution and filter-sterilize through a 0.2-μm syringe filter. Store solutions at –20°C in 1 mL aliquots. Alternatively, silver thiosulfate (STS) may be used in place of $AgNO_3$ (*see* **Note 5**).
8. Zeatin Mixed Isomers (10 mg/mL stock): Prepare solution by dissolving zeatin in a few drops of 1 *M* HCl. Make up to concentration with distilled water. Filter-sterilize through a 0.2-μm syringe filter. Store solutions in 1-mL aliquots at –20°C.
9. 2,4-Dichlorophenoxyacetic acid, 2,4-D (1 mg/mL stock): Prepare solution by dissolving 2,4-D in a few drops of 1 *M* NaOH. Make up to concentration with distilled water. Filter-sterilize through a 0.2-μm syringe filter. Store solutions at –20°C in 2-mL aliquots (*see* **Note 6**).
10. Picloram (1 mg/mL stock): Prepare solution by dissolving in a few drops of 1 *M* NaOH. Make up to concentration with distilled water. Filter-sterilize through a 0.2-μm syringe filter. Store solutions in 2-mL aliquots at –20°C (*see* **Note 6**).

2.2.2. Culture Media

1. L7-V liquid medium (2X concentrated stock): 200 mL L7 macrosalts (10X), 2 mL L microsalts (1000X), 20 mL ferrous sulfate/chelate (Sigma F-0518, 100X), 0.4 g myoinositol, 80 mL 3AA (25X), 60 or 180 g maltose (*see* **Note 7**). Make up to 1 L with distilled water. Adjust pH to 5.7. Filter-sterilize through a 0.2-μm medium filter and store at 4°C.
2. R liquid medium (2X concentrated stock): 200 mL L7 macrosalts (10X), 2 mL L microsalts (1000X), 20 mL ferrous sulfate/chelate (Sigma F-0518, 100X), 10 mL vitamins/inositol (200X), 60 g maltose. Make up to 1 L with distilled water. Adjust pH to 5.7. Filter-sterilize through a 0.2-μm medium filter, and store at 4°C.
3. Callus induction medium: To prepare 800 mL, mix 400 mL L7-V liquid medium (2X), 0.4-1.6 mL of 1 mg/mL 2,4-D stock solution for wheat culture or 3.2 mL of 1 mg/mL Picloram stock solution for tritordeum culture, 0.4 mL 20 mg/mL $AgNO_3$ stock, and 400 mL Agargel (10 g/L) or 400 mL agarose (8 g/L). Pour medium into sterile tissue-culture Petri dishes (*see* **Notes 8–10**).
4. Regeneration medium (RZ): To prepare 800 mL, mix 400 mL R liquid medium (2X), 80 μL 2,4-D stock, 0.4 mL zeatin stock, 0.4 mL $AgNO_3$ stock, 400 mL Agargel, or agarose. Pour medium into sterile Petri dishes or into Magenta boxes (*see* **Notes 9** and **10**).
5. Regeneration medium (RO): To prepare 800 mL, mix 400 mL R liquid medium (2X) and 400 mL Agargel or agarose. Pour medium into sterile Petri dishes or into Magenta boxes (*see* **Note 10**).

2.2.3. Solutions for DNA Delivery and Selection

1. Gold particle mixture: Suspend 40 mg gold powder (*see* **Note 11**) in 1 mL of absolute ethanol, and sonicate for 2 min until particles are clearly dispersed. Centrifuge briefly (5 s), and discard supernatant. Repeat ethanol wash twice. Resuspend gold in 1 mL of sterile distilled water, and sonicate for 2 min. Centrifuge for 5 s, and discard supernatant. Repeat the process once, and resuspend finally in 1 mL sterile distilled water. Aliquot in 50-µL vol/Eppendorf tube, mixing between aliquots to ensure an equal distribution of particles. Store at –20°C.
2. Calcium chloride (2.5 *M* stock): Prepare solution and filter-sterilize through a 0.2-µm syringe filter. Store at 4°C in 55-µL aliquots.
3. Spermidine solution (0.1 *M* stock): Spermidine is supplied as a 0.92 g/mL density solution. Take 15.8 µL of solution, and add sterile distilled water to give a total volume of 1 mL. Store at –20°C in 25-µL aliquots. Do not reuse thawed aliquots (*see* **Note 12**).
4. Plasmid DNA (1 µg/µL stock): Prepare plasmid DNA using a Qiagen purification kit. Resuspend DNA pellet in sterile distilled water or TE buffer (10 m*M* Tris and 1 m*M* EDTA, pH 8.0), and adjust concentration as appropriate (*see* **Note 13**).
5. Selection agents: Glufosinate ammonium (10 mg/mL stock), Bialaphos (10 mg/mL stock), geneticin disulfate (G418) (50 mg/mL stock), paromomycin sulfate (100 mg/mL stock). Prepare solutions, and filter-sterilize through a 0.2-µm syringe filter. Store solutions at –20°C in aliquots (*see* **Note 6**).

2.2.4. Histochemical GUS Assay Solutions

1. Sodium-phosphate buffer, pH 7.0 (0.5 *M* stock): Prepare 1 *M* Na_2HPO_4 and 1 *M* NaH_2PO_4. Mix 58 mL 1 *M* Na_2HPO_4 and 42 mL 1 *M* NaH_2PO_4. Make up to 200 mL with distilled water. Store solution at –20°C in 20-mL aliquots.
2. Potassium ferricyanide (50-m*M* stock): Prepare solution, and store at –20°C in 10-mL aliquots.
3. Potassium ferrocyanide (50-mM stock): Prepare solution, and store at –20°C in 10-mL aliquots.
4. X-Gluc buffer: To prepare 100 mL, dissolve 50 mg X-Gluc (5-bromo-4-chloro-3-indolyl-β-D-glucuronide, cyclohexylammonium salt) in 1 mL methyl cellusolve or dimethylformamide. Add the following solutions: 20 mL 0.5 *M* sodium-phosphate buffer, pH 7.0, 1 mL each of 50 m*M* potassium ferricyanide and potassium ferrocyanide and 77 mL distilled water. Filter-sterilize through a 0.2-µm syringe filter and store at –20°C in 2-mL aliquots.

2.3. Additional Requirements

1. 1 *M* and 5 *M* NaOH or KOH and HCl.
2. 70% Absolute ethanol.
3. 10% Domestos commercial bleach (*see* **Note 15**).
4. Sterile water.
5. Culture room conditions: 16-h photoperiod, 25 ± 1°C.

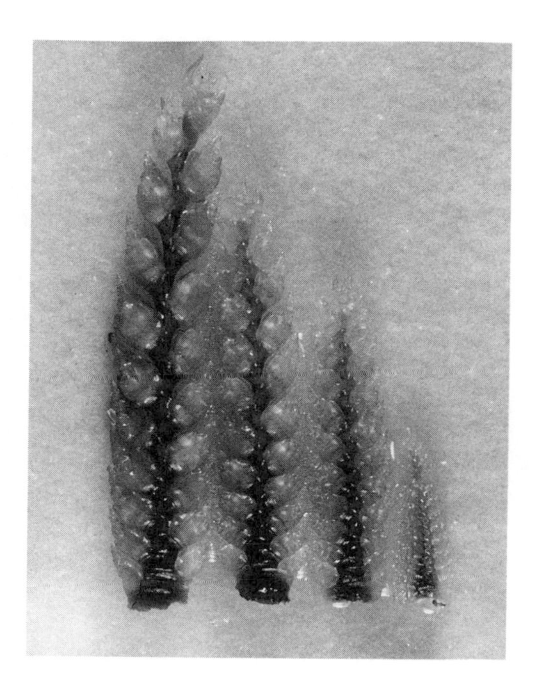

Fig. 1. Wheat inflorescences ranging in length from <0.5–1.0 cm ready for dissection.

3. Methods

3.1. Collection and Dissection of Inflorescences

1. Harvest tillers containing inflorescences ranging in length from 0.5–1.0 cm (*see* **Note 14** and **Fig. 1**) by cutting below the inflorescence-bearing node (the last node of the tiller). Trim the tillers to approx 8- to 10-cm length, and seal the upper end with Nescofilm.
2. Sterilize the tiller sections by rinsing in 70% ethanol for 5 min, then 30 min in 10% Domestos (*see* **Note 15**) with gentle agitation, and rinse twice with sterile distilled water.
3. Dissect inflorescences under sterile conditions by cutting away the whorled leaves surrounding the inflorescence with a scalpel blade. Cut inflorescence into transverse sections of approx 1-mm length (*see* **Note 16**).

3.2. Culture of Inflorescences

3.2.1. Culture of Inflorescence Explants for Somatic Embryogenesis and Plant Regeneration

1. Plate the pieces of inflorescence (1-mm explants) on callus induction medium. Culture approx 10–15 explants/9-cm Petri dish.
2. Keep cultures in the dark at 25 ± 1°C for 3–4 wk.

Fig. 2. Regenerating callus from inflorescence after 3 wk in regeneration medium.

3. Transfer cultures to regeneration medium (RZ) and culture further for 3–4 wk at 25 ± 1°C under 16 h photoperiod (*see* **Note 17** and **Fig. 2**).
4. On succeeding rounds of subculture, small plantlets are transferred to regeneration medium (RO) for further development prior to transfer to soil (*see* **Note 18**). Incubate cultures under the same conditions specified above.
5. Transfer plantlets to 2:1 soil–perlite mixture once a good root system is established, and keep initially in a propagator for acclimation for 1–2 wk. Grow the plants in a controlled environment chamber or under greenhouse conditions as outlined in **Subheading 2.1.** Repot the plants when necessary.

3.2.2. Culture of Inflorescence Explants for Particle Bombardment

1. To prepare for particle bombardment, place up to 30 explants at the center of a Petri dish containing callus induction medium, covering a target area of 2.5 cm in diameter.
2. Preculture the explants in the dark at 25 ± 1°C for 1–6 d prior to bombardment (*see* **Note 19**).

3.3. Particle Bombardment

3.3.1. DNA–Gold Precipitation

1. Sonicate a 50-µL aliquot of the gold solution for 1 min for deagglomeration of particles (*see* **Note 20**).
2. Add 5 µL (5 µg) of plasmid DNA, and mix gently. If using more than one plasmid, i.e., for cotransformation, the amounts of DNA are added in equimolar quantities (for bombardment of controls, *see* **Note 21**)

3. Place 50 µL of $CaCl_2$ and 20 µL of spermidine into an Eppendorf tube lid, and mix well. Close the lid, and tap the solution down into the gold–DNA mixture (*see* **Note 22**). Pulse centrifuge for 5 s to pellet DNA-coated particles. Discard supernatant.
4. Add 150 µL of absolute ethanol to wash the particles (*see* **Note 23**). Pulse centrifuge for 5 s to pellet particles. Discard the supernatant and resuspend particles in 85 µL of absolute ethanol (*see* **Note 24**).

3.3.2. DNA Delivery

1. Gently mix the coated particles, and immediately take a 5-µL aliquot. Drop the mixture onto the center of a macrocarrier, and allow to dry (*see* **Note 25**).
2. Place target cultures (prepared as explained in **Subheadings 3.1.** and **3.2.**) into the chamber.
3. Operate the particle gun according to the instructions supplied with the Bio-Rad PDS 1000/He gun using the following conditions: He acceleration pressure 650–1100 psi, vacuum 28 in. Hg, 2.5 cm gap distance (distance between rupture disk and macrocarrier), 5.5 cm target distance (distance between stopping screen and target plate), and 0.8 cm stopping plate aperture (*see* **Note 26**).

3.3.3. Transient GUS Expression Assay

1. Two days after bombardment, randomly select a few pieces of tissues for a histochemical GUS assay. Immerse the tissues in X-Gluc buffer with 0.1% v/v Triton X-100.
2. Incubate overnight at 37°C and then for 2 d at 26°C prior to assessment. Assess blue staining (*see* **Note 27**).
3. After initial incubation, explants may be stored in 70% ethanol or in 1 *M* TE buffer with a pH higher than 7.

3.3.4. Selection of Cultures for Recovery of Transgenic Plants

1. After bombardment (*see* **Subheading 3.3.2.**), spread the explants evenly over the surface of the culture medium in 2–3 9-cm Petri dishes (not exceeding 10-15 explants/plate).
2. Culture explants on callus induction medium for 3-4 wk as explained in **Subheading 3.2.1.**
3. Transfer explants to regeneration medium (RZ) containing the appropriate selection agent, and incubate as usual for 3-4 wk (*see* **Note 28**).
4. Transfer cultures to RO medium containing selective agent, and culture for a further 3-4 wk until surviving plantlets show signs of resistance.
5. Transfer plantlets to pots, and grow them as explained in **Subheading 3.2.1.**

4. Notes

1. The following elite wheat cultivars are also responsive; Consort, Hereward, Rialto, Riband Soissons, Cadenza, Imp, Avans, Canon.

2. We also grow donor plant materials under greenhouse conditions, but these plants are subjected to seasonal changes. Our conditions for growing tritordeum and wheat plants are:
   a. Soil composition: 75% L&P fine-grade peat, 12% screened sterilized loam, 10% 6 mm screened, lime free grit, 3% medium-grade vermiculite, 3.5 kg Osmocote/m^3 soil (slow-release fertilizer, 15-11-13 NPK plus micronutrients), 0.5 kg PG mix/m^3 soil (14-16-18 NPK granular fertilizer plus micronutrients).
   b. Light: growth rooms provided with HQI lamps 400W (Osram) and glasshouses with natural light supplemented with sodium lamps 400 W (Son-T).
   c. Pest control: sulfur (Thiovit) spray every 4–5 wk and biological control using *Amblyseius caliginosus* for the control of thrips.
3. Autoclave settings are 121°C, 15 psi for 20 min.
4. Agargel and agarose are used at end concentrations of 0.5 and 0.4%, respectively. Shake solutions very well before and just after autoclaving to avoid lumps. Agarose is only used in media supplemented with the selection agent paromomycin, which precipitates out of solution in Agargel-solidified media.
5. We prepare STS by mixing 0.1 *M* $AgNO_3$ and 0.1 *M* $Na_2S_2O_3$ at 1:4 ratio. Add $AgNO_3$ stock into $Na_2S_2O_3$ slowly to avoid precipitation. STS is used freshly made at a final concentration of 30 μ*M* (equivalent to 10 mg/L). $AgNO_3$ and $Na_2S_2O_3$ solutions are air- and photo-sensitive. Store filter-sterilized solutions in the dark at 4°C.
6. Picloram, 2,4-D, glufosinate ammonium, Bialaphos, G418, and paromomycin are toxic substances. Check chemical purities of glufosinate ammonium (Greyhound, UK) and Bialaphos (Meiji Seika Kaisha LTD, Tokyo) after every purchase. Calculations to achieve accurate concentrations may need to be adjusted accordingly.
7. Use 60 and 180 g maltose in the callus induction medium for tritordeum and wheat cultures, respectively. We find that 9% maltose in the callus induction medium is optimal for regeneration in wheat.
8. The callus induction medium is supplemented with 0.5–2 mg/L 2,4-D and 4 mg/L Picloram for wheat and tritordeum culture, respectively. The optimal 2,4-D concentration varies for different wheat genotypes.
9. Silver nitrate is not essential for tritordeum, but does not have any negative effect. The addition of $AgNO_3$ in the induction medium has been seen to be beneficial in most wheat cultures, although it has to be tested for each new genotype.
10. The gelling agent stock solution is melted in a microwave oven. Mix all the media components, and finally add the gelling agent. To minimize condensation in the plates, do not pour the medium until it has cooled to approximately 50°C. Pour approx 40 mL of medium/9-cm tissue-culture Petri dish or 80 mL/Magenta box. Therefore, 800 mL medium will make approx 20 plates or 10 Magenta boxes.
11. The sizes and sources of gold particles we use are: 0.6-μm particles (Bio-Rad, CA) and 0.4 to 1.2-μm particles (Heraeus, Karlsruhe, Germany). Bio-Rad particles are used for wheat transformation, but either type is used for tritordeum.

12. Spermidine solution is very hygroscopic and oxidizable. We suggest storing the original spermidine solution in 15.8-μL aliquots at –80°C. Each aliquot will be used to make up 1 mL of 0.1 *M* spermidine stock.
13. We use a number of combinations of scorable and selectable marker genes cobombarded with genes of interest. Generally, we use the uidA gene as scorable marker, and bar and neo genes as selectable markers (*see* also **Note 28**).
14. Inflorescences are located above the last node of a tiller. Therefore, by cutting under this node, the part of the tiller harvested will contain the inflorescence. The upper end of the tiller is sealed to prevent the penetration of sterilizing agent into the inner region where the explant is located. Inflorescences of 0.5–1.0 cm are generally most responsive, but there is genotypic variation. It takes an element of familiarity with the plant material to be able to harvest tillers with the right age/size of inflorescence.
15. Domestos is a chlorine-based commercial bleach containing 4.5 μg/L free chlorine. While sterilizing, continuously agitate the tillers by placing on a shaker.
16. Regeneration is inhibited in wheat if explants are <1 mm size.
17. Silver nitrate is not essential in the regeneration medium for tritordeum cultures.
18. It takes about three to four rounds of subculture (9–12 wk) on regeneration medium before plantlets are of sufficient size to be potted out. Wheat plants may need more than one round of regeneration on RZ medium.
19. Preculture allows the tissues to recover from isolation before being subjected to bombardment. This should also allow for any contamination to be detected prior to bombardment. In wheat, the preculture period conditions the tissues for bombardment, since the high-maltose concentration in the medium (9%) partially plasmolyses the cells. Partial plasmolysis is apparently important for the tissues to withstand particle bombardment. Preculture of explants for 1–6 d prior to bombardment increases the number of transgenic plants produced in tritordeum. Preculture of wheat explants for 1 d appears optimal.
20. There is evidence that oversonication may result in particle aggregation.
21. For bombardment controls, i.e., particle bombardment without DNA, 5 μL of sterile distilled water are added in place of plasmid DNA.
22. Precipitation onto the gold particles occurs very rapidly. The Ca^{2+} ions from $CaCl_2$ act to bind the DNA onto the gold particles. Therefore, for a uniform precipitation, DNA–gold and $CaCl_2$-spermidine mixture should be mixed well before they are combined.
23. Resuspend as many of the particles as possible by scraping them off the side of the tube with a pipet tip and breakup any aggregation. It is important to make sure that the gold particles are fully resuspended at this stage, since aggregations are more difficult to deal with later.
24. Seal Eppendorf tube lids with Nescofilm to minimize evaporation of ethanol, and keep on ice until required. However, it is advisable to use the gold–DNA preparation immediately.
25. Allow the particles to dry slowly on a nonvibrating surface in a 6-cm Petri dish. Only a few macrocarriers should be loaded with gold at any one time, so that they

are used when recently dried. Examine macrocarriers microscopically prior to bombardment to determine the uniformity and spread of particles, discarding any that have agglomerated. A "sand-dune" appearance of particles is observed as a consequence of too rapid ethanol evaporation.

26. We use acceleration pressures of 650 or 900 psi for wheat and 1100 for tritordeum. Wheat tissues are more sensitive to high bombardment pressures.
27. The same procedure may be used to assay GUS expression in tissues of putatively transformed plants, such as leaves, roots, and flowers. For GUS assays on green tissues, chlorophyll should be removed after staining by incubating the tissues in absolute ethanol. The length of the incubation time depends on the age of the leaf and should last until the chlorophyll has been totally extracted. It may be necessary to replace the ethanol several times.
28. We generally use the selectable marker genes bar and neo for transformation. The bar gene confers resistance to the herbicides Basta (glufosinate ammonium/PPT) and Bialaphos. The neo gene confers resistance to the antibiotics G418 and paromomycin. We use the following concentrations for selection: 2–4 mg/L gluphosinate ammonium, 3–5 mg/L Bialaphos, 25–50 mg/L G418, and 50 mg/L paromomycin.

References

1. Chen, T. H., Lam L., and Chen, S. C. (1985) Somatic embryogenesis and plant regeneration from culturing young inflorescences of *Oryza sativa* L. (rice). *Plant Cell Tissue Organ Cult.* **4,** 51–54.
2. Rout, J. R. and Lucas, W. J. (1996) Characterization and manipulation of embryogenic response from in-vitro-cultured immature inflorescences of rice (*Oryza sativa* L.). *Planta* **198,** 127–138.
3. Pareddy, D. R. and Petolino, J. F. (1990) Somatic embryogenesis and plant regeneration from immature inflorescences of several elite inbreds of maize. *Plant Sci.* 67, 211–219.
4. Ozias-Akins, P. and Vasil, I. K. (1982) Plant regeneration from cultured immature embryos and inflorescences of *Triticum aestivum* L. (wheat): Evidence for somatic embryogenesis. *Protoplasma* **110,** 95–105.
5. Maddock, S. E., Lancaster, V. A., Risiott, R., and Franklin, J. (1983) Plant regeneration from cultured immature embryos and inflorescences of 25 cultivars of wheat (*Triticum aestivum*). *J. Exp. Bot.* **34,** 915–926.
6. Eapen, S. and Rao, P. S. (1985) Plant regeneration from immature inflorescence callus cultures of wheat, rye and triticale. *Euphytica* **34,** 153–159.
7. Redway, F. A., Vasil, V., Lu, D., and Vasil, I. K. (1990) Identification of callus types for long-term maintenance and regeneration from commercial cultivars of wheat (*Triticum aestivum* L.) *Theor. Appl. Genet.* **79,** 609–617.
8. Sharma, V. K., Rao, A., Varshney, A., and Kothari, S. L. (1995) Comparison of developmental stages of inflorescence for high requency plant regeneration in *Triticum aestivum* L. and *T. durum* Desf. *Plant Cell. Rep.* **15,** 227–231.

9. George, L., Eapen, S., and Rao, P. S. (1989) High frequency somatic embryogenesis and plant regeneration from immature inflorescence cultures of two Indian cultivars of sorghum (*Sorghum bicolor* L. Moench). *Proc. Indian Acad. Sci.* **99,** 405–410.
10. Wen, F. S., Sorensen, E. L., Barnett, F. L., and Liang, G. H. (1991) Callus induction and plant regeneration from anther and inflorescence culture of Sorghum. *Euphytica* **52,** 177–181.
11. Thomas, M. R. and Scott, K. J. (1985) Plant regeneration by somatic embryogenesis from callus initiated from immature embryos and immature inflorescences of *Hordeum vulgare. J. Plant. Physiol.* **121,** 159–169.
12. Nagarathna, K. C., Shetty, S. A., Harinarayana, G., and Shetty, H. S. (1993) Selection for downy mildew resistance from the regenerants of pearl millet. *Plant Sci.* **90,** 53–61.
13. Vishnoi, R. K. and Kothari, S. L. (1996) Somatic embryogenesis and efficient plant regeneration in immature inflroescence culture of *Setaria italica* (L.) Beauv. *Cereal Res. Commun.* **24,** 291–297.
14. Xu, Z. H., Wang, D. Y., Yang, L. J., and Wei, Z. M. (1984) Somatic embryogenesis and plant regeneration in cultured immature inflorescences of *Setaria italica. Plant Cell. Rep.* **3,** 149–150.
15. Linacero, R. and Vazquez, A. M. (1990) Somatic embryogenesis from immature inflorescences of rye. *Plant Sci.* **72,** 253–258.
16. Barcelo, P., Vazquez, A., and Martin, A. (1989) Somatic embryogenesis and plant regeneration from tritordeum. *Plant Breeding* **103,** 235–240.
17. Barcelo, P., Hagel, C., Becker, D., Martin, A., and Lörz, H. (1994) Transgenic cereal (tritordeum) plants obtained at high efficiency by microprojectile bombardment of inflorescence tissue. *Plant J.* **5,** 583–592.
18. Marcinska, I., Dubert, F., and Biesaga-Koscielniak, J. (1995) Transfer of the ability to flower in winter wheat via callus tissue regenerated from immature inflorescences. *Plant Cell Tissue Organ Cult.* **41,** 285–288.
19. Martin, A. and Sanchez-Monge, E. (1982) Cytology and morphology of the amphiploid *Hordeum chilense* X *Triticum turgidum* conv. durum. *Euphytica* **31,** 261–267.

8

Cryopreservation of Rice Tissue Cultures

Erica E. Benson and Paul T. Lynch

1. Introduction

The biotechnological improvement of rice is largely dependent on the maintenance of dedifferentiated cultures as either callus and/or suspension cultures. For example, the production of transgenic plants of rice (*Oryza sativa* L.) either by direct DNA uptake into protoplasts or by particle bombardment is dependent on embryogenic callus or cell suspension cultures from which fertile plants can be regenerated ***(1,2)***. However, over time the morphogenic competence of dedifferentiated rice cultures declines ***(3)***. Therefore, new cultures have to be regularly initiated and characterized in order to maintain a constant supply of embryogenic cells. This approach is highly problematic, particularly with Indica, Varietal Group 1 ***(4)***, rice varieties ***(5)***. Cryopreservation of embryogenic cells provides a more efficient means of ensuring a constant supply of competent cells for genetic manipulation. The recovery of embryogenic rice cultures after cryogenic storage that were capable of plant regeneration has been reported by several groups ***(6–9)***. Embryogenic callus, and more commonly, suspension cultures from a range of different rice varieties have been cryopreserved , including Indica (Varietal Group 1) varieties ***(9)***, Japonica (Varietal Group 6) varieties ***(7,10)***. Transgenic rice suspension cultures have also been successfully recovered from cryogenic storage ***(6)***.

Suspension cultures of rice have been cryopreserved using controlled-rate cooling ***(6,7)*** and vitrification procedures ***(11,12)***. Protocols based on the controlled-rate freezing of cells normally requires preculture in culture medium containing osmotically active compounds (for example, mannitol) prior to cryopreservation. Preculture can significantly enhance the postthaw recovery of plant cells ***(13)***. This effect is not solely owing to dehydration effects ***(14)***, but can also involve the activation of genes coding for factors that protect plant

From: *Methods in Molecular Biology, Vol. 111: Plant Cell Culture Protocols*
Edited by: R. D. Hall © Humana Press Inc., Totowa, NJ

cells from environmental stresses ***(15–17)***. After preculture, the cells are treated with cryoprotectants (total concentration 1–2 *M*), which results in moderate cell dehydration ***(18)***. Further dehydration occurs during slow freezing. Ice crystal formation in the external medium leads to an increase in solute concentration and, therefore, the osmotic removal of water from the cells. On further ultrarapid cooling in liquid nitrogen, the cell contents vitrify ***(19)***. Although plant cells can tolerate extracellular ice crystal formation, such ice formation can be damaging, for example, owing to the mechanical stress to cells caused by the growth of extracellular ice crystals ***(20)***. The more recently developed vitrification procedures ***(21,22)*** involve cryopreservation strategies, which prevent ice nucleation and the subsequent growth of ice crystals. They are based on a combination of cryoprotective mechanisms, including severe cell dehydration at nonfreezing temperatures as a result of exposure to concentrated nonpenetrating cryoprotectants (total concentration 5–8 *M*), followed by rapid freezing leading to vitrification ***(23)***. The application of high concentrations of penetrating cryoprotectants, for example, dimethyl sulfoxide (DMSO), increases the viscosity of the intracellular solutes, and as a result, ice nucleating events are inhibited. The avoidance of ice crystal formation in vitrification procedures and the lack of a need for a controlled-rate freezer are generally considered to be advantagous ***(23)***. However, this must be counterbalanced by the need to use high concentrations of potentially toxic chemicals in the vitrification solutions. Glasses formed during vitrification are not stable and can destabilize on rewarming, during which ice nucleation can occur. It is thus critical to ensue rapid rewarming. Therefore, in vitrification, it is the rewarming rate that is critical rather than cooling parameters.

This chapter presents a series of protocols based on published methods used to cryopreserve rice suspension and callus cultures. In the case of suspension cultures, controlled-rate freezing and vitrification procedures are described. Factors influencing postthaw regrowth and methods of determining postthaw cell viability are also presented in the context of protocol development.

2. Materials

To maintain sterility of the rice cultures, all appropriate manipulations should be performed in a laminar flow bench, using aseptic techniques and sterile materials.

2.1. General Requirements

1. Liquid nitrogen and ice.
2. Small bench-top dewar of 2- to 5-L capacity.
3. Long-term storage dewar with an appropriate drawer or cane inventory system.
4. Polypropylene cryovials, 2 mL with graduation marks and label, or 1- to 0.5-mL polypropylene straws. Cryovials can be obtained, for example, from Sigma and straws (type 101, cattle, nonsterilized) from Instruments de Medicine Veterinaire, l'Aigle, France.

5. Small spatula, sieves with 45-µm nylon mesh, aluminum cryocanes, Pasteur pipets, forceps, ice bucket, 100- and 250-mL conical flasks.
6. A water bath set at 45°C. (*see* **Note 1**).
7. Rice callus and suspension cultures grown on appropriate culture medium, e.g., AA2 ***(24)*** or modified R2 ***(9)***.
8. Safety equipment: including, cryogloves, face shield.

2.2. Controlled-Rate Freezing Requirements (Cell Suspensions)

1. Controlled-rate freezer able to cool at a rate of –1°C/min. However, alternative cooling methods may suffice (*see* **Note 2**).
2. Liquid preculture medium: rice culture medium, e.g., AA2 ***(24)*** or modified R2 ***(9)***, both containing 60 g/L mannitol (*see* **Note 3**).
3. Cryoprotectant solution: glycerol, 46 g/L, DMSO, 39 g/L, sucrose 342.3 g/L, and proline 10.0 g/L made up in liquid standard culture medium. (*see* **Note 4**).
4. Postthaw recovery medium: rice culture medium containing 4.0 g/L agarose (*see* **Notes 3** and **5**) in 9.0-cm Petri dishes (25 mL of media/dish). The media onto which the cells will be placed immediately after thawing should be overlaid with two filter paper disks (Whatman No. 1, 5.5-cm diameter), one on top of the other.

2.3. Controlled-Rate Freezing Requirements (Callus Cultures)

1. Controlled-rate freezer able to cool at a rate of –1°C/min (*see* **Note 2**).
2. Full-strength cryoprotectant solution: glycerol, 46 g/L, DMSO, 39 g/L, sucrose 342.3 g/L, and proline 3.3 g/L made up in liquid standard culture medium, such as modified R2 medium ***(8)*** (*see* **Note 4**).
3. Half-strength cryoprotectant solution, prepared in full-strength culture medium, but at half the above concentrations.
4. Water bath set at 30°C (*see* **Note 1**).
5. Ammonium-ion-free liquid rice culture medium.
6. Postthaw culture media: semisolid rice culture medium (*see* **Notes 3** and **5**) in Petri dishes.

2.4. Vitrification Requirements

2.4.1. Method A, Based on Huang et al. (12)

1. Preculture medium A—liquid rice culture medium (e.g., AA2) containing 60 g/L sucrose.
2. Preculture medium B—liquid rice culture medium containing 72.9 g/L sorbitol.
3. PVS2 vitrification solution ***(25,26)*** comprising rice preculture medium B to which is added: 30% (v/v) glycerol, 15% (v/v) ethylene glycol, and 15% (v/v) DMSO (*see* **Note 4**).
4. A 25% (v/v) solution of PVS2 diluted with preculture medium B.
5. Washing solution comprising 1.2 *M* sorbitol made up in liquid rice culture medium.
6. Semisolid rice culture medium (e.g., AA2 medium containing 0.8% w/v agar) containing 40% (w/v) soluble starch.
7. Semisolid rice culture medium.

2.4.2. Method B, Based on Watanabe and Steponkus (11)

1. Loading solution 126.3 g/L DMSO in liquid rice culture medium, e.g., Linsmaier and Skoog medium, LS *(27)*, held at 20°C.
2. Vitrification solution, 445 g/L DMSO and 187 g/L sorbitol in rice culture medium, held at 0°C (*see* **Note 4**).
3. Unloading solution, 91.1 g/L sorbitol in liquid rice culture medium.
4. Semisolid rice culture medium (*see* **Note 5**).

2.5. Selection of Tissue for Freezing

For the successful cryopreservation of rice tissue cultures, it is vital to select the most appropriate tissue. Generally, an actively growing area of callus cultures and cell suspensions in the exponential phase of growth should be selected. Factors including the embryogenic potential of the cells *(7)*, cell aggregation *(10)*, and cultivar *(9)*, have been shown to influence postthaw rice cell regrowth significantly.

2.6. Requirements for the Assessment of Postthaw Recovery

2.6.1. Fluorescein Diacetate (FDA)

1. Stock solution of FDA 0.1% (w/v) in acetone, stored in the dark at 4°C.
2. Pasteur pipets, test tube, slides and cover slips.
3. Liquid standard culture medium.
4. UV microscope with 490-nm excitation and 530-nm barrier filters.

2.6.2. Triphenyl Tetrazolium Chloride (TTC)

1. Reagent mixture: 0.6% (w/v) TTC in 0.5 *M* Na_2HPO_4/KH_2PO_4 buffer, pH 7.4, supplemented with 0.05% (v/v) Tween 80. The buffer can be stored in a refrigerator for several months. Do not use if the solution appears cloudy. The reagent mixture should be prepared immediately prior to use.
2. Distilled water.
3. 95% (v/v) Ethanol.
4. Pipets (10-mL), Pasteur pipets, cuvets, test tubes.
5. Boiling water bath.
6. Spectrophotometer capable of measuring absorbance at 490 nm.

3. Methods

3.1. Controlled-Rate Freezing (Cell Suspensions), Based on Lynch et al. (7)

1. At the standard subculture time, transfer the cells to be cryopreserved to preculture medium using the standard culture protocol.
2. Three to 4 d after subculture harvest the cells on 45-μm nylon mesh filters and transfer to cryovials (0.75 mL/vial).

3. Prelabel the vials, and to each vial add 0.75 mL of filter-sterilized, chilled (on ice) cryoprotectant mixture. Immediately seal the vial, and shake to ensure the cells and the cryoprotectant mixture are well mixed and place on ice.
4. Mount the vials onto labeled aluminium cryocanes, and incubate on ice for 1 h.
5. Shake the ice and water off the canes and vials. If necessary, dry with tissue paper and transfer to the controlled-rate freezer (e.g., a Planer Kryo 10 Series III; Planer BioMed, Sunbury, Middlesex, UK) precooled to 0°C. Place the controlled-rate freezer's sample chamber temperature probe into a vial containing 1.5 mL of chilled cryoprotectant mixture. This provides a temperature profile of a "sample" during freezing. Start the freezing program (*see* **Note 6**).
6. Immediately after the controlled-rate freezer programme is completed remove the vials from the freezer sample chamber and plunge them into a dewar of liquid nitrogen. This should be located next to the freezer. After 5–10 min, the vials can be transferred to a long-term storage dewar, and their position logged into an appropriate inventory system (*see* **Note 7**).
7. After storage, plunge the vials into a water bath at 45°C (*see* **Note 1**). Complete thawing normally takes approx 2 min Remove the vials from the water bath as soon as all the ice has melted, and transfer to a laminar flow bench (*see* **Notes 7** and **8**).
8. Dry the vials with tissue paper.
9. Aspirate the cryoprotectants with a Pasteur pipet and place the cells on the top of a double layer of filter paper (Whatman no. 1, 5.5-cm diameter) overlaying the postthaw recovery medium (*see* **Note 5**).
10. Seal the dishes with Nescofilm (Nippon Shoji Kaisha Ltd., Osaka, Japan), and incubate in the dark at 28°C, or as appropriate for the rice culture being used.
11. After 3 d, transfer the cells and filter paper to a fresh dish of postthaw recovery medium (*see* **Note 5**).
12. Postthaw cell regrowth is normally visible to the naked eye within 5 d of thawing. Sufficient cell regrowth to allow reinitiation of suspension cultures normally occurs within 25 d of thawing.
13. Suspension cultures can be reinitiated using the same techniques used to initiate new nonfrozen cultures. It has been noted that cultures initiated from frozen cells often develop more quickly than new, nonfrozen cultures and can be in a state suitable for protoplast isolation within 28 d after initiation *(7)*.

3.2. Controlled-Rate Freezing (Callus Cultures), Based on Cornejo et al. (8)

1. Select callus with a compact nodular appearance from cultures 4- to 8-mon old, which have been subcultured every 4 wk.
2. Place individual calli into vials containing 1 mL of 1/2 strength chilled cryoprotectant solution, and incubate on ice for 15 min.
3. Remove the cryoprotectant solution, replace with an equal volume of full strength chilled cryoprotectant solution and incubate on ice for a further 15 min.

4. Mount vials on aluminum canes, and transfer to a controlled-rate freezer with its sample chamber at 4°C. Start the freezing program, cool to –25°C at -1°C/min, and hold at –25°C for 30 min.
5. Immediately after the controlled-rate freezer program is completed, remove the vials from the freezer sample chamber, and plunge them into a dewar of liquid nitrogen. After 5–10 min, the vials can be transferred in to a long-term storage dewar and appropriate inventory system (*see* **Note 7**).
6. After storage, plunge the vials into a water bath at 30°C (*see* **Note 1**). Complete thawing normally takes approx 2 min. Remove the vials from the water bath as soon as all the ice has melted, transfer to a laminar flow bench, and place on ice (*see* **Notes 7** and **8**).
7. Gradually dilute the cryoprotectant solution with ice-cold ammonium ion-free culture medium.
8. Remove the thawed calli from the vials and transfer to semisolid culture medium ***(9)***. Seal the Petri dishes with Nescofilm, and incubate in the dark at 27°C, or as appropriate for the rice culture being used.
9. Transfer to fresh culture medium every 3 wk.
10. One to 2 mon after thawing, remove the regrowing callus from nongrowing sections placed on fresh culture medium, and maintain for a further 3–6 wk until actively growing callus lines have been established.

3.3. Vitrification

3.3.1. Method A, based on Huang et al. (12)

1. Using standard culture procedures, 11 d prior to freezing, transfer the suspension cells into preculture medium A, and incubate under normal conditions.
2. Using standard culture procedures, 1 d prior to freezing, transfer the suspension cells into preculture medium B, and incubate under normal conditions.
3. Transfer 3 mL of cell suspension culture into a 10-mL centrifuge tube, add 6 mL of 25% (v/v) PVS2 solution, and incubate at 20°C for 10 min.
4. Remove the 25% (v/v) PVS2 solution, resuspend the cells in 3 mL of 100% (v/v) chilled PVS2 solution, and incubate on ice for 7.5 min.
5. Using a wide-bore Pasture pipet, transfer 0.75-mL aliquots of the suspended cells (at a final packed-cell volume of 40%) into polypropylene straws (straws surface sterilized with ethanol).
6. Quickly seal each straw by drumming sealing powder (obtained from Instruments de Medicine Veterinaire) into the open end. Ensure the powder is wetted by the suspension in the straw. Alternatively, the straws can be sealed with a hot pair of forceps or a sealing machine.
7. Plunge straws into liquid nitrogen. After 5–10 min, the samples can be transferred into a long-term storage dewar and appropriate inventory system (*see* **Note 7**).
8. Rapidly thaw the cells by placing the straws in a water bath at 37°C.
9. Introduce 1 mL of washing solution into each warmed straw, and incubate for 2 min.
10. Remove the supernatant, replace with 1 mL of washing solution, and incubate the cells for 25 min at 25°C.

11. Drain the cells from the cut end of each straw onto semi-solid rice culture medium containing 40% (w/v) starch. Incubate at 26°C for 2d.
12. Transfer the cells onto semisolid AA2 medium, and incubate at 26°C. Cell growth can be observed within 4 d after thawing.

3.3.2. Method B, Based on Watanabe and Steponkus (11)

1. Allow the cells to settle in cultures 3–4 d after subculture, and drain off the culture medium.
2. To 500 µL of settled cells add 2.5 mL of loading solution, and incubate at 20°C for 20 min.
3. Remove the loading solution, and place the cells on ice. Resuspend the cells in 2.5 mL of chilled vitrification solution.
4. Transfer 500-µL aliquots of the cell suspension in to polypropylene straws, and plunge in to liquid nitrogen. The total time the cells were exposed to vitrification solution prior to freezing is 4 min. After 5–10 min, the vials can be transferred to a long-term storage dewar with an appropriate inventory system (*see* **Note 7**).
5. Store straws in liquid nitrogen.
6. After storage, thaw the cells by warming the straws in the air for 10 s followed by 10 s in an ethyl alcohol bath at 20°C (*see* **Notes 1**, **7**, and **8**).
7. Immediately after thawing, transfer the straw to a laminar flow bench, cut both ends and expel the contents into 8 mL of unloading solution. Allow the cells to settle, remove the supernatant, and replace with 2 mL of fresh unloading solution. Maintain at 20°C for 30 min.
8. Transfer 100 µL of settled cells onto 4 mL of semisolid culture medium in a 3.5-cm Petri dish. Seal the dishes with Nescofilm. Incubate cultures at 25°C under fluorescent light (2000 lx) or in the dark at a temperature appropriate for the rice cell line used.
9. Within 3–4 wk of thawing, sufficient cell regrowth should have occurred to allow the reinitiation of suspension cultures.

3.4. Assessment of Postthaw Recovery

The presence of postthaw cell regrowth can be determined by regular (at 7-d intervals) examination of the thawed cells under a binocular microscope. Such examinations can be combined with an assessment of morphological features, such as color and wetness, of the thawed cells, which can be important indicators of cryoinjury *(3)*. Growth can be quantified by fresh and/or dry weight determinations, and the growth expressed in terms of % change after thawing or % difference as compared with nonfrozen controls. In combination with such observations it is also useful to assess the viability of the thawed cells using techniques ,such as FDA *(28)* staining or TTC reduction *(29)* as described below. Viability determinations can provide a useful indication of potential postthaw growth and can be used to ascertain the effects of different postthaw culture conditions *(3)*. However, such determinations should not be made in isolation, since apparently high cell viability (determined for example by TTC reduction) soon after thawing is not always a prelude to postthaw cell growth *(3)*.

3.4.1. FDA, Based on Widholm (28)

1. Place 1–2 drops of the FDA stock solution into 10 mL of liquid culture medium and mix well.
2. Add 1–2 drops of the diluted stain to the cell sample mounted on a microscope slide.
3. Leave for a few minutes before placing the cover slip over the sample.
4. Count the number of cells in a field of view under bright field, and then count the number of fluorescent cells in the same field of view with UV illumination.

$$\% \text{ Viability} = (\text{number of fluorescent cells/total number of cells}) \times 100 \quad (1)$$

Determine the viability in at least five fields of view to calculate an overall sample mean viability.

3.4.2. TTC, Based on Steponkus and Lamphear (29)

1. Take 100 mg fresh weight cell samples, place in a test tube, and add 3.0 mL of TTC reagent solution.
2. Infiltrate reagent solution under vacuum for 10 min.
3. Incubate samples in the dark overnight at 25°C.
4. Remove the TTC reagent and wash the cells by resuspending in distilled water.
5. Remove the distilled water.
6. Resuspended the cells in 7 mL of 95% ethanol.
7. Extract the formazin color complex by placing the samples in boiling water until all the ethanol had been driven off. To avoid the buildup of ethanol fumes in the laboratory, this stage should be performed in a fume cupboard.
8. Resuspend cooled samples in 10 mL of ethanol, vortex, and allow the cells to settle out. Settling of the cells can be assisted by centrifuging the samples at 80*g* for 5 min.
9. Measure the absorbance of the supernatant at 490 nm , against an ethanol blank. Viability can be expressed as absorbance at 490 nm/100 mg or /g fresh weight, or as a percentage of an unfrozen control. Sample replication is advisable to take into account variation between cell sampling procedures.

4. Notes

1. To avoid the cryovials moving around in the circulation of the water bath, and to reduce the potential for microbial contamination, place jars of sterile water in the bath, which should be allowed to equilibrate to the temperature of the water bath. To thaw, the cryovials are plunged into the sterile water in the jars.
2. Controlled-rate freezing has generally been achieved using computer programmable cooling systems manufactured, for example, by Planer Products Ltd. and CryoMed. These machines provide a very accurate and reproducible means of freezing rice cultures. However, controlled-rate freezers are expensive (usually

over $12,000). Less sophisticated alternatives using standard laboratory freezers have been successfully used to cryopreserve rice cultures ***(9)***. Specifically, vials are incubated for 2 h at –25°C prior to plunging into liquid nitrogen, or incubated for 1 h at –25°C followed by 1 h at –70°C prior to plunging into liquid nitrogen.

3. AA2-based culture medium must be filter-sterilized. To prepare semisolid AA2 based medium, mix equal quantities of double-strength AA2 medium (filter-sterilized), with melted double-strength agarose (Sigma Type 1) 8.0 g/L in water (autoclaved), and immediately pour into Petri dishes.
4. Cryoprotectants must be of a high purity or, in the case of DMSO, spectroscopically pure. After preparing the cryoprotectant mixture, adjust the pH to 5.8. Cryoprotectants should be prepared immediately prior to use, chilled on iced water, and filter-sterilized. Owing to the high viscosity of vitrification solutions, autoclaving is the only practical method of sterilization.
5. Postthaw recovery medium tends to be based on the media the rice cells are normally maintained in, for example, AA2 ***(7)***, modified R2 ***(9)***, or LS ***(10)***. However, the presence of ammonium ions in the culture medium immediately after thawing may be detrimental to postthaw cell growth ***(7,30)***, as is the presence of fructose as the carbon source in the culture medium ***(3)***. The addition of the surfactant Pluronic F-68 (0.01% w/v) to postthaw culture medium has been shown to promote postthaw rice cell growth ***(31)***. Supplementing the culture medium with activated charcoal for the first 3 d postthaw can enhance postthaw cell growth ***(18)***.

   Modifications of the pregrowth and recovery medium by the addition of the iron chelating agent desferrioxamine can enhance the recovery of cryopreserved rice cell cultures ***(32)***. Thus, short-term applications of the drug (3 d exposure) before and after cryopreservation, at concentrations within the range of 0.5–10 mg/L can improve recovery potential. This is further enhanced if cations are removed from the culture medium during the period of application. Desferrioxamine has been marketed as the drug Desferral by Ciba Geigy and can be obtained from Ciba Laboratories, Horsham UK. Its mode of action involves the reduction of oxidative stress via the removal of cations, which enhance the Fenton reaction and hydroxyl radical production. Rice cell cultures are known to undergo oxidative damage during cryopreservation ***(33)***.
6. Where controlled-rate freezers have been employed to freeze rice cultures, the freezing programmes most widely employed are based on that of Withers and King ***(34)***. This specifically involves a hold for 10 min at 0°C, cooling at –1°C/min to –35°C, and holding at –35°C for 35 min.
7. Samples can be stored either under liquid nitrogen or in liquid nitrogen vapor, with the level of liquid nitrogen maintained below the lowest vials. Under the latter conditions, no liquid nitrogen enters the vials, reducing the chances of vials exploding when thawed. Losses of liquid nitrogen by evaporation are less in these systems. However, significantly the temperature of the uppermost vials may tend to oscillate and rise above –110°C. This can cause devitrification and ice crystal growth in the cells, which will result in the loss of cell viability.

8. Care should be taken when thawing vials or straws, since any liquid nitrogen in them will rapidly evaporate, and this can lead to either vials or straws rupturing explosively. Therefore, appropriate safety precautions should be observed, including wearing a face shield.
9. To reduce photo-oxidation damage to the cells *(35)*, all manipulations during the recovery phase are carried out in a nonilluminated laminar flow bench.

References

1. Kothari, S. L., Davey, M. R., Lynch, P. T., Finch, P. T., Finch, R. P., and Cocking, E. C. (1991) Transgenic rice, in *Transgenic Plants,* vol. 2 (Kung, S. D. and Wu, R., eds.), Butterworths, London, pp. 3–20.
2. Cao, Duan, J., X., McElroy, D., and Wu, R. (1992) Regeneration of herbicide resistant transgenic rice plants following microprojectile-mediated transformation of suspension culture cells. *Plant Cell Rep.* **11,** 586–591.
3. Lynch, P. T. and Benson, E. E. (1991) Cryopreservation, a method for maintaining the plant regeneration capacity of rice cell suspension cultures, in *Proceedings of the Second Rice Genetic Symposium*, IRRI, Los Banos, pp. 321–332.
4. Glaszmann, J. C. (1987) Isozymes and classification of Asian rice varieties *Theor. Appl. Genet.* **74,** 21–30.
5. Hodges, T. K., Peng, Lyznik, J., L. A., and Koetje, D. S. (1991) Transformation and regeneration of rice protoplasts, in *Rice Biotechnology* (Khush, G. S. and Toenniessen, G. H., eds.), CAB International, Wallingford, pp. 157–174.
6. Meijer, E. G. M., van Iren, F., Schrijnemakers, E., Hensgens, L. A. M., van Zijderveld, M., and Schilperoort, R. A. (1990) Retention of the capacity to produce plants from protoplasts in cryopreserved line of rice (*Oryza sativa* L.). *Plant Cell Rep.* **10,** 171–174.
7. Lynch, P. T., Benson, E. E., Jones, J., Cocking, E. C., Power, J. B., and Davey M. R. (1994) Rice cell cryopreservation: the influence of culture methods and the embryogenic potential of cell suspensions on post-thaw recovery. *Plant Sci.* **98,** 185–192.
8. Cornejo, M-J, Wong, V. L., and Blechi, A. E. (1995) Cryopreserved callus - A source of protoplasts for rice transformation. *Plant Cell Rep.* **14,** 210–214.
9. Jain, S., Jain, R. K., and Wu, R. (1996) A simple and efficient procedure for cryopreservation of embryogenic cells of aromatic Indica rice varieties. *Plant Cell Rep.* **15,** 712–717.
10. Watanabe, K., Kawai, F., and Kanamorii, M. (1995) Factors affecting cryoprotectability of cultured rice (*Oryza sativa* L.) cells—cell wall and cell aggregate size. *Cryo Lett.* **16,** 147–156.
11. Watanabe, K. and Steponkus, P. L. (1995) The vitrification of *Oryza sativa* cells, L. *Cryo Lett.* **16,** 255–262.
12. Huang, C. N., Wang, J. H., Yan, Q. S., Zhang, X. Q., and Yan, Q. F. (1995) Plant regeneration from rice (*Oryza sativa* L.) embryogenic suspension cells, cryopreserved by vitrification. *Plant Cell Rep.* **14,** 730–734.

13. Riberio, R. C. S., Jekkel, Z., Mulligan, B. J., Cocking, E. C., Power, J. B., and Davey, M. R. (1996) Regeneration of fertile plants from cryopreserved cell suspensions of *Arabidopsis thaliana* (L.) Heynh. *Plant Sci.* **115,** 115–121.
14. Chen, T. H. H., Kartha, K. K., Constabel, F., and Gusta, L. V. (1984) Cryopreservation of alkaloid-producing cell-cultures of periwinkle (*Catharanthus roseus*). *Plant Physiol.* **75,** 726–731.
15. Skriver, K. and Mundy, J. (1990) Gene expression in response to abscisic acid and osmotic stress. *Plant Cell* **2,** 503–512.
16. Watanabe, K., Sato, F., Yamada, Y., Kawai, F., Kanamori, M., and Mitsuda, H. (1992) Characterisation of polypeptides in cultured rice cells differing in cryoprotectability. *J. Plant Physiol.* **136,** 443–447.
17. Chandler, P. M. and Robertson, M. (1994) Gene expression regulated by abscisic acid and its relation to stress tolerance. *Ann. Rev. Plant Physiol. Plant Mol. Biol.* **45,** 113–141.
18. Scrijnemakers, E. W. M. and van Iren, F. (1995) A two-step or equilibrium freezing procedure for the cryopreservation of plant cell suspensions, in *Cryopreservation and Freeze-Drying Protocols* (Day, D.J and McLellan, M. R., eds.), Humana Press, Totowa, NJ, pp. 103–111.
19. Meryman , H. T. and Williams, R. J. (1985) Basic principles of freezing injury in cells; natural tolerance and approaches to cryopreservation, in *Cryopreservation of Plant Cells and Organs* (Kartha K. K., ed.), CRC, Boca Raton, pp. 13–48.
20. Grout, B. W. W. (1995) Introduction to the *in vitro* preservation of plant cells, tissues and organs, in *Genetic Preservation of Plant Cells In Vitro* (Grout, B. W. W., ed.), Springer-Verlag, Berlin, pp. 1–12.
21. Uragami, A., Sakai, A., Nagai, M., and Takahashi, T. (1989) Survival of cultured cell and somatic embryos of *Asparagus officinalis* cryopreserved by vitrification. *Plant Cell Rep.* **8,** 418–421.
22. Langis, R., Schnabel, B., Earle, E. D. and Steponkus, P. L. (1989) Cryopreservation of *Brassica campestris* L. cell suspensions by vitrification. *Cryo Lett.* **10,** 421–428.
23. Reinhoud, P. J., Uragami, A., Sakai, A. and Van Iren, F. (1995) Vitrification of plant cell suspensions, in *Cryopreservation and Freeze-Drying Protocols* (Day, D.J and McLellan, M. R., eds.), Humana Press, Totowa, NJ, pp. 113–120.
24. Abdullah, R., Cocking, E. C., and Thompson, J. A. (1986) Efficient plant regeneration from rice protoplasts through somatic embryogenesis. *Bio/Technology* 4, 1087–1093.
25. Uragami, U. (1991) Cryopreservation of asparagus (*Asparagus officinalis* L.) cultured *in vitro*. Res. Bull. Hooaido Nat. Agric. Expt. Stat. **156,** 1–37.
26. Sakai, A., Kobayashi, S., and Oiyama, I. (1990) Cryopreservation of nucellar cells of navel orange (*Citrus sinensis* Osb. var. *brasiliensis* Tanaka) by vitrification. *Plant Cell Rep.* **9,** 30–33.
27. Linsmaier, E. M. and Skoog, F. (1965) Organic growth factor requirements of tobacco tissue cultures. *Physiol. Plant.* **8,** 100–127.

28. Widholm, J. M. (1972) The use of florescein diacetate and phenosafranine for determining viability of cultured plant cells. *Stain Technol.* **47,** 189.
29. Steponkus, P. L. and Lamphear, F. O. (1967) Refinement of the triphenyl tetrazolium chloride method of determining cold injury. *Plant Physiol.* **42,** 1423–1426.
30. Kuriyama, A., Watanabe, K., Ueno, S., and Mitsuda, H. (1989) Inhibitory effect of ammonium ions on recovery of cryopreserved rice cells. *Plant Sci.* **64,** 231–235.
31. Anthony, P., Jelodar, N. B., Lowe, K. C., Power, J. B. and Davey, M. R. (1996) Pluronic F-68 increases the post-thaw growth of cryopreserved plant cells. *Cryobiol*ogy **33,** 508–514.
32. Benson, E. E., Lynch, P. T., and Jones, J. (1995) The use of the iron chelating agent desferrioxamine in rice cell cryopreservation: a novel approach for improving recovery. *Plant Sci.* **110,** 249–258.
33. Benson, E. E., Lynch, P. T., and Jones, J. (1992) The detection of lipid peroxidation products in cryoprotected and frozen rice cell; consequences for post freeze recovery. *Plant Sci.* **85,** 107–114.
34. Withers, L. A. and King, P. J. (1980) A simple freezing unit and cryopreservation method for plant cell suspensions. *Cryo Lett.* **1,** 213–220.
35. Benson, E. E. and Noronha-Dutra, A. A. (1988) Chemiluminescence in cryopreserved plant tissue cultures: the possible involvement of singlet oxygen in cryoinjury. *Cryo Lett.* **9,** 120–131.

9

Noncryogenic, Long-Term Germplasm Storage

Ali Golmirzaie and Judith Toledo

1. Introduction

Germplasm conservation is an important goal in the preservation of genetic diversity. Conventional methods, such as field and seed conservation, are used for most crops, although nonconventional methods, such as tissue culture and cryopreservation, have been used to support those methods.

Tissue culture can be used to maintain plant material, thus safeguarding it against losses caused by adverse environmental changes, pests, and diseases during growth under field conditions. For in vitro establishment, the choice of a particular medium depends on the plant species, but in general, the medium should contain minerals, a carbon source, vitamins, and low concentrations of growth regulators (**Table 1**). In this medium, plants can grow in vitro for 2–3 mon, when it is then necessary to make subcultures to fresh media. This procedure increases the amount of labor required, especially when working with a large collection.

In vitro maintenance of plants for longer periods of time, without subculture, is done through growth rate reduction, in which modifications to environmental conditions or changes to some media components are made. Examples are reducing growth temperature, reducing light intensity, using growth regulators, limiting mineral supply, adding osmotic stressants, or combining any of these procedures.

1.1. Growth Temperature

Temperature ranges established for culture rooms are similar to field temperatures for a given crop. To minimize the growth rate of temperate species, we can reduce room temperature to near zero; for tropical crops, a moderate reduction in normal growth temperature is applicable ***(1,2)***.

From: *Methods in Molecular Biology, Vol. 111: Plant Cell Culture Protocols*
Edited by: R. D. Hall © Humana Press Inc., Totowa, NJ

Table 1
Components of Culture Media of Potato Included with the Murashige and Skoog Mineral Salts Mixture and Vitamins (*11*, and *see* Appendix) (International Potato Center, 1997)

Ingredients, mg/mL	Propagation medium	Conservation medium	Tuberization medium
Gibberellic acid	0.1		
Sucrose%	2.5	2.5	8.0
Sorbitol%		4.0	
Benzyl amino purine			5.0
Chlorocholine chloride			500
Agar%		8.0	
Phytagel%	3.5	3.5	
pH	5.6	5.6	5.6

1.2. Light Intensity

In vitro plants use sugars as a carbon source through heterotrophic absorption in the culture vessel. Although these plants still maintain their photosynthetic ability (autotrophic absorption), this can be restricted by low CO_2 concentrations. By using low light intensity, we can reduce the support of carbon obtained autotrophically, which then results in delayed growth ***(3)***. For in vitro conservation of potato plantlets, low light intensity is used along with other stressants ***(4)***.

1.3. Growth Regulators

Abscisic acid (ABA) reduces the overall growth rate of in vitro plants. Although it is used for several crops, physiological changes or mutations can appear, which can threaten germplasm genetic stability ***(2–5)***.

1.4. Nutrient Supply

Limiting the nutrient supply can delay plant growth by decreasing the amount of specific minerals, such as nitrogen or magnesium. This is done for various crops ***(4,6)***. However, symptoms of nutrient deficiency can appear, which lead to severe stress, thus reducing plantlet viability.

1.5. Osmotic Stressants

Minerals dissolved in water are introduced into cells through differences in osmotic pressures (inside and outside the plant cell). The inclusion of sugar in media increases the osmotic potential, thus reducing the uptake of minerals by cells. As a consequence, plant growth is delayed. Sugar alcohols (as mannitol

Table 2
Number of Accessions of *Solanum* Species Maintained Under Long-Term Conditions from 1–5 Years (Unpublished Data)

Solanum species	1	2	3	4	5	Total
S. andigena	1	569	666	1555	70	2861
S. stenotomum	0	57	54	129	5	245
S. tuberosum	0	38	24	68	8	138
S. phureja	0	37	37	66	9	149
S. chaucha	0	23	23	61	1	108
S. goniocalyx	0	15	6	31	1	53
S. juzepczukii	0	6	3	8	0	17
S. ajanhuiri	0	6	3	8	0	17
Others	0	184	125	280	18	607

and sorbitol) are used extensively in germplasm conservation. Osmoticums enter the cell slowly and produce osmotic effects without being metabolized ***(1,2,7)***.

A combination of osmotic stressants and low temperatures increases the length of time between subcultures. For the in vitro conservation of potato, good results have been obtained by combining osmotic stressants, low temperatures, and low light intensity ***(8)***. Most potato accessions can be maintained for up to 4 yr without subculture under these conditions (**Table 2**) (**Fig. 1**).

1.6. Long-Term Maintenance Through In Vitro Tubers

Microtuber production is an alternative method for clonal multiplication in tuberous species. Microtubers of potato have been developed at CIP for a wide range of genotypes for the purpose of international germplasm distribution, potato seed production, and as an alternative to germplasm conservation. Microtubers are produced in 2–3 mon and can be stored at 10°C for 10 mo ***(9)*** after harvest. Additionally, tuber dormancy can be controlled by environmental changes ***(9,10)***, or sprout growth can be retarded by storage of sprouted tubers embedded in conservation medium, thus permitting another means of long-term conservation.

2. Materials

1. Potato plantlets growing on propagation medium.
2. Culture media (**Table 1**) contain Murashige and Skoog salts (***11***; *see* Appendix) (Gibco 10632–016) and Murashige and Skoog vitamins (***11***; *see* Appendix) (Sigma M7150); growth regulators are included in the media before autoclaving; agar or phytagel is included after adjusting the pH. The media are autoclaved at 121°C for 20 min at 15 psi before use.

Fig. 1. Potato plantlets after 4 yr of storage at 6–8°C in conservation media.

3. Growth regulators: Gibberellic acid (Sigma G7645), 6-benzylaminopurine (Sigma B5898), chlorocholine chloride (Sigma C4049).
4. Sucrose grade II (Sigma S5391).
5. Phytagel (Sigma P8169) or agar (Sigma A1296).
6. Surgical knife handle #11 with blade #15.
7. Stainless-steel forceps, 18 mm.
8. 25 × 150 mm pyrex test tubes, containing 12 mL of culture media.
9. 100X 15-mm Petri dishes containing two sheets Whatman no. 1 filter paper.
10. Bacti-incinerator for instruments.
11. Laminar flow cabinet for aseptic manipulation.
12. 18–22°C culture room (air conditioner with thermostat control), 3000 lx, 16 h light.
13. 6–8°C culture room (air conditioner with thermostat control), 1000 lx, 16 h light.

3. Methods

3.1. *In Vitro Propagation by Single Nodes*

1. Under aseptic conditions, take out in vitro plantlets obtained from buds growing on propagation media from the culture tube, and place on a Petri dish (*see* **Notes 1** and **2**).

2. Slice the plantlets into pieces or nodes (micropropagation). Transfer each node into a glass tube 18 × 150 mm with propagation medium (**Table 1**), 3 nodes/tube *(11)*.
3. Transfer to incubation room. After 2–3 mo, plants with more than 10 nodes will be obtained. Incubation room conditions are 18–22°C, 16 hr light, 3000 lx.

3.2. Transfer to Slow-Growth Conditions

1. Cut plants growing on propagation medium at 18–22°C into single nodes, and place in tubes of 25 × 150 mm containing conservation medium (**Table 1**), approx 5 nodes/tube and 5 tubes/accession (*see* **Note 3**).
2. Grow plantlets for 2 3 wk at 18–22°C (until roots and the first two nodes are obtained), after which, transfer to a cool room at 6°C, 16 hr light, 1000 lx (*see* **Note 4**).
3. Make evaluations of stress response weekly for the first few months, and then make monthly evaluations (*see* **Notes 5**, **6**, and **7**).
4. When plantlets show glassiness, or become yellowish or brownish, replace with new nodes. Sometimes some explants produce this effect, whereas other nodes grow normally.
5. When plants begin to die and produce symptoms, such as stems turning yellow at the bottom, media show empty bubbles, or when plants reach the top of the tube, they must be subcultured. Under this evaluation, most accessions of the potato collection were maintained for 4 yr (**Table 2**; *see* **Note 8**).

3.3. Subculture After Slow Maintenance

1. Transfer plants to propagation medium when the maintenance period is finished in order to recover normal growth. During the first month, nodes will show slow growth. However, normal growth will be recovered during the next few months. Incubation room conditions for this period are the same as for the propagation room. If stress symptoms persist, a second micropropagation (**Subheading 3.1.**) is necessary (*see* **Notes 1** and **3**) .
2. Place the micropropagated plants obtained on conservation medium (**Subheading 3.2.**).

3.4. Long-Term Maintenance Through in Vitro Tubers

1. Propagate in vitro plantlets in magenta boxes using propagation medium.
2. After 3 wk of growth, add the liquid tuberization medium over the plants growing in the magenta boxes, and place in a darkroom for 2–4 mo.
3. Microtubers appear from the second month and continue growing for the next few months.
4. When the microtubers have taken on a rounded form and the connection with the stolon is tight, they are ready for harvesting.
5. Store microtubers either in Petri dishes with a wet filter paper or embedded in conservation medium.

4. Notes

1. Plantlets for slow-growth conservation must come from young cultures that have good physiological status in order to obtain the best effect of osmotic stress and therefore a longer period of conservation.
2. Maintenance of large collections involves the use of numbers or names to identify accessions. Databases (Foxpro for Windows, MS Access queries, or MS SQL-Server) with specific information can be useful for management, because they provide a source of labels to print when necessary.
3. When plantlets grown in conservation medium are subcultured onto new conservation medium after their growth period, viability in nodes is reduced and plantlets have low vigor. In this case, we recommend a source of plantlets growing on propagation medium.
4. The culture growth room for longer periods of plantlet conservation has to be maintained in the best aseptic conditions as possible by avoiding sources of contamination (dust, dirt, or contaminated material) and cleaning the room frequently (using bactericides, fungicides, and acaricides). At the same time, a restricted area has to be considered to prevent trespassing. To ensure the stability of culture room growth conditions, we recommend using an electronic alarm system connected to a control panel in a secure room.
5. During growth, the lateral buds are stimulated to grow and produce several branches that during the first year, look like a bush at the bottom of the tube. After the first year, each branch continues growing toward the top of the tube. Aerial roots are normally present and develop profusely over time. Leaves are normally very small, and stems are thick.
6. Slow growth can be affected by a reduction in osmotic stress produced when plants take up mannitol, thus accumulating the substance in tissues or using it as a source of carbon. This has been reported in wheat (embryo culture), rape (seedling culture), and potato (stem culture) ***(12)***.
7. There is no difference between dirty (virus-infected) and clean (virus-free) plantlets growing on conservation medium.
8. The genetic stability of plantlets growing in slow-growth conditions has to be monitored; there are some indications that low temperatures (5–10°C) might cause biochemical lesions during seed storage and cryopreservation. In this case, morphological comparisons and molecular techniques should be developed to compare field and in vitro accessions.

References

1. Dodds, J. H. and Roberts, L. W. (1985) *Experiments in Plant Tissue Culture,* 2nd ed. Cambridge University Press, Cambridge, NY.
2. George, E. F. and Sherrington, P. D. (1984) *Plant Propagation by Tissue Culture.* Exegetics Ltd., Edington, Wetsbury, Wilts, England.
3. Hughes, K. W. (1981) In vitro ecology: exogenous factors affecting growth and morphogenesis in plant culture system. *Environ. Exp. Botany* **21,** 281–288.

4. Lizarraga, R., Huaman, Z., and Dodds, J. H. (1989) In vitro conservation of potato germplasm at the International Potato Center. *Am. Potato J.* **66,** 253–276.
5. Wescott, R. J. (1981) Tissue culture storage of potato germplasm. 2. Use of growth retardants. *Potato Res.* **24,** 343–352.
6. Wescott, R. J. (1981) Tissue culture storage of potato germplasm. 1. Minimal growth storage. *Potato Res.* **24,** 331–342.
7. Thompson, M. R., Douglas, T. J., Obata-Sasamoto, H., and Thorpe, T. A. (1986) Mannitol metabolism in cultured plant cells. *Physiol. Plant.* **67,** 365–369.
8. Espinoza, N., Lizárraga, R., Sigueñas, C., Buitrón, F., Bryan, J., and Dodds, J. H. (1992) *Tissue Culture: Micropropagation, Conservation, and Export of Potato Germplasm.* CIP Research Guide 1. International Potato Center, Lima, Peru.
9. Estrada, R., Tovar, P., and Dodds, J. H. (1986) Induction of in vitro tubers in a broad range of potato genotypes. *Plant Cell Tiss. Org. Cult.* **7,** 3–10.
10. Tovar, P., Estrada, R., Schilde-Rentschler, L., and Dodds, J. H. (1985). Induction of in vitro potato tubers. *CIP Circular 13* **(4):1–4**. International Potato Center, Lima, Peru
11. Murashige, T. and Skoog, F. (1962) A revised medium for rapid growth and bio-assays with tobacco tissue culture. *Physiol. Plant.* **15,** 473–497.
12. Lipavska, H. and Vreugdenhil, D. (1996) Uptake of mannitol from the media by in vitro grown plants. *Plant Cell, Tiss. Org. Cult.* **45,** 103–107.

III

Plant Propagation In Vitro

10

Micropropagation of Strawberry via Axillary Shoot Proliferation

Philippe Boxus

1. Introduction

Elite strawberry plants can only be multiplied by a vegetative method. This explains the great dissemination of parasites, such as viruses, mycoplasmas, nematodes, soil fungi, and tarsonems, which are transmitted directly or indirectly to the descendants. The impossibility of efficiently fighting against such parasites, the introduction of nonresistant cultivars, and international trade have rendered the problem more and more acute. However, most strawberry-producing countries have drawn up a program of certification that takes these various parasites into account.

After heat therapy or meristem culture, healthy plants are isolated in screened cages, far from any source of contamination. The soil is disinfected by fumigation. Every mother plant is indexed, and daughter plants are then planted in the open air in well-isolated and disinfected plots. They remain under very strict phytosanitary observation. The plants produced in this way constitute registered stock, which is considered virus-free, though it is impossible to give absolute evidence of this. This traditional method of multiplying healthy plants requires a minimum delay of 3 yr before a healthy cultivar is at the producer's disposal.

Massive production of strawberry plants in vitro, introduced in 1974 *(1)*, offered an interesting improvement to this very strict and very slow scheme. Therefore, within five years, European growers had adopted this method. The technical process is very simple. The presence of 1 mg/L benzylaminopurine (BAP), an active cytokinin, in the basal culture medium induces an extensive development of axillary dormant buds. This axillary branching continues

From: *Methods in Molecular Biology, Vol. 111: Plant Cell Culture Protocols*
Edited by: R. D. Hall © Humana Press Inc., Totowa, NJ

from one subculture to the next, but as soon as the transfer is made onto a BAP-free basal medium, the branching stops, and rooted strawberry plantlets develop.

The first industrial application of strawberry micropropagation was reported in Germany *(2)*. About 10 years later, the annual production of micropropagated strawberry plants reached 7.5 million. This high-quality material, used as stock mother plants, produces a large portion of the billion conventional strawberry plants annually produced in the world (CEE, Cost 87, January 1988). In the US alone, in 1985, 350,000 plants were raised in vitro, whereas 700 million plants were produced in the field *(3)*.

Numerous field trials, made to compare fruit production of healthy plants orginating from tissue culture with standard material, did not show any statistical differences *(4–7)*. However, some sporadic occurrences of abnormal fruit setting were observed on specific cultivars. A hyperflowering habit entailed malformations and small fruits, although this trouble was only present on specific cultivars after numerous subcultures. Therefore, it was decided in several countries to limit to 10 the number of subcultures for each mericlone *(6)*. Moreover, legislation in some countries rejected or limited the use of this new methodology. In The Netherlands, for instance, only the everbearing cultivars could be propagated in vitro.

Although the limitation of subculture number was very effective in controlling the fruit quality of strawberry microplants, we were obliged to wait for Jemmali et al.'s researchs to understand the reasons for this hyperflowering behavior *(8)*. Indeed, they observed a new kind of bud during the proliferation phase. Named "stipular buds," they appeared at a specific position on the stipule (**Fig. 1**). After isolation and transfer onto a proliferation medium, they showed a very high propagation rate in vitro, and a hyperflowering habit in production fields. Jemmali et al. identified a higher free BAP content in these buds *(9)*.

It is possible to limit the occurrence of these stipular buds by decreasing the BAP concentration from 1 mg/L to 0.5 or 0.25 mg/L. Moreover, the number of subcultures has to be strictly limited to 10. Under these conditions, we have a very high-quality stock material, which is well adapted to the new intensive crop system.

These observations underline the risks introduced by adventitious regeneration in a mass propagation system. However, on the contrary, the axillary branches developed from preformed axillary buds are genetically and phenotypically stable, as was already underlined *(10)* and again confirmed in our very recent trials on strawberry plants *(11)*.

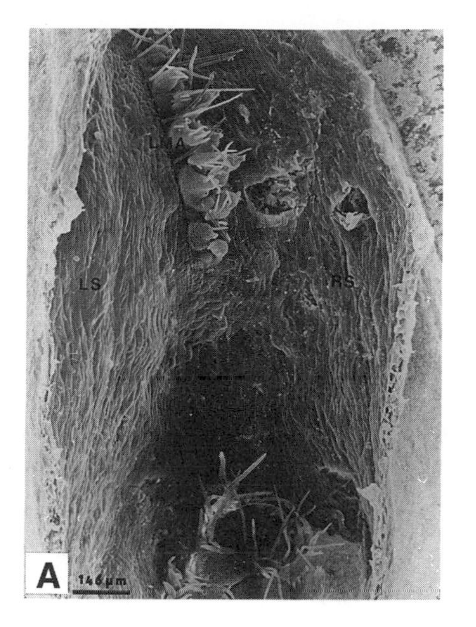

Fig. 1. Neoformation of stipular buds on cv 'Gorella' subcultured on a proliferation medium containing 1 mg/L BAP. **(A)** The buds present a fasciation, parallel to the petiole axis. **(B)** The stipular bud develops just on the middle of the two stipule tips. LS: left stipule, RS: right stipule, SB: stipular bud, AB: axillary bud, PA: petiole axis.

Table 1
Stock Solutions for Strawberry Culture Medium (Knop Macronutrients) ***(12)***

$Ca(NO_3)_2$	100 g/L
KNO_3	100 g/L
$MgSO_4 \cdot 7H_2O$	100 g/L
KH_2PO_4	100 g/L

2. Materials

1. Plant material must be selected in a field production. From the selected mother plants, collect preferentially the youngest runners (*see* **Notes 1** and **2**).
2. Reagents for plant material sterilization: alcoholic solution 96%, calcium hypochlorite 5% (w/v, *see* **Note 3**).
3. Binocular microscope and dissecting tools, razor blades and handles, scalpels with disposable blades, and fine forceps adapted for meristem excision (12 cm long).
4. Stock mother solutions: prepared with chemicals, quality pro analysis (*see* **Note 4**)
   a. Macronutrients: the four salts are stored at room temperature in liquid solution (100 g/L) (**Table 1**).

Table 2
Stock Solution for (1000X) Murashige and Skoog Micronutrients *(13)*

Salts	MS solution, mg/L end concentration	1 L MS: stock solution 1,000X, mg/L
$MnSO_4 \cdot H_2O$	16	16,000
$ZnSO_4 \cdot 7H_2O$	8.6	8,600
H_3BO_3	6.2	6,200
KI	0.83	830
$Na_2MoO_4 \cdot 2H_2O$	0.25	250
$CuSO_4 \cdot 5H_2O$	0.025	25
$CoCl_2 \cdot 6H_2O$	0.025	25

Table 3
Stock Solution for (1000 x) Murashige and Skoog Vitamins and Amino Acids *(12)*

Vitamins or amino acid	MS solution, mg/L end concentration	1 L MS: stock solution 1,000X, mg/L
Nicotinic acid[a]	0.5	500
Pyridoxine, HCl	0.5	500
Glycine	2.0	2000
Thiamine, HCl	0.1	100
Meso-inositol	100.0	100,000

[a]Nicotinic acid must be dissolved by gentle heating.

b. Micronutrients: strength 1000X, stored at +4°C (**Table 2**).
c. Vitamins: strength 1000X , stored at +4°C (**Table 3**).
d. Growth substances:
   i. Cytokinin BAP strength 1000X, dissolved in a few drops HCl 32%, then make up to volume with distilled water, stored at +4°C in dark bottle.
   ii. Auxin, indolyl butyric acid (IBA), strength 1000X, dissolved in 1 *M* KOH, stored at +4°C in dark bottle.
   iii. Gibberellic acid, GA3, strength 1000X, dissolved in water, stored at +4°C.

5. Agar and glucose of good quality (*see* **Note 5**).
6. Glassware:
   a. Tissue-culture tubes: 16 x 75 mm with screw cap for meristem culture, and 25 x 150 mm with loose-fitting caps (kaputs, Belco) for the first subcultures.
   b. Tissue-culture jars, 0.5 L, diameter 10 cm, with a glass lids (*see* **Note 6**).
   c. Precut food polythene film (23 x 23 cm) to seal the lids of the jars.

Table 4
Composition of Strawberry Culture Media *(1)*

	Meristem culture medium	Propagation medium	Rooting medium
Macronutrients	Knop	Knop	Knop
Micronutrients	Knop	Knop	Knop
Vitamin mixture	MS	MS	MS
Hormonal substances (mg/L)			
BAP	0.1	0.5	—
IBA	1.0	1.0	1.0
GA_3	0.1	0.1	—
Glucose (g/L)	40.0	40.0	40.0
Agar (g/L)	5.2	5.2	5.2.
Activated charcoal (g/L)	—	—	0.5
pH	5.6	5.6	5.6

MS: Murashige and Skoog; BAP: benzylaminopurine; IBA: indol butyric acid; GA_3: gibberellic acid

7. Horizontal laminar air flow hood for sterile transfers.
8. Sterile disposable Petri dishes (90–100 x 15 mm), fine but long forceps (25 cm) to divide clusters of axillary shoots.
9. Draft-free growth room with air conditioning, temperature between 22 and 25°C. The photoperiod is fixed at 16–8 h/d and night, with a light intensity of 30–50 µmol/m^2/s, diffused by daylight TL lamps. The control of relative humidity is only necessary in very humid countries. The growth room shelves (0.75- or 1.50-m width) are metal in order to prevent an excessive heat inside the jars resulting from the greenhouse effect during the light period (*see* **Note 7**).
10. Greenhouse facilities or plastic tunnels must be available during the five or six weaning weeks. The strawberry plantlets are normally very resilient. Therefore, heating or cooling equipment is not obligatory.

3. Methods

After media preparation and the sterilization of the plant material, all subsequent steps should be carried out in a flow cabinet under aseptic conditions.

3.1. Preparation of Culture Medium

1. Prepare the culture media from mother stock solutions and distilled water. Except for the hormonal balance, the media have the same compositions for the three propagation stages (**Table 4**).
2. Prepare culture media in a gauged flask, and before adding the agar and the glucose, adjust the pH to 5.6 with 1 *M* KOH or 1 *N* HCl solutions.
3. After boiling, the medium is distributed, 2.5 mL in the small test tubes, 15 mL in the large ones and 140 mL in the 0.5-L jars. Finally the material is

sterilized by autoclaving at 110°C for 20–30 min, depending on the size of your autoclave.

4. After autoclaving, the jars or test tubes are kept in a clean room. They are ready to use 1 d later (*see* **Note 8**).

3.2. Initiation Phase or Meristem Tip Culture

1. The position of the meristem tip on the mother plant does not influence the plant propagation ability. However, excision is easier from young runner tips than from well-developed plants (*see* **Note 9**).
2. Disinfect the samples by washing with water and weak detergent. Then, dip for 2 min in an absolute alcohol solution. Afterward, dip the samples for 10–20 min in a 5% calcium hypochloride solution. Finally, the samples are transferred to a washing bath filled with sterile water (*see* **Note 10**).
3. The samples are excised one by one using fine forceps and pieces of razor blade. The explant size should be <0.3 mm to guarantee a maximum sanitary state of the produced plants. They are aseptically placed in test tubes (16 x 75 mm) containing 2.5 mL meristem culture medium.
4. The cultures are placed on the shelves in the growth room.
5. About 1 mo later, the excised meristem tips have developed into a rosette of leaflets. They must be transferred to a propagation medium to start the mericlonal propagation (*see* **Note 11**).

3.3. Mericlonal Propagation

1. As soon as the first rosette of leaflets has developed to a size of a few millimeters, transfer to a 25 x 150 culture tube sealed by kaputa (loose-fitting caps), containing 15 mL propagation medium. Each culture tube is labeled to keep the mericlones separate.
2. The racks with the culture tubes are placed in the light on the shelves in the growth room for 6–8 wk.
3. Within 3 or 4 wk after this transfer, two or three new buds will appear at the base of the petioles of the oldest leaves. These young axillary buds grow very quickly and, in turn, produce new axillary buds. After 6–8 wk in the presence of cytokinin (BAP 0.5 mg/L), the axillary buds invade the entire culture tube. The initial plantlet is transformed into a cluster of buds, often without any callus. These clusters may contain 15–25 small buds. Each has several short petioles, ended by a very small unifoliate leaf. No roots are present on these buds.
4. At this stage, the clusters of buds may be aseptically separated and transferred onto a fresh medium to continue the clonal propagation. Every 6–8 wk, one isolated bud will produce 15–25 new axillary buds (*see* **Note 12**).
5. During the second division of the clusters, only a small fraction of the isolated buds is transferred to proliferation medium to continue the propagation. The remainder is placed on rooting medium to be stored at +2°C as mother stock plants ("germplasm") or to produce rooted plantlets to be used for virus indexing (see **Subheading 3.4.**).

6. Economically, the division within isolated buds is not very easy and takes a lot of time. Therefore, after one or two transfers on proliferation medium, the clusters are now divided more roughly into four to five small tufts of buds. These are transferred onto fresh proliferation medium, 7 explants/0.5 L jar (*see* **Note 12**).
7. The jars are placed on the shelves in the light. About 4 wk later in the presence of 0.5 mg/L cytokinin, new large clusters are formed and may be divided further (*see* **Note 13**).
8. To maintain a high proliferation rate, the transfers must be done immediately after the appearance of the first roots. In this way, the proliferation rate is about four- to fivefold per month, i.e., several million from one meristem in only 1 yr (*see* **Note 14**).

3.4. Rooting Phase

1. To obtain rooted plantlets, the clusters are divided into small tufts of three well-developed buds. These explants are transferred to a BAP-free medium containing 0.5 g/L activated charcoal. Each 0.5-L jar contains 15 explants (*see* **Note 15**).
2. Placed on the lighted shelves in the growth room, the first roots appear 7–10 d later, and the first true leaves (trifoliate leaves) develop (*see* **Note 16**).
3. Within 5–6 wk, well-rooted plantlets occupy the whole jar. These have developed 4- to 8-cm high leaves and 3- to 7-cm long roots. They are now ready for weaning (*see* **Note 17**).
4. Under **Subheading 3.3., step 5**, the possibility of creating a germplasm with a part of the clusters obtained from the second or third subculture was considered. Toward this aim, isolated buds are placed in 25 x 150 mm test tubes, closed with kaputs and sealed with food polythene film. After 4–5 wk in the light, the test tubes containing well-developed, rooted plantlets can be stored in complete darkness in a cold room at +2°C. These can then be stored in this way for 2 or 3 yr. These cold-stored plantlets could then be used to start a new propagation line. They will replace the material coming from meristem culture. Of course, in this case, it will be necessary to take the total number of subcultures into account to avoid the risk of deviant plants emerging (see **Subheading 1.**).
5. Under **Subheading 3.3., step 5**, it was also recommended to root, during the first steps of the propagation process, a minimum of two plants, which must serve for virus indexing. As the propagation line is pursued during the indexing tests, it is evident that all the production should be destroyed in the case of virus contamination (*see* **Note 18**).

3.5. Weaning Phase

1. Rooted plantlets are removed from the jars coming directly either from the growth room or the cold room.
2. Wash under tap water to eliminate a maximum of gelled agar from the roots.
3. The small tufts of clean rooted plantlets are separated into in single plantlets with the help of fine forceps (*see* **Note 19**).

Table 5
Field Performance, Runner Production in Italy, Year 1979 *(4)*

Cultivar	Propagation[a]	Number of runners	% runners >8 mm
'Gorella'	TC	62.0	71.7
	standard	39.3	79.9
'Belrubi'	TC	26.2	88.6
	standard	10.8	83.4
'Aliso'	TC	146.3	54.2
	standard	89.5	61.5

[a]TC : micropropagation in tissue culture

4. After grading and counting, the wetted single plantlets are stored in small plastic boxes and placed at +2°C until the transfer to the soil (*see* **Note 20**).
5. The rooted plantlets are acclimatized in trays filled with a peat mixture substrate with a separation of 5 x 5 cm. They are covered with a plastic sheet. The optimum ambient temperature is 22°C/15°C day and night. Strawberry plantlets can survive more drastic temperatures without damage if they are covered by a plastic sheet. However, long periods of low temperature (±5°C) or high temperature (±32°C) must be avoided. Covering with a milky white plastic sheet can protect the plantlets from direct intense solar radiation.
6. Immediately after transplantation, the plantlets are watered with a solution containing a broad spectrum fungicide to avoid damage from fungi (*see* **Note 21**).
7. After the watering with the fungicide, the plastic sheet is firmly sealed. Every 2 or 3 d, the state of the plants must be controlled. As soon as there is need for a fungicide treatment, repeat.
8. After 2 or 3 wk, when the plants develop new leaves, the plastic sheet can progressively be lifted up.
9. Generally, 4 wk after soil transfer, the first runners come out. Then young plants can be planted in the nurseries for the production of runners (*see* **Note 22**).

3.6. Field Behavior

1. Runner production: The main characteristics of micropropagated strawberry plants in the field are a flatter habit of the nonflowering plants and a very large number of stolon initiations. At the beginning of the summer, the stolons are shorter and finer than those formed on standard plants. At the end of the growing season, the runner plants obtained from micropropagated plants are identical to the conventional ones. Only the number of runners per mother plant can differentiate the two origins (**Table 5**). This ability to produce larger quantities of runners disappears during the second year of production. This increased runner production is very important for the everbearing cultivars, which are generally poor producers of runners.

Table 6
Field Performance: Fruit Production in France, cv. "Gorella" (Mean of 3 Years, 1984–1986) *(6)*

	Southeast		Southwest	
	Commercial grade, g/plant	Mean weight, g/fruit	Commercial grade, g/plant	Mean weight, g/fruit
Standard	837	18.2	510	17.5
New clones[a]	892	18.0	602	17.6
Old clones[b]	700	16.1	418	14.8

[a]New clones: 'Gorella' clones with <10 subcultures in vitro.
[b]Old clones: 'Gorella' clones with >30 subcultures in vitro.

Table 7
Fruit Production: Year 1988, cv. "Rapella," Wilhelminadorp and Horst (The Netherlands) *(7)*

	Production kg/acre, class I and II
First runner progeny of TC plants	278 A[a]
Second runner progeny of TC plants	281 A
Standard plants	244 B

[a]Different letters indicate statistical differences at $p = 0.05$.

2. Fruit production: The comparisons made in several countries, with different varieties, do not show any significant difference between the fruit production from micropropagated or standard strawberry plants, providing that mericlones with an high number of subcultures have been avoided (**Tables 6** and **7**).

4. Notes

1. To be sure to start the clonal propagation with the true cultivar, it is better to select the material in a field production. Ask the help of a good farmer, a specialist of the chosen variety, to select the best mother plants in the field. Some laboratories keep the mother plants in screenhouses for several years, and always use the same plants for the meristem excision. This method seems not as reliable as the previous one.
2. If possible, collect ±20 young runner tips, removed from several mother plants. Give a number to each mother plant, and keep the runner tips of each mother plant in separate bags. As soon as possible, place the bags in the vegetable compartment of a refrigerator. The samples can be stored easily for 2 or 3 wk before excision.
3. Calcium hypochlorite is very hygroscopic and must be stored under dry conditions. The 5% calcium hypochlorite solution must be prepared fresh before use. This solution can be used only for 2 or 3 d.

4. All the mother stock solutions are prepared from proanalysis chemical reagents dissolved in distilled water. Preparation of large quantities is possible. In this case, distribute the mother stock solution into different containers. Sterilize the mineral mother stock solutions by autoclaving, freeze the mother stock solutions of vitamin, and growth substances. Before use, always control that there is no precipitation in the concentrated mother solutions.
5. Use a microbiological-grade agar and sugar.
6. Some laboratories use other glass containers, named "Meli" jars. These have a polythene lid and a content of 350 mL.
7. The control of relative humidity is important in countries with very humid and hot seasons, monsoon seasons for instance. If not controlled, the risks of mould and bacterial contaminations are too high. In spite of the air regulation in the room, a temperature gradient exists in most growth rooms. Generally, the upper shelves are warmer than the lower ones. Too high-temperature in the jars results from a high ambient temperature, but also from the greenhouse effect, which can cause an increase of 6–8°C in the jars in comparison with the ambient air. Therefore, it is important to place the jars immediately on a metal shelf. This has the ambient temperature and will eliminate the excessive temperature from the jars.
8. After autoclaving, cool the jars as early as possible to avoid hydrolysis of the culture medium. This can occur in summer during hot days, especially when the media contain little agar. In very hot countries, it will be necessary to increase the agar concentration.
9. To excise the meristems, the very young runner tips are the best plant material. This material is collected in a fruit production field in June, before the runners have formed leaves and roots. This very young material has fewer hairs surrounding the meristematic zone and results in little contamination. When meristems are removed from dormant plants or from cold stored plants, a higher percentage of the cultures will be contaminated. When only crowns of adult plants are available, avoid deep crowns or those from cold stored plants, because these are frequently contaminated by bacteria.
10. After sterilization with calcium or sodium hypochloride, it is not necessary to rinse the explants several times with sterile water, because chloride ions are not toxic in the culture medium. If after 1 mo the excised meristem has not developed a rosette of leaflets, this explant can be subcultured for another month onto the same meristem culture medium.
11. At each transfer, carefully clean the explants by removal of all necrotic parts, callus, and roots to avoid oxidation or diffusion of phenolic compounds into the fresh culture medium.
12. Dip the explants deeply into the medium to assure a good contact between the dormant axillary buds and the medium. This will enlarge the number of axillary shoots developed during the subculture. There is no risk of asphyxia provided that only a portion of the explant is out the medium.
13. The varietal sensitivity to the temperature is variable. The most resilient cultivars should be placed on the upper shelves. A general yellowing of the plantlets dur-

ing the proliferation phase is characteristic of too high a temperature. In this case, change the level of the shelves or decrease the air temperature, but take care, because at a temperature that is too low, the plants will not grow fast enough.

14. At each transfer, take the number of subcultures into account to be sure to stop at the 10th subculture. Dip the explants correctly into the medium. Take care not to have too hard a medium. Control that the mother jar, which must be divided, is not contaminated by fungus or bacteria. For these latter infections, it is necessary to inspect visually the jars by transparency. If there is any doubt, e.g., presence of white cloud at the base of the explants or some turbidity in the condensation water at the surface of the medium, discard this mother jar. Some bacteria are very difficult to eliminate, even by flaming! Finally, always transfer before or just at the appearance of the first roots. If this is not possible, e.g., no media or no manpower available, immediately place the jars in the cold room at +2°C to stop development. This last point is very important to maintain the high rate of multiplication.
15. The introduction of activated charcoal in the rooting medium improves the growth of the roots a little, but also, the black color of the medium is an excellent identification label. Some laboratories prefer MS medium *(12)* to Knop medium *(13)* as macronutrients during the rooting phase. Indeed, on this medium, the plantlets will grow faster. However, these MS plantlets are softer and also more susceptible during the weaning phase.
16. Plantlets in rooting phase are more sensitive to an excess temperature. In this case, the roots stop their growth and develop a black tip. As soon as the jars are placed at a lower temperature, the roots start to grow again.
17. To extend the period of work in the laboratory, it is possible to root strawberry plantlets some weeks before the normal weaning time. Rooted plantlets can be stored at +2°C for 4–6 mo without any damage, but in the case of long storage periods, place the rooted plantlets as early as possible in the cold room to avoid oxidations or phenol production.
18. Very often, the propagation of new mericlones is pursued during the time required for virus indexing. However, until now, the percentage of virus elimination has been about 100%, so the risk of being obliged to destroy all the production for sanitary reasons is very low.
19. In general, each 0.5-L jar produces 45 to 60 rooted plantlets, 3–7 cm high. Some cultivars with very fine buds can give, per jar, higher numbers of rooted plantlets. However, they are so thin that it is better to keep 2 or 3 buds together to facilitate their acclimatization.
20. The little rooted plantlets are stored in 12 x 8 x 4 cm plastic boxes covered by a see-through lid. These plantlets are moist when they are introduced in these boxes. Also, after a few days, check the condensation water inside and eliminate the excess water to keep the material in good condition for 1 mo.
21. Thiram (TMTD) 0.5–1.0% is a very effective fungicide, without any phytotoxicity for control of *Botrytis* and *Pythium*, the most dangerous fungi during the acclimatization phase. Avoid use of very freshly sterilized substrates, which are too easily invaded by unfavorable germs.

22. Transfer the plants to the nursery field as soon as possible to avoid a halt in the development of the young runners. Severe stress, induced by a poor transfer from the greenhouse to the field, can even induce flowers under long day length and, consecutively, can stop runner production for several weeks.

References

1. Boxus, Ph. (1974) The production of strawberry plants by *in vitro* micropropagation. *J. Hort. Sci.* **49,** 209–210.
2. Westphalen, H. J. and Billen, W. (1976) Erzugung von Erdbeerpflanzen in grossen Mengen durch Sprosspitzenkultur. *Erwerbsobstbau* **18,** 49, 50.
3. Swartz, H. J. and Lindstrom, J. T. (1986) Small fruit and grape tissue culture from 1980 to 1985 Commercialization of the technique, in *Tissue Culture as a Plant Production System for Horticultural Crops* (Zimmerman, R. H., Griesbach, R. J., Hammerschlag, F. A., and Lawson, R.H, eds.), Martinus Nijhoff, Dordrecht, pp. 201–220.
4. Damiano, C. (1980) Strawberry micropropagation, in *Proceedings of the Conference on Nursery Production of Fruit Plant Through Tissue Culture: Applications and Feasibility*. USDA-SEA, Agricultural Research Results, ARR-NE. Beltsville, MD, pp. 11–22.
5. Boxus, Ph. (1985) Comportement du fraisier issu de micropropagation *in vitro*. *Fruit Belge* **410,** 106–110.
6. Navatel, J. C., Varachaud, G., Roudeillac, P., and Bardet, A. (1986) Multiplication *in vitro*: comportement agronomique de plants de fraisiers issus de pieds-mères produits par micropropagation par rapport au matériel classique. Synthèse de trois années d'expérimentation. *Infos-CTIFL* **21,** 2–8.
7. Dijkstra, J. (1988) Samenvatting onderzoekresultaten vollegrondsaardbeien (1987) Internal Report (personal communication).
8. Jemmali, A., Boxus, Ph., and Kinet, J. M. (1992) Are strawberry arising from adventitious stipule buds also true to type? *Acta Hort.* **319,** 171–176.
9. Jemmali, A., Boxus, Ph., Kevers, Cl., and Gaspar, Th. (1995) Carry-over of morphological and biochemical characteristics associated with hyperflowering of micropropagated strawberries. *J. Plant Physiol.* **147,** 435–440.
10. d'Amato, F. (1977) Cytogenetics of differentiation in tissue and cell cultures, in *Applied and Fundamental Aspects of Plant Cell, Tissue and Organ Culture* (Reinert, J. and Bajaj, Y. P. S., eds.), Springer Verlag, Berlin, pp. 344–357.
11. Jemmali, A., Robbe, A., and Boxus, Ph. (1997) Histochemical events and other fundamental differences between axillary and adventitious stipular shoots in micropropagated strawberry. *Third Intern. Strawberry Symposium, Veldhoven (The Netherlands). Acta Hort.* **439,** 341–345.
12. Knop, W. (1965) Quantitative Untersuchungen uber die Ernahrungsprozesse der Pflanzen. *Landwirstsch. Vers. Stn.* **7,** 93–107.
13. Murashige, T. and Skoog, F. (1962) A revised medium for rapid growth and bioassays with tobacco tissue culture. *Physiol. Plant* **15,** 473–497.

11

Meristem-Tip Culture for Propagation and Virus Elimination

Brian W. W. Grout

1. Introduction

The essence of meristem-tip culture is the excision of the organized apex of the shoot from a selected donor plant for subsequent in vitro culture. The conditions of culture are regulated to allow only for organized outgrowth of the apex directly into a shoot, without the intervention of any adventitious organs *(1–3)*. The excised meristem tip is typically small (often <1 mm in length) and is removed by sterile dissection under the microscope, as in the potato example detailed in this chapter (**Fig. 1**). The explant comprises the apical dome and a limited number of the youngest leaf primordia, and excludes any differentiated provascular or vascular tissues. A major advantage of working with such a small explant is the potential that this holds for excluding pathogenic organisms that may have been present in the donor plants from the in vitro culture (*see below*). A second advantage is the genetic stability inherent in the technique, since plantlet production is from an already differentiated apical meristem and propagation from adventitious meristems can be avoided *(3–9)*. Shoot development directly from the meristem avoids callus tissue formation and adventitious organogenesis, ensuring that genetic instability and somaclonal variation are minimized. If there is no requirement for virus elimination, then the less demanding, related technique of shoot-tip culture may be more expedient for plant propagation. In this related procedure the explant is still a dissected shoot apex, but a much larger one that is easier to remove and contains a relatively large number of developing leaf primordia. Typically, this explant is between 3 and 20 mm in length, and development in vitro can still be regulated to allow for direct outgrowth of the organized apex.

From: *Methods in Molecular Biology, Vol. 111: Plant Cell Culture Protocols*
Edited by: R. D. Hall © Humana Press Inc., Totowa, NJ

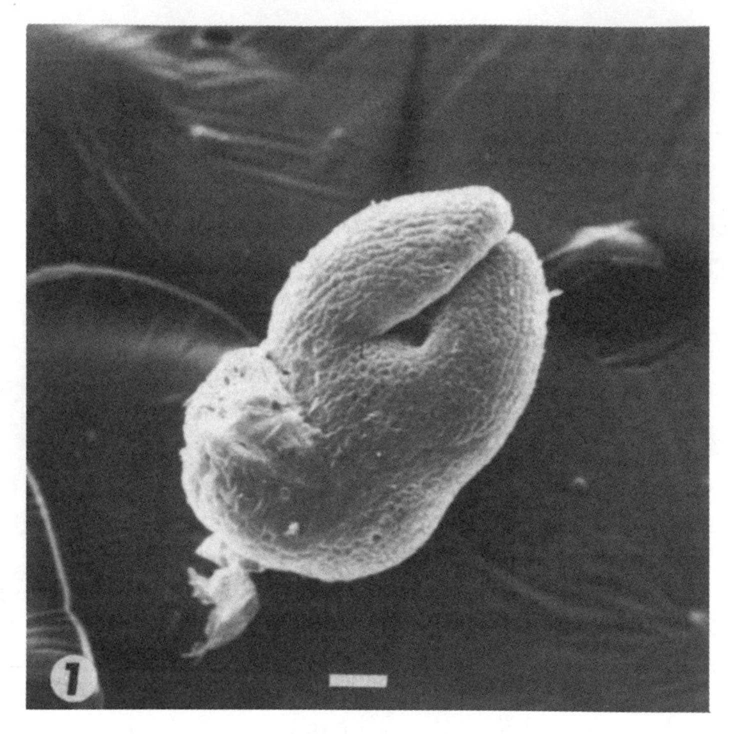

Fig. 1. A freshly excised meristem tip from an axillary bud of the potato *Solanum tuberosum*. The two smallest emergent leaf primordia are present. Scale bar represents 50 μ*M*.

The axillary buds of in vitro plantlets derived from meristem-tip culture may also be used as a secondary propagule. When the in vitro plantlet has developed expanded internodes, it may be divided into segments, each containing a small leaf and an even smaller axillary bud (**Fig. 2**). When these nodal explants are placed on fresh culture medium, the axillary bud will grow directly into a new plantlet, at which time the process can be repeated. This technique adds a high propagation rate to the original meristem-tip culture technique, and together the techniques form the basis of micropropagation, which is so important to the horticulture industry (*1–4*; *see* **Table 1**).

It is possible, however, that callus tissue may develop on certain portions of the growing explant, particularly at the surface damaged by excision (**Fig. 3**). The only acceptable situation under such circumstances is where the callus is slow-growing and localized, and the callus mass and any organized development on it can be excised at the first available opportunity. Studies from the author's laboratory on cauliflower plants regenerated from wound callus and floral meristems in vitro showed normal phenotypes for all the plants produced by

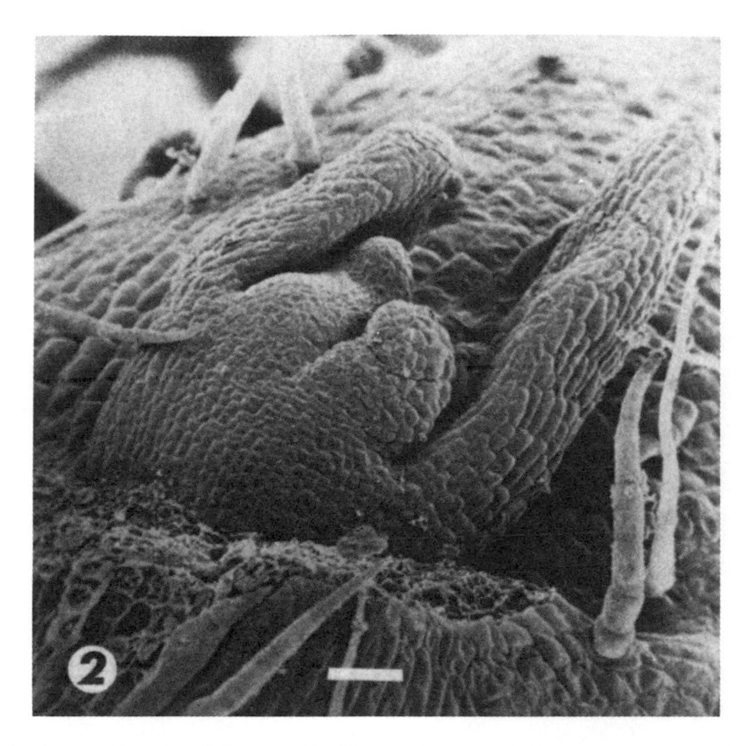

Fig. 2. An axillary bud from a freshly excised nodal segment, taken from a meristem-derived potato plantlet in vitro. A number of small leaf primordia are apparent, emphasizing the effectively normal structure of this very small axillary organ. Scale bar represents 50 μ*M*.

Table 1
The Propagation Potential Inherent in the Meristem-Tip Culture Technique[a]

Time from culture initiation, mo	No of nodes available[b] for further subculture, equivalent to plantlet number
3	5
4	25
5	125
6	625
7	3125
8	15,625

[a]The data are for potato species, *Solanum curtilobum,* and were calculated from propagation rates achieved in the author's laboratory.
[b]Assumes 5 nodes available/plantlet.

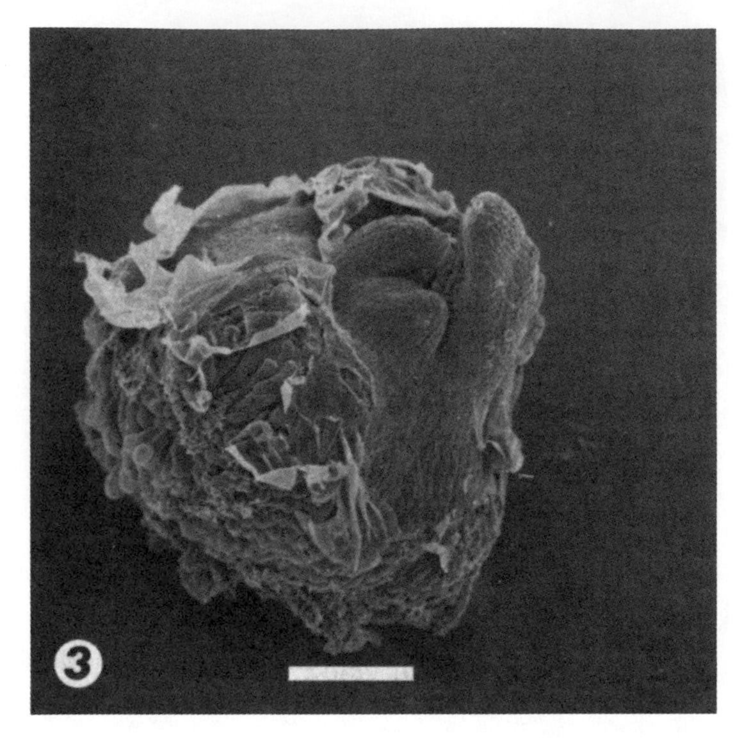

Fig. 3. A meristem-tip culture of potato 6 d after initiation. A distinct swelling can be seen at the wound surface, and larger callus cells are beginning to proliferate from it. Scale bar represents 500 μ*M*.

meristem culture, whereas those of callus origin included rosette and stunted forms, and both glossy and serrated leaves as examples of phenotypic deviance.

These abnormal plants, derived from callus tissue, and normal plants from meristem culture showed the same DNA levels, as measured by scanning microdensitometry, but showed considerable differences when the isozymes of acid phosphatase was examined by gel electrophoresis. This suggests a range of unacceptable variation at the gene level, introduced by the adventitious origin of the shoots.

A further advantage of meristem culture is that the technique preserves the precise arrangement of cell layers necessary if a chimeral genetic structure is to be maintained. In a typical chimera, the surface layers of the developing meristem are of differing genetic background, and it is their contribution in a particular arrangement to the plant organs that produces the desired characteristics, e.g., the flower color in some African violets. As long as the integrity of the meristem remains intact and development is normal in vitro, then the chimeral pattern will be preserved. If, however, callus tissues are allowed to

Table 2
The Relationship of Explant Size to Effective Virus (WCMV) Elimination in Clover[a]

Explant size, mm	No. explants taken into culture	No. of plants established in soil	No. virus-free plants
<0.6	90	18	18
0.6–1.2	113	45	19
1.3–1.8	190	102	25
1.9–2.4	158	88	11
2.5–3.0	174	92	11

[a]Data from ***(15)***.

form and shoot proliferation, subsequently, is from adventitious origins, then there will be a risk that the chimeral layers of the original explant may not all be represented in the specially required form in the adventitious shoots.

The technique of meristem culture may be exploited in situations where the donor plant is infected with viral, bacterial, or fungal pathogens, whether or not symptoms of the infection are expressed. The basis of eradication is that the terminal region of the shoot meristem, above the zone of vascular differentation, is unlikely to contain pathogenic particles. If a sufficiently small explant can be taken from an infected donor and raised in vitro, then there is a real possibility of the derived culture being pathogen-free. Such cultures, once screened and certified, can form the basis of a guaranteed disease-free stock for further propagation ***(9–11)***. The meristem-tip technique can be linked with heat therapy to improve the efficacy of disease elimination, or antiviral, chemotherapeutic agents may be investigated ***(9–14)***.

Whatever variants of technique are employed for virus eradication, the key to success is undoubtably the size of the explant. The smallest explants are those that typically, will be the least successful during in vitro culture, but will produce the highest proportion of virus-free material when entire plants are reared in the glasshouse or field. This is clearly illustrated by attempts at the elimination of WCMV from *Trifolium pratense* (**Table 2**; ***15***).

If meristem culture alone is not successful in producing any virus-free plants, then temperature-stress treatment of donor plants and/or the use of antiviral agents will have to be considered. There are no prescriptive, global methodologies for these treatments, and an empirical study will be required. Heat therapy relies on the growth of the donor plants at elevated temperatures, typically 30–37°C and may involve a treatment of several weeks. Where in vitro cultures provide the donor material, it is possible to heat treat the cultures also. A number of factors may contribute to the absence of virus particles from the

Table 3
The Effect of Extended Low-Temperature (2–4°C) Treatment on the Elimination of Hop Latent Viroid from Hop Plants[a]

	Duration of cold treatment, mo	Propagation of viroid-free plants
Variety 1	6	0/4
	17	4/23
Variety 2	21	1/14
	9	3/12

[a]Adapted from ***(14)***.

meristems of treated plants, including reduced movement of the particles to the apical regions, a thermally induced block on viral RNA synthesis, and inactivation of virus particles. It is useful to note that extended, low-temperature treatment of donor material may also be effective ***(14)***. In attempts to eliminate hop latent viroid, the low-temperature (2–4°C) treatment of parent plants was only effective if extended beyond 6 mo (**Table 3**). Particularly when subjecting donor plants to temperature stresses, it is extremely important to understand the physiology and developmental behavior of the plants concerned, so that maximum stress can be applied while maintaining an acceptable pattern of growth. A good knowledge of the taxonomy of the plant under investigation is also valuable when selecting likely treatments from the literature, since close, or distant, relatives can be identified.

Antiviral chemicals can be used as additives in the culture medium ***(11,12)***, and one of the most widely used is ribavarin, also known as virazole. This compound is a guanosine analog with broad-spectrum activity against animal viruses and appears also to be active against plant virus replication in whole plants. Increasing concentrations of ribavirin and increasing length of culture incubation in the presence of the compound typically increase the effectiveness of virus elimination, but slowed growth and phytotoxity may be evident at high concentrations.

The major advantages of meristem culture are that it provides:

1. clonal propagation in vitro with maximal genetic stability;
2. the potential for removal of viral, bacterial, and fungal pathogens from donor plants;
3. the meristem tip as a practical propagule for cryopreservation and other techniques of culture storage;
4. a technique for accurate micropropagation of chimeric material; and
5. cultures that are often acceptable for international transport with respect to quarantine regulations.

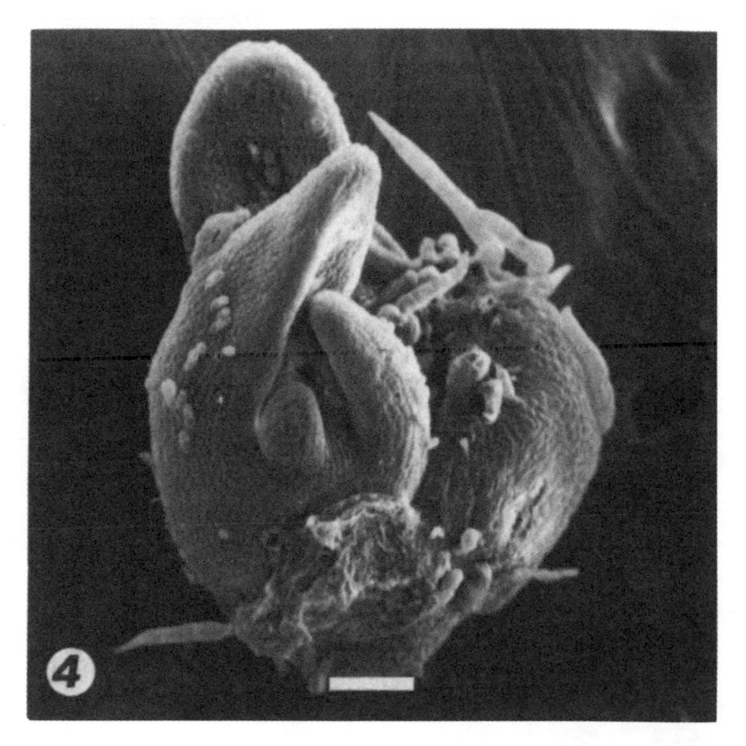

Fig. 4. A meristem-tip culture of potato 10 d after initiation. When excised, this explant had only two obvious leaf primordia, and has both developed and extended considerably during the culture period. Scale bar represents 500 μ*M*.

The technique of meristem culture, with optional procedures for virus elimination, is detailed below for potato species (*see* **Fig. 4**). The steps described are broadly applicable to a wide range of plant subjects, although formulations of media and reagents may need to be altered.

2. Materials

1. Glasshouse-grown potato plants provide the stem sections with axillary buds to be dissected. These are raised in an insect-proof facility where possible, and maintained on capillary matting on raised benches so that overhead watering is avoided. The donor plants can be subjected to heat treatments for virus eradication, involving growth at 33°C for 4–6 wk.
2. Wide-necked test tubes make suitable culture vessels for incubating meristem cultures, capped with purpose-bought tops, or aluminum foil, and sealed after inoculation with a Parafilm strip.
3. The culture vessels contain 2.5 mL of Murashige and Skoog medium (***16*** and *see* Appendix) solidified with 0.8–1.0% w/v agar and supplemented with 0.2 mg/L

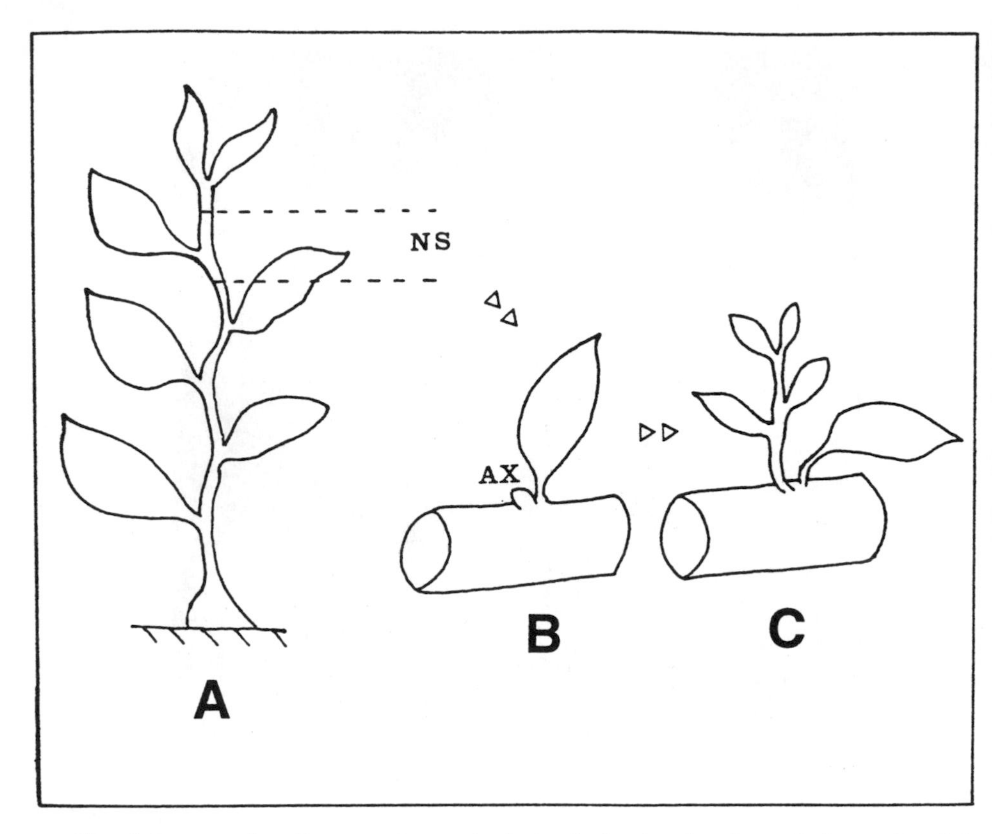

Fig. 5. Propagation from meristem-tip derived plantlets by the technique of nodal culture. **(A)** A plantlet showing extension growth in vitro. The nodal segment to be excised is indicated (NS). **(B)** The excised nodal segment as it is transferred onto fresh culture medium, showing the axillary bud (AX) that will be responsible for subsequent growth. **(C)** The pattern of development of a successful nodal segment culture, showing extension growth of the new plantlet.

1-naphthalene acetic acid and 0.5 mg/L gibberellic acid, pH 5.8 (*see also* ***17–20***). Alternatively, liquid media may be used in conjunction with paper bridges or fiber supports in the culture vessel. If required for virus elimination, ribovarin at 200 μM is filter-sterilized and added to the medium before solidification.

4. Nodal cultures, derived from in vitro plantlets, are excised as detailed in **Fig. 5** and cultured in identical vessels to the excised meristems. The growth medium is the same as that for meristem culture without growth hormone supplements.
5. Presterilization of the excised stem segments is carried out using an absolute ethanol dip for 30 s with a subsequent sterilization using 5% v/v sodium hypochlorite with a drop of Tween 80 (added as a wetting agent). This is particularly important when the tissue surfaces are waxy or coated with epidermal hairs. Sterile distilled water for rinsing is also required.

4. A dissection microscope with a magnification of at least 15× and mounted in a laminar flow cabinet is required. A piece of expanded polystyrene covered in white plastic film and taped to the microscope stage is ideal for holding sterilized stem segments in place for dissection, using sterilized pins. The tips of 12-gage hypodermic needles are used to carry out the dissection.
5. A growth room is needed that provides a controlled environment in which to incubate the cultures. Lighting at 4000 lx at culture level with a 16-h photoperiod, provided by warm white fluorescent tubes, and a constant temperature of 25 ± 1°C provides optimal culture conditions.

3. Methods

1. Select a suitable donor plant, in this case, any of the *Solanum tuberosum* ssp *tuberosum* types (*see* **Note 1**), following any desired temperature pretreatments. Excise stem segments containing at least one node from the donor plant.
2. Remove mature and expanding foliage to expose the terminal and axillary buds. Cut donor segments to 4-cm lengths, and presterilize by immersion in absolute ethanol for 30 s.
3. Sterilize by immersing the donor tissues in the sodium hypochlorite solution, with added detergent, for 8 min (*see* **Note 2**).
4. Following surface sterilization, rinse the tissues three times in sterile distilled water.
5. Mount the stem segment on the stage of the dissection microscope, and use the tips of hypodermic needles to dissect away progressively smaller, developing leaves to expose the apical meristem of the bud, with the few youngest of the leaf primordia (*see* **Note 3**).
6. Excise the explant tissue that should comprise the apical dome and the required number of the youngest leaf primordia (*see* **Note 4**).
7. After excision, the explant is transferred directly onto the selected growth medium, and the culture vessel is closed.
8. Transfer the completed meristem-tip culture to the growth room.
9. If the explant is viable, then enlargement, development of chlorophyll, and some elongation will be visible within 7–14 d (**Fig. 4**; *see* **Note 5**).
10. Maintain the developing plantlet in vitro until the internodes are sufficiently elongated to allow dissection into nodal explants.
11. To prepare nodal explants, remove the plantlets from the culture vessels under sterile conditions, and separate into nodal segments (**Fig. 5**). Each of these is transferred directly onto fresh growth medium to allow axillary bud outgrowth. Extension of this bud should be evident within 7–14 d of culture initiation (*see* **Note 6**).

4. Notes

1. Donor plants should be selected for general health and vigor. Donor tissues should be taken from young, actively growing stems to ensure the best chance of success. However, if donor plants have been subjected to temperature stresses as a pretreatment, then allowances must be made for apparent foliar injury and slowed

growth. Using donor plants that have been glasshouse-raised and only watered from below will help minimize infection problems.

2. Be careful. It is often easy to oversterilize during surface sterilization and lethally damage tissues and, therefore, the meristems in an attempt to eradicate surface pathogens.
3. It is only with practice that reproducible, high levels of culture success will be achieved, since early attempts at dissection will often result in damaged tissue. Hypodermic needles used as dissecting tools often need to be discarded after two to three meristem excisions.
4. It seems that the presence of leaf primordia is essential for successful culture growth, so removal of too small an explant may restrict success.
5. A consequence of dissection may be the production in the tissue of toxic, oxidized polyphenolic compounds ***(3)***. These are more prevalent in cultures of woody species. Their effects may be minimized by rapid serial transfer of the excised meristem tip to fresh medium as soon as significant browning of the medium occurs, or by reducing light available to the culture in the early stages of development. Attempts can be made to reduce oxidation of the secreted polyphenolic compounds by incorporating antioxidants into the growth medium. These might include ascorbic acid, dithiothreitol, and polyvinyl pyrrolidone.
6. Media are commonly species-specific with regard to the formulation of the required growth regulators and their concentrations, and may also be specific for the other organic and inorganic constituents. The specificity may even extend to the cultivar/race level with respect to optimal growth. For potato, a range of successful media for different varieties and species, at a range of ploidy levels, have been published ***(17–20)***. Reference sources can provide an appropriate growth medium to begin an empirical study, using the formulation for the closest taxonomic relative if no medium is detailed for the particular plant under investigation.

References

1. Murashige, T. (1978) The impact of plant tissue culture on agriculture, in *Frontiers of Plant Tissue Culture* (Thorpe, T. A., ed.), IAPTC, Calgary, pp. 15–26.
2. Murashige, T. (1978) Principles of rapid propagation, in *Propagation of Higher Plants Through Tissue Culture* (Hughes, K. W., Henke. R., and Constantin., eds.), US Department of Energy CONF-7804111, US Technical Information Center, Washington, DC, pp. 14–24.
3. Hu, C. Y. and Wang, P. J. (1984) Meristem, shoot-tip and bud culture, in *Handbook of Plant Cell Culture* (Evans, D. A., Sharp, W. R., Ammirato, P. V., and Yamada, Y., eds.), Macmillan, New York, pp. 177–277.
4. Debergh, P. C. and Zimmerman, R. H. (eds.) (1991) *Micropropagation. Technology and Application.* Kluwer Academic Publishers, p. 484.
5. Murashige. T. (1974) Plant propagation through tissue culture. *Ann. Rev. Plant Physiol.* **25,** 135–166.
6. Ancora, G., Belli-Donini, M. L., and Cozzo, L. (1981) Globe artichoke plants obtained from shoot apices through rapid in vitro micropropagation. *Sci. Hort.* **14,** 207–213.

7. Reisch, B. (1984) Genetic variability in regenerated plants, in *Handbook of Plant Cell Culture,* vol. 1, (Evans, D. A., Sharp, W. R., Ammirato, P. V., and Yamada, Y., eds.), Macmillan, New York, pp. 748–769.
8. Scowcroft, W. R., Bretell, I. R. S., Ryal, S. A., Davies, P., and Pallotta, M. (1987) Somaclonal variation and genomix flux, in *Plant Tissue and Cell Culture* (Green, C. F., Somers, D. A., Hackett, W. P., and Bisboev, D. D., eds.), A. Liss, pp. 275–288.
9. Vanzaayen, A., Vaneijk, C., and Versluijs, J. (1992) Production of high quality, healthy ornamental crops through meristem culture. *Acta Bot. Neerl.* **41,** 425–433.
10. Walkey, D. G. A. (1980) Production of virus-free plants by tissue culture, in *Tissue Culture Methods for Plant Pathologists* (Ingram, D. S. and Helgeson, J. P., eds.), Blackwell Scientific, UK, pp. 109–119.
11. Kartha, K. K. (1986) Production and indexing of disease-free plants, in *Plant Tissue Culture and Its Agricultural Applications* (Withers, L. A. and Alderson, P. G., eds.), Butterworths, UK, pp. 219–238.
12. Long, R. D. and Cassells, A. C. (1986) Elimination of viruses from tissue cultures in the presence of antiviral chemicals, in *Plant Tissue Culture and Its Agricultural Applications* (Withers, L. A. and Alderson, P. G., eds.), Butterworths, UK, pp. 239–248.
13. Lim, S., Wong, S., and Goh, C. (1993) Elimination of Cymbidium mosaic virus and Odontoglossum Ringspot Virus from orchids by meristem culture and thin section culture with chemotherapy. *Ann. Appl. Biol.* **122,** 289–297.
14. Adams, A. N., Barbara, D. J., Morton, A., and Darby, P. (1996) The experimental transmission of hop latent viroid and its elimination by low temperature treatment and meristem culture. *Ann. Appl. Biol.* **128,** 37–44.
15. Dale, P. J. and Cheyne, V. A. (1993) The elimination of clover diseases by shoot tip culture. *Ann. Appl. Biol.* **123,** 25–32.
16. Murashige, T. and Skoog, F. (1962) A revised medium for rapid growth and bioassay using tobacco tissue cultures. *Physiol. Plant.* **15,** 473–497.
17. Westcott, R. J., Henshaw, G. G., Grout, B. W. W. and Roca, W. M. (1977) Tissue culture methods and germplasm storage in vitro. *Acta Hort.* **78,** 45–49.
18. Miller, S. A. and Lipschutz, L. (1984) Potato, in *Handbook of Plant Cell Culture,* vol. 3 (Ammirato, P. V., Evans, D. A., Sharp, W. R., and Yamada, Y., eds.), Macmillan, pp. 291–326.
19. Henshaw, G. G., O'Hara, J. F. and Westcott, R. J. (1980) Storage of potato germplasm, in *Tissue Culture Methods for Plant Pathologists* (Ingram, D. S. and Helgeson, J. P., eds.), Blackwell, pp. 71–78.
20. George, R. A. T. (ed.) (1986) *Technical Guideline or Seed Potato Micropropagation and Multiplication.* Food and Agriculture Organization of the United Nations, Rome, Italy.

12

Clonal Propagation of Orchids

Brent Tisserat and Daniel Jones

1. Introduction

Methods for clonal propagation of the two major morphological groups of orchids, i.e., sympodials and monopodials, are presented. The first group, sympodials, includes such genera as *Cymbidium, Cattleya, Dendrobium,* and *Oncidium.* They are characterized by a multibranching rhizome that can supply an abundance of axillary shoots for use as explants. They were among the first orchids to be successfully propagated, and techniques for their in vitro initiation (i.e., establishment) and subsequent proliferation are well established ***(1–8)***. The second group, monopodials, include *Phalaenopsis* and *Vanda,* and are characterized by a single, unbranched axis that possesses few available axillary shoots for use as explants. Significantly different in their morphologies, these two groups require different approaches to explant selection and subsequent culturing. The successful large-scale micropropagation of monopodials is, in fact, a relatively recent achievement ***(9)*** and is the culmination of a wide variety of studies using different media compositions and supplements ***(10–20)***.

Since the introduction of orchid meristem culture in 1960, a divergence in orchid micropropagation techniques developed based on using two distinctly different culture schemes. In one method, the initial explant is induced into (and is subsequently maintained in) an undifferentiated callus or protocorm-like body (PLB) state. PLBs are spherical tissue masses that resemble an early stage of orchid embryo development. Proliferation occurs via PLB multiplication, and differentiation into plantlets is permitted only after the desired volume of callus is achieved. The alternate method minimizes the role of callus, and encourages the differentiation of cultures into plantlets early in the procedure. Consequently, proliferation is accomplished by the induction of axillary shoots from plantlets derived from the original explant.

From: *Methods in Molecular Biology, Vol. 111: Plant Cell Culture Protocols*
Edited by: R. D. Hall © Humana Press Inc., Totowa, NJ

This chapter describes four methods of propagation: one callus and one axillary shoot technique for each of the two morphological groups of orchids. Advantages and disadvantages associated with these two methods will be discussed and possible variation pointed out (*see* **Subheading 4.**).

2. Materials

1. Disinfesting solution: Use a solution of 1% sodium hypochlorite (household bleach diluted with distilled water) with 0.1% Tween 80 emulsifier. Prepare fresh solution for each treatment, and use within the hour.
2. Postdisinfestation holding solution: Prepare a 100-m*M* phosphate buffer with 8.76 m*M* sucrose. Adjust the pH to 5.8, and autoclave for 15 min at 1.05 kg/cm^2 and 121°C.
3. Culture media: Use Murashige and Skoog inorganic salts (*see* Appendix) with: 8.76 m*M* sucrose, 0.003 m*M* thiamine-HCl, 0.005 m*M* pyridoxine-HCl, 0.008 m*M* nicotinic acid, 0.555 m*M* myo-inositol, and 100 mg/L casein hydrolysate. This basic formulation is supplemented as indicated in **Table 1**. An initiation medium (IM) and a differing proliferation medium (PM) are required for each method. The proliferation steps of **Subheadings 3.3.** and **3.4.** necessitate the preparation of a series of media (**Table 1**) to determine optimum yields. The pH of all solutions is adjusted to 5.7 with 1.0 *N* HCl or NaOH before being dispensed into 25 × 150 mm culture tubes (for agar media) or 250-mL flasks (for liquid media) in 25-mL aliquots. Culture tubes and flasks, capped with appropriate closures, are autoclaved for 15 min at 1.05 $kg \cdot cm^{-2}$ and 121°C.
4. Culture conditions: Cultures receive 15–25 $\mu mol/m^2/s$ from wide-spectrum fluorescent tubes for 16 h daily. Temperature is maintained at 27°C.
5. Sympodial orchid explants (*see* **Table 2**). Developing shoots ($<^1/_2$ their mature size) are severed from the rhizome (**Fig. 1**).
6. Monopodial orchid explants (*see* **Table 2**): Inflorescences (preferably with flowers still intact) should be severed near their point of emergence from leaf bases.

3. Methods

3.1. Clonal Propagation of Sympodials by Callus Production

1. Leaves surrounding developing shoots are removed, exposing underlying lateral buds and shoot tips. Though the form of the leaves may vary from one sympodial genus to another, the basic structure of the developing shoot is as depicted in **Fig. 1A**.
2. Excised shoots are sterilized for 20 min in disinfesting solution.
3. Replace disinfesting solution with buffered holding solution. Tissues may remain in this solution while they await final trimming.
4. Excise explants for culturing by maneuvering the blade of a scalpel behind the lateral buds. Remove the buds by cutting approx 1 mm into the underlying stem tissue. The shoot tip is removed by cutting just below the base of the elliptical leaf primordia. All dissecting operations are performed while the tissues are immersed in the holding solution. Final trimming of explants involves the removal

Table 1
Supplements to the Basal Medium Employed in the Four Orchid Micropropagation Methods[a]

Methods	Media	Supplements		
		BA, μ*M*	NAA, μ*M*	Agar, mg/L[b]
3.1.	IM	7.83	1.07	—
	PM	0.78	13.43	6000
3.2.	IM	7.83	1.07	—
	PM	11.75	2.68	6000
3.3.	IM	19.58	5.37	6000
	PM	0.39	2.68	6000
		0.39	5.37	6000
		0.39	8.05	6000
		3.92	2.68	6000
		3.92	5.37	6000
		3.92	8.05	6000
3.4.	IM	19.58	5.37	6000
	PM	7.84	0.54	6000
		15.67	0.54	6000
		31.34	0.54	6000
		7.84	5.37	6000
		15.67	5.37	6000
		31.34	5.37	6000
3.1.–3.4.	RM	—	—	6000

[a]IM, initiation medium; PM, proliferation medium; RM, rooting medium; BA, benzyladenine; NAA, naphthaleneacetic acid.

[b]Agar Type: Bacto-agar, Difco Laboratories, Detroit, MI.

Table 2
Some Commonly Encountered Monopodial and Sympodial Orchid Genera

Monopodials		
Aerides	*Cyrtorchis*	*Renanthera*
Aerangis	*Doritis*	*Rhynchostylis*
Angraecum	*Neofinetia*	*Vanda*
Ascocentrum	*Phalaenopsis*	*Vandopsis*
Sympodials		
Brassavola	*Cymbidium*	*Masdevalia*
Brassia	*Dendrobium*	*Miltonia*
Broughtonia	*Encyclia*	*Odontoglossum*
Cattleya	*Epidendrum*	*Oncidium*
Cochlioda	*Laelia*	*Sophronitis*

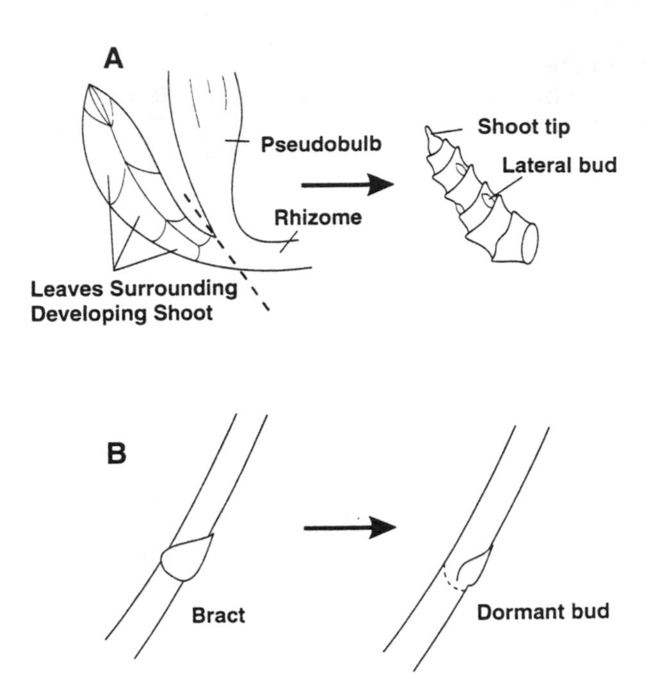

Fig. 1. Orchid explant selections. **(A)** Shoot explant selection for sympodial orchids. Dotted line indicates point of removal of developing shoot (left). Excised shoot showing exposed lateral buds and shoot tip used for culturing (right). **(B)** Inflorescence stem explant selection for monopodial orchids. Explant with original bract intact (left). Dotted line indicates location where bract was formerly attached to the stem (right).

of one or two of the tunicate leaf scales of the lateral buds or leaf primordia surrounding the shoot tips. Finally, carefully remove any fibrous stem tissue that may have adhered to the lateral buds or shoot tips when dissected from the stem.

5. Place explants in flasks containing liquid IM (**Table 1**), and agitate on a gyrotory shaker at 100 rpm.
6. Within a few days, explants should turn green and begin to swell. After 4–6 wk, the explants should be large and plump. At this time, remove the explants from culture, lightly scar with two or three superficial scalpel incisions, and place on PM (**Table 1**). If, after the initial 4–6 wk, any cultures appear to have undergone little or no growth, reculture to fresh IM, and observe for suitability for transfer to PM in another few weeks. In addition, if at any time the medium of an initiation culture begins to discolor (i.e., darken because of phenolic compound accumulation), it should be replaced with fresh medium immediately (*see* **Note 1**).
7. On transfer to PM, explants begin to produce callus or PLBs on their surface and from the incisions made before transfer. Large PLB clusters should be broken up and subcultured to fresh PM to enhance multiplication. Continue sectioning and subculturing callus masses until the desired volume of PLBs is achieved.

8. To induce differentiation into plantlets with roots, transfer PLBs to the rooting medium (RM) in **Table 1**.

3.2. Clonal Propagation of Sympodials by Axillary Shoot Formation

1. Follow **steps 1–5** as described in **Subheading 3.1.** Step 6 is executed using **Subheading 3.2.** PM instead of **Subheading 3.1.** PM (**Table 1**).
2. When transferred to PM, cultures will start to produce well-differentiated axillary shoots and/or callus masses, which quickly differentiate into shoots. From this point onward, most shoots should give rise to additional axillary shoots without passing through a callus stage. Shoot clumps should be separated to enhance proliferation, and individual shoots subcultured to fresh PM. This step is repeated until the desired number of plants is obtained (*see* **Note 2**).
3. To establish a root system, shoots are transferred to RM (**Table 1**).

3.3. Clonal Propagation of Monopodials by Callus Production

1. Cut the inflorescence into sections, each piece possessing one dormant node with approx 2 cm of the stem attached above and below the node (*see* **Note 3**).
2. With fine-tipped forceps, remove the bracts surrounding the dormant buds (**Fig. 1B**). All traces of the bracts, including any old, papery, or scared tissues that may be present at their bases, should be removed before disinfestation.
3. Wash nodal sections by agitation in dilute household detergent for 5 min.
4. Sterilize in disinfesting solution for 20 min. Agitate frequently.
5. Replace disinfesting solution with buffered holding solution. Agitate vigorously to ensure deactivation of the remaining sterilant.
6. With a scalpel, recut the two ends of the nodal sections, making sure that each fresh cut is at least 5 mm below the margins created by the bleached (damaged) areas at each end of the section.
7. Place nodal sections on **Subheading 3.3.** IM so that the bud is just above the medium surface. Within a week, buds turn green and begin to swell. After 12–16 wk, explants possess multiple shoots with well-developed leaves. Occasionally, this shoot cluster may also possess peripheral callus masses that contain newly developing plantlets in various stages of development (**Fig. 1B**).
8. If callus masses described in **step 7** are present on the explant, they should be severed from the rest of the node culture, sectioned into smaller pieces, and transferred to the PM series of **Subheading 3.3.** (**Table 1**).
9. Cultures not possessing visible callus are treated as follows: Sever the nodal plantlets from the stem, being sure to leave at least a 3–5-mm stub of the nodal base still attached to the stem. Severed shoots may be used for **Subheading 3.4.** (**Fig. 2**). Reculture the stem section, devoid of its nodal shoots, on fresh IM. The cut surface and periphery of the nodal stub will produce callus, which can be handled as described in **step 8** (**Fig. 2**).
10. The PM (from **Table 1**) deemed optimum for a particular callus should be employed for further subculturing (*see* **Notes 4–6**). Continue sectioning callus clusters and transferring to fresh PM until the desired volume of callus is achieved.

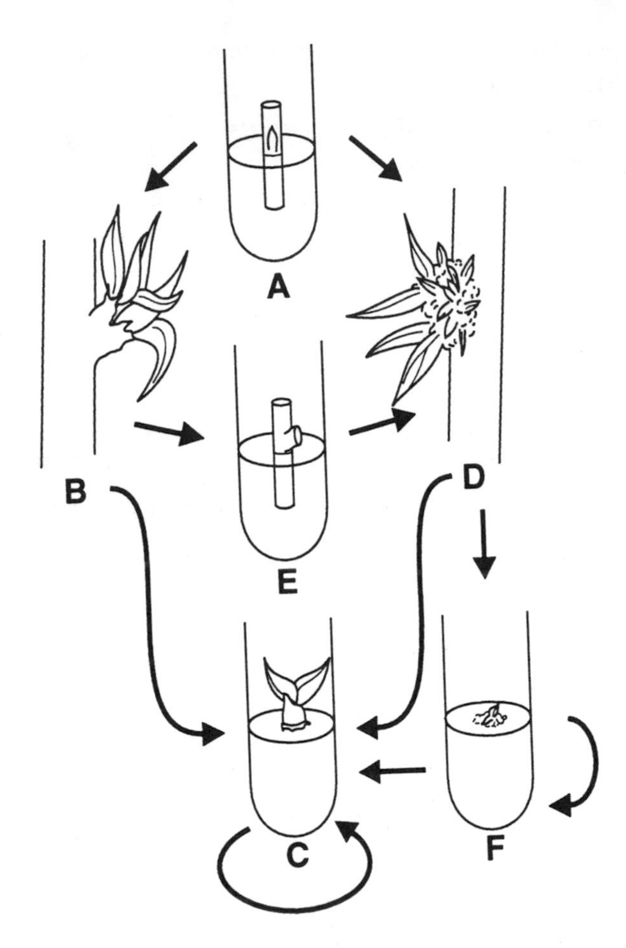

Fig. 2. Steps in the micropropagation of monopodial orchids. **(A)** Newly initiated inflorescence node culture. **(B)** Multiple Shoots developing from a node. **(C)** An individual shoot separated from the shoot cluster derived from either B or D. **(D)** Multiple shoots, with accompanying callus, developing from node. **(E)** Recultured inflorescence node with shoots removed to induce callus from the remaining stub. **(F)** Isolated callus and shoots obtained from culture D.

11. To induce differentiation into plantlets with roots, callus or PLBs should be transferred to RM (**Table 1**).

3.4. Clonal Propagation of Monopodials by Axillary Shoot Formation

1. Follow s**teps 1–7** as described for **Subheading 3.3.**
2. Multiple nodal shoots are separated, so that each plantlet has a sufficient base to support its leaves above the medium surface (**Fig. 2**). Culture plantlets on the entire PM series in this section (**Table 1**).

3. The medium giving the highest axillary shoot yield for a particular plant is used for all subsequent proliferation. Continue separating shoot clusters into individual plantlets and subculturing on fresh PM until the desired number of plantlets is achieved (*see* **Note 7**).
4. To establish a root system, shoots are transferred to RM (**Table 1**).

4. Notes

1. Exudation of phytotoxic phenolics from wounded orchid tissue poses a serious problem for in vitro culture. The most effective way of overcoming this problem in orchids is frequent media replenishment. Inclusion of activated charcoal into the media may also be beneficial. It is important that culture vigor not bc compromised by phenolic accumulation, because once attenuated, the productivity of a culture is difficult to restore.
2. It should be apparent that the IM formulations of all four methods are closely allied to the PM formulations of **Subheadings 3.2.** and **3.4.** The primary reason for indirectly approaching even the callus proliferation techniques by first employing a shoot proliferation analog is to achieve greater initial explant survival. The presence of cytokinins in the IM improves the survival rate of explants, probably owing to the antisenescence properties of the cytokinins.
3. In monopodial propagation, it is important that inflorescences be young and healthy. Ideally, flowers at the apex should still be fresh. Older inflorescences, or ones from which the flowers have faded, will give adequate, but not optimum results.
4. In monopodial callus propagation, loss of chlorophyll, and an accompanying decline in growth, may be observed in some cultures. In such instances, transfer of thc callus to PM lacking sucrose can often restore color and growth to cultures. All subsequent proliferation steps should, therefore, employ sucrose-free media. When differentiation into plantlets is desired, use sucrose-containing RM (**Table 1**).
5. Generally, callus propagation tends to be swifter than axillary shoot propagation. However, culturing orchids for extended periods in an undifferentiated callus state can result in significant occurrences of mutation. Genetic aberration has long been accepted as a risk in callus proliferation. However, as the demand for orchid clones increases, the problem of mutation in callus is only likely to worsen. Conversely, axillary shoot proliferation is very resistant to genetic aberration, and its development was largely in response to increasing concern over the occurrence of mutation ***(3,5)***.
6. Though NAA and BA perform satisfactorily, experiments to substitute other growth regulators for them should be beneficial. In particular, cytokinins, such as isopentenyladenine and zeatin riboside (the latter being very expensive), enhance orchid micropropagation. Also, numerous other growth regulators, such as thidiazuron, have also been employed that show promise in orchid micropropagation ***(19)***.
7. Both axillary shoot methods presented are capable of producing approx 1000 individual plants in the first year of culture, and as many as 100,000 by the end of the second year. The two callus methods described are capable of producing 100,000 or more plants within the first year of culture.

References

1. Churchill, M. E. (1973) Tissue culture of orchids. *New Phytol.* **72,** 161–166.
2. Lindermann, E. G. P. (1970) Meristem culture of *Cattleya. Am. Orchid Soc. Bull.* **39,** 1002–1004.
3. Morel, G. (1960) Producing virus free *Cymbidiums. Am. Orchid Soc. Bull.* **29,** 495–497.
4. Morel, G. (1964) A new means of clonal propagation of orchids. *Am. Orchid Soc. Bull.* **33,** 473–478.
5. Sagawa, Y. (1966) Clonal propagation of *Cymbidiums* through shoot meristem culture. *Am. Orchid Soc. Bull.* **35,** 188–192.
6. Mujib, A. and Jana, B. K. (1994) Clonal propagation of *Dendrobium* madame pompadour through apical meristem. *Adv. Plant Sci.* **7,** 340–346.
7. Scully, R. (1967) Aspects of meristem culture in the *Cattleya* alliance. *Am. Orchid Soc. Bull.* **36,** 103–108.
8. Wimber, D. (1963) Clonal multiplication of *Cymbidiums* through tissue culture of the shoot meristem. *Am. Orchid Soc. Bull.* **32,** 105–107.
9. Kuhn, L. (1982) Orchid propagation in the EYMC. *Marie Selby Bot. Gard. Bull.* **9,** 32–34.
10. Griesbach, R. J. (1983) The use of indoleacetylamino acids in the in vitro propagation of *Phalaenopsis* orchids. *Sci. Hort.* **19,** 363–366.
11. Intuwong, O. (1974) Clonal propagation of *Phalaenopsis* by shoot tip culture. *Am. Orchid Soc. Bull.* **43,** 893–895.
12. Huang, L. C. (1984) Alternative media and method for *Cattleya* propagation by tissue culture. *Am. Orchid Soc. Bull.* **53,** 167–170.
13. Rotor, G. (1949) a method of vegetative propagation of *Phalaenopsis* species and hybrids. *Am. Orchid Soc. Bull.* **18,** 738–739.
14. Sagawa, Y. (1961) Vegetative propagation of *Phalaenopsis* stem cuttings. *Am. Orchid Soc. Bull.* **30,** 803–809.
15. Scully, R. (1966) Stem propagation of *Phalaenopsis. Am. Orchid Soc. Bull.* **35,** 40–42.
16. Lay, M. and Fan, F. (1978) Studies on the tissue culture of orchids. *Orchid Rev.* **Oct,** 308–310.
17. Tanaka, M. (1977) Clonal propagation of *Phalaenopsis* by leaf tissue culture. *Am. Orchid Soc. Bull.* **46,** 733–737.
18. Zimmer, K. and Pieper, W. (1978) Clonal propagation of *Phalaenopsis* by excised buds. *Orchid Rev.* **July,** 223–227.
19. Ernst, R. (1994) Effects of thidiazuron on in vitro propagation of *Phalaenopsis* and *Doritaenopsis* (Orchidaceae). *Plant Cell Tissue Organ Cult.* **39,** 273–275.
20. Tokuhara, K. and Mii, M. (1993) Micropropagation of *Phalaenopsis* and *Doritaenopsis* by culturing shoot tips of flower stalk buds. *Plant Cell Re.* **13,** 7–11.

13

In Vitro Propagation of Succulent Plants

Jill Gratton and Michael F. Fay

1. Introduction

Maintenance of collections of succulent plants can be problematic, because many of these species are very susceptible to rots caused by bacteria and fungi. Rooting and establishment of cuttings can also be difficult. Tissue-culture techniques have been applied to a wide range of succulents, and the relevant literature has recently been reviewed ***(1,2)***. Here we describe methods for the micropropagation of cacti and other succulents that have been developed at Kew for overcoming the problems mentioned above.

The techniques have been used most successfully with members of *Asclepiadaceae,* healthy plants having been produced from small pieces of unhealthy tissue in the genera *Caralluma, Huernia, Stapelia,* and *Ceropegia.* Some other genera, notably *Hoodia,* have, however, proven intractable using the methods described here. Succulent species of *Aizoaceae* and *Crassulaceae* have also been propagated, but the methods will not be described here, since they are closely based on those used for *Asclepiadaceae.*

Cacti have proven more difficult. This is probably largely owing to the structure of the plants, with the meristematic areas being embedded in the areole tissue. Removal of woody spines can damage this tissue or allow it to be damaged by the sterilant. Successes have, however, been achieved, and include species of *Opuntia* and *Mammillaria.* Succulent *Euphorbia* spp. are also difficult, but several species have been micropropagated. Species with woody spines seem to be particularly problematic.

Micropropagation can be of particular value in the propagation of rare and endangered species, especially where viable seeds are not available. Bulk propagation and subsequent distribution of those species that are endangered by overcollection may alleviate pressure on the wild populations.

From: *Methods in Molecular Biology, Vol. 111: Plant Cell Culture Protocols*
Edited by: R. D. Hall © Humana Press Inc., Totowa, NJ

2. Materials

1. Media: Murashige and Skoog (MS) medium (*3* and *see* Appendix) is obtained as a ready-mixed powder, without sucrose, growth regulators, or agar. It is made up at the standard concentration with 30 g/L sucrose and a range of growth regulators, which are added before autoclaving. These are detailed in the **Subheading 3.**

   Adjust the pH of the medium to 5.6–5.8 using dilute NaOH or HCl as appropriate. Dissolve 9 g/L of agar in the medium using a microwave oven (2 min for each 250 mL of medium in a 1000-W microwave oven), and stir thoroughly. This method was developed using Sigma purified agar (Product No. A7002). Dispense the medium into the required vessels, and sterilize in the autoclave for 15 min at 121°C and 1.05 kg/cm^2. If the medium is not to be used immediately, store the vessels in plastic bags at 4°C in the dark.
2. Sterilization: An ethanol *(95%)* dip is used as a pretreatment (10–30 s) when the tissues are infested with pests. Sodium hypochlorite (BDH, 10–14% available chlorine) is used as the sterilant, diluted to 3–10% v/v in deionized water (0.3–1.4% available chlorine), with the addition of a few drops of Tween 80/L.
3. Compost: Open, free-draining composts are used, since these have been shown to promote rooting and to decrease losses from rotting off during the weaning stage. The most frequently used compost is composed of three parts fine loam, two parts coir, one part sharp grit, six parts montmorillonite (particle size up to 4 mm), and six parts Perlite (3- to 6-mm particle size). To this compost, slow-release fertilizer is added as required. Calcined montmorillonite possesses good absorption properties. The montmorillonite and Perlite make the compost considerably more open, thus aiding establishment.

3. Methods

The techniques described here were developed at Kew in an attempt to make the best possible use of what little material is often available. The basic principle on which the method is based is the induction of shoot formation from dormant meristems in the areole of cacti or from axillary buds in other succulents, using media containing cytokinins, alone or in combination with low concentrations of auxins.

3.1. Initial Preparation of Tissue

1. Excise and discard dead or rotting material.
2. If the material is infested with pests, dip it in 95% ethanol for 10–30 s, and then wipe it with a soft tissue or paintbrush.
3. Remove any soil adhering to the plant material by washing under running water or agitating in water containing a few drops of Tween 80. A fine paintbrush can be useful for dislodging debris trapped by hairs or spines.
4. Remove any damaged tissue, taking care not to damage the underlying tissues.

5. Woody spines and hairs found in cacti and succulent *Euphorbia* spp. are difficult to surface-sterilize effectively. They are therefore removed, using watchmaker's forceps, fine-pointed scalpel blades, or hypodermic needles (*see* **Note 1**). Great care must be taken not to damage the tissue underneath the spine, since the meristematic area often lies very close to the base of the spines. Where the spines are less woody, they can be cut back as close as possible to the base, rather than being totally removed.

3.2. Surface Sterilization

1. The pieces of plant tissue should be left as large as possible for surface sterilization, to minimize damage to healthy tissue by the sterilant. Where possible, a range of sterilization times and sterilant concentrations are used, but this depends on the amount of tissue available. Sodium hypochlorite (0.3–1.4% available chlorine in deionized water) is used as the sterilant, with a few drops of Tween 80/L. Sterilization times range from 5 to 15 min. The most commonly used regime is a 5% dilution of BDH sodium hypochlorite (0.5–0.7% final available chlorine) plus Tween 80 for 10 or 15 min. Surface sterilization is improved if the beaker is put onto a magnetic stirrer set at a speed high enough to keep the plant tissue submerged, but not so high that it will damage the plant tissue.
2. Rinse the material three times in sterile deionized water.

3.3. Preparation of the Surface Sterile Material for Culture

The size of the explant used for culture varies greatly, depending on the structure of the plant and the extent to which any rot present has advanced. The general procedures for each of the groups of plants are described below.

3.3.1. Asclepiadaceae

In asclepiads, axillary buds are found in the axils of the vestigial leaves and are often visible to the naked eye. Remove these buds with some surrounding tissue, using a scalpel and forceps. The first cut should be made directly below the vestigial leaf and the second as far above the bud, since the structure of the plant will allow. A vertical cut made behind these cuts will separate the required explant from the main body of the plant (**Fig. 1**).

In species where the stems are very ridged and the vestigial leaves are very close together, make vertical cuts between the ridges as far away from the buds as possible. In these species, we have found it better to use pieces of tissue with several buds for initiating cultures, rather than trying to separate out the individual buds.

In *Ceropegia* spp., nodal sections are taken. Where there are two buds present at the node, make a vertical cut through the node, and culture the buds separately.

Fig. 1. *Caralluma* micropropagation. A shoot ready for explant excision (left, approx ×1.5), a prepared explant with two dormant buds (middle, approx ×4), and an established proliferating shoot culture (right, approx ×3).

3.3.2. Cactaceae

In cacti where the areoles are widely spaced, remove them with 3–5 mm of the surrounding tissue. In *Mammillaria* and other genera with tubercles, remove the areoles with as much of the tubercle tissue as possible. Where the areoles are close together, make vertical incisions into the tissue, taking care not to cut right through, and excise small pieces consisting of several areoles.

3.3.3. Euphorbiaceae

These are cut up into explants similar to those from asclepiads, after removal of spines.

3.4. Culture

1. Place the individual explants in boiling tubes containing 10–15 mL of medium, with the cut surface of the explant in contact with the medium, and the bud or areole facing upward. Where sufficient tissue is available, explants are cultured on a range of semisolid media based on MS with added growth regulators. The most commonly used growth regulator combinations are 1 or 2 mg/L benzylaminopurine (BAP) + 0.1 mg/L naphthylacetic acid (NAA) (*see* **Note 2**).
2. Maintain the cultures in a culture room at 22–25°C with a 16-h photoperiod (light intensity 1000–4200 lx) (*see* **Note 3**).

3. Examine the cultures after a few days for signs of contamination. Where no more material is readily available, it has proven possible in some cases to carry out a second surface sterilization if contamination has occurred. Solutions of sodium hypochlorite (0.5–1.0% available chlorine) have been used for 5–30 min, depending on the type of material, but it is difficult to give more specific instructions.
4. After a variable length of time on the media described above, axillary buds on some of the explants will begin to grow out (*see* **Note 4**). When these have grown sufficiently (to 2–3 cm in asclepiads), cut them into sections, each with one or a few buds, and place them onto fresh medium in 1/2- or 1-lb honey jars. These sections and the stump on the original explant should continue to produce lateral shoots, thus establishing a proliferating system (**Fig. 1**). Continue subculturing in the same way at 1- to 2-mo intervals until sufficient shoots have been obtained.
5. When this stage is reached, rooting in vitro is attempted. Place excised shoots onto a range of semisolid media based on MS ± growth regulators. The most commonly used media for rooting are MS with: (a) no growth regulators; or (b) 0.01, 0.1, 0.25, 0.5, or 1.0 mg/L NAA. Alternative media are detailed in **Note 5**.
6. When the plantlets are well rooted, tease them out of the medium gently, and wash them carefully under running water to remove traces of agar.
7. Pot the plantlets in 2-in square pots with 1–4 plants/pot, using the compost described in **Subheading 2.** (*see* **Note 6**). Place the pots on a bed of sand with under heating (if available) in a dry intermediate glasshouse (21°C day, 16°C night). Where rotting off is a problem, fungicide drenches have proven useful.
8. Plants can take up to 6 mo to become fully established and begin to grow vigorously.

4. Notes

1. In cacti, the removal of spines can result in damage to the dormant buds located in the areole. In particularly sensitive species, sterilization with 0.1% mercuric chloride for 2–10 min can obviate the necessity to remove spines prior to sterilization. As a result, the sensitive bud tissue is not exposed to damage during the subsequent treatment with sodium hypochlorite.

   Mauseth and Halperin *(4)* recommended singeing the spines on cacti instead of removing them. This has been tried with little success.

   A major problem with removing spines can be that the newly exposed tissues are very susceptible to damage by the sterilant. Spines in some species of *Euphorbia* can be removed after sterilization. Where there is very little tissue available, the cut ends can be sealed with paraffin wax, thus minimizing the damage.
2. With some plants, it has proven necessary to use media with higher cytokinin concentrations to initiate proliferating cultures. The most frequently used concentrations are 5 or 10 mg/L BAP with 0.1 or 0.5 mg/L NAA.
3. During the initiation process with *Asclepiadaceae,* compact callus sometimes forms at the base of explants. Remove this at each subculture. If the callus threatens to take over the culture, transfer the tissue to a medium with lower growth regulator concentrations.

4. The pathway for cacti described in the **Subheading 3.** is the ideal, but in certain species of cacti (e.g., some *Mammillaria* spp.), a callus is sometimes formed before shoots are regenerated. Vitrification can also occur; techniques for overcoming this that can be tried include:
   a. Incorporation of 1–3 g/L activated charcoal.
   b. Use of media of lower ionic strength, e.g., half-strength MS ($^1/_2$MS).
   c. Use of media with higher agar concentration (e.g., 1.4%).
   d. Allowing the medium to dehydrate partially before transferring the cultures.
5. Where rooting will not take place on any of the media listed above, the following variations have been used with some success:
   a. Indoleacetic acid (IAA) can be used at the same concentrations in place of NAA.
   b. NAA can be used at higher concentrations (5–20 mg/L).
   c. $^1/_2$MS ± NAA or IAA can be used.
6. Where dehydration is a problem during weaning, steps must be taken to stop the plantlets from dying. Keep the plants under a clear propagator lid until they have become acclimatized to the lower humidity. Gradually lower the humidity by opening the vents in the lid, and eventually remove the lid altogether. The use of open, free-draining composts is very important at this stage.

   With particularly difficult species, we have had some success using an intermediate stage between agar and pot culture. Sterilize the standard compost in honey jars, and then add sterile water or mineral salt solutions (e.g., $^1/_2$MS without sucrose or agar). Then "pot" the plantlets in this, and seal the vessels as normal. When the plantlets appear to have become established, transfer the jars to the glasshouse, gradually loosen the tops over a few days, and then pot the plantlets normally.

References

1. Fay, M. F. and Gratton, J. (1992) Tissue culture of cacti and other succulents—a literature review and report of micropropagation at Kew. *Bradleya* **10,** 33–48.
2. Fay, M. F., Gratton, J., and Atkinson P. J. (1995) Tissue culture of succulent plants—an annotated bibliography. *Bradleya* **13,** 38–42.
3. Murashige, T. and Skoog, F. (1962) A revised medium for rapid growth and bioassays with tobacco tissue cultures. *Physiol. Plant* **15,** 473–497.
4. Mauseth, J. D. and Halperin, W. (1975) Hormonal control of organogenesis in *Opuntia polyacantha (Cactaceae). J. Am. Bot.* **62,** 869–877.

14

Micropropagation of Flower Bulbs

Lily and Narcissus

Merel M. Langens-Gerrits and Geert-Jan M. De Klerk

1. Introduction

For most bulbous crops, artificial (vegetative) propagation methods have been developed, such as scaling (lily), scooping (hyacinth), and chipping (narcissus). Because the speed of these methods is often low, introduction of newly bred cultivars (either produced by conventional breeding or by genetic modification) or of pathogen-free bulbs (produced by meristem culture) requires a long period of time. In tulip, for which no artificial propagation method exists, this can even take 20–25 yr. Micropropagation considerably shortens this period. Furthermore, because of the large number of propagation cycles in the field, conventionally produced bulbs may become easily infected. Micropropagation produces starting material that is completely or predominantly pathogen-free.

The main difference between micropropagation of bulbs and other crops is that bulblets and not microcuttings are produced. In contrast to microcuttings, bulblets mostly do not require a rooting treatment and acclimatization, but they do require breaking of dormancy. It should be noted that in almost all protocols, at least one step involves adventitious regeneration of a shoot meristem. Many if not all sports (mutated plants) are chimeras, and since the chimeric structure is lost in adventitious regeneration, the characteristic for which a sport has been selected may very well get lost when these protocols are used.

The protocols for micropropagation of flower bulbs consist of six stages:

1. Mother plant preparation: Storage conditions of the bulbs may determine the ability to regenerate. To avoid contamination, healthy undamaged bulbs are used. Initial contamination can be strongly reduced by a hot water treatment: 1 h at 54°C in narcissus *(1)* and 1 h at 43°C in lily *(2)*.

From: *Methods in Molecular Biology, Vol. 111: Plant Cell Culture Protocols*
Edited by: R. D. Hall © Humana Press Inc., Totowa, NJ

2. Initiation: Usually, explants are excised from scales or flower stems. From these explants, adventitious shoots/bulblets regenerate in vitro. Apical and axillary buds may also be used and may be the preferable explant, since they give better results in virus elimination and also have better propagation rates than adventitious buds. In iris, bulblets originating from apical or axillary buds show better performance for a number of years after planting (M. van Schadewijk, personal communication).
3. Propagation: Both shoots and bulblets can be used for further adventitious or axillary propagation. Shoots have higher propagation rates than bulblets. In hyacinth, e.g., 2–3 new bulblets regenerate/bulblet, whereas 10–40 new shoots regenerate/shoot.
4. Bulb growth: In the final propagation cycle, bulblets should be produced. Bulblets are firm structures and can be handled easily. Shoots planted in soil sometimes do not form a bulb or form aberrant bulbs *(3)*. Conditions stimulating bulb growth include high sucrose concentration, absence or low concentrations of growth regulators, darkness, charcoal, and moderate temperatures. Bulblets should be sufficiently large (about 100 mg) to achieve good growth after planting. Large bulblets of lily grow faster than small ones *(4)*.
5. Dormancy breaking. For fast and uniform sprouting, and for optimal growth, bulblets need a temperature treatment to break dormancy *(5,6)*. This requires either several weeks at low temperatures (2–9°C) or in some species, e.g., iris, a temperature treatment of 4 wk at 30°C *(3)*. The optimal temperature and duration for tissue-culture-derived bulblets are the same as for conventionally produced ones. Some tissue-culture conditions, in particular, temperature, sugar concentration, and duration of the culture cycle, affect the dormancy level and leaf formation *(7–9)*. Dormancy development is prevented by addition of fluridone, an inhibitor of abscisic acid synthesis *(10)*.
6. Planting: For planting, usually the same steamed potting composts can be used as for conventionally propagated bulbs. During the first year, growth may be in a glass house. Bulb growth is considerably enhanced by planting directly in soil, probably because of the lower soil temperature (hyacinth and iris, A. F. L. M. Derks, personal communication). Growth in an aphid-proof glass house or gauze house is essential when virus infection has to be avoided.

The protocols developed for propagation in vitro are often similar to conventional propagation protocols. For the five main bulbous crops, tulip, lily, narcissus, hyacinth and iris, satisfactory micropropagation protocols exist, except for tulip *(11,12)*. In this chapter, protocols for lily (straightforward) and narcissus (more complicated) will be described. A protocol for iris can be found in ref. *(3)* and for hyacinth in ref. *(13)*.

2. Materials

2.1. Lily

1. Clean, healthy bulbs (*see* **Note 1**).
2. 70% (v/v) Alcohol, sterile water.

3. Sodium hypochlorite solution (1% available chlorine) + 0.03 ÷ Tween 20.
4. Containers with medium: the medium consists of MS salts ***(14)***, 3% sucrose (w/v), 100 mg/L myo-inositol, 0.4 mg/L thiamine-HCl, 0.27 μ*M* naphthaleneacetic acid (NAA), 0.6% (w/v) agar, and a pH of 6.0 before autoclaving. This medium can be used both for initiation and propagation (*see* **Note 2**). Bulb growth in the final step can be on the same medium, but with 6% sucrose.

2.2. Narcissus

1. Clean, healthy bulbs (*see* **Note 3**).
2. 70% (v/v) Alcohol, sterile water.
3. Sodium hypochlorite solution (1% available chlorine) + 0.03 ÷ Tween 20.
4. Containers with medium: basal medium consists of MS salts ***(14)***, 3% sucrose (w/v), 100 mg/L myoinositol, 0.4 mg/L thiamine-HCl, 0.54 μ*M* NAA, 4.4 μ*M* benzyladenine (BAP), 0.6% (w/v) agar, and a pH of 6.0 before autoclaving. The basal medium can be used for initiation and propagation. Bulb growth can be induced on the same medium, but with 9% sucrose, without BAP and with 5 g/L activated charcoal. Rooting takes place on basal medium without BAP.

3. Methods

3.1. Lily

1. Remove roots.
2. Wash bulbs thoroughly with water.
3. Remove brown or damaged outer scales (*see* **Note 4**).
4. Peel off clean and healthy-looking scales. Remove brown tips. Since regeneration is best at the basal side of the scale, take care not to damage this part.
5. Rinse scales for about 30 s with 70% alcohol.
6. Sterilize scales for 30 min in sodium hypochlorite.
7. Rinse three times with sterile water.
8. Cut explants of about 7 × 7 mm (*see* **Note 5**; **Fig. 1**).
9. Place explants with abaxial side on medium (**Fig. 1**).
10. Explants can be cultured at 20–25°C. In the dark, larger bulblets are obtained than in the light (16 h/d, 30 $\mu mol/m^{-2}/s^{-1}$).
11. After about 10 wk, regenerated bulblets can be used for further propagation. Individual scales of the bulblets are used as explant; large scales can be cut in two or three parts. This step can be repeated every 10 wk.
12. After the final propagation cycle, larger bulblets can be obtained by subculturing single bulblets on medium with high- (6%) sucrose concentration for a number of weeks or months. This treatment is often used to "store" the bulblets until planting time in the spring.
13. Before planting, bulblets receive a cold treatment of several (6–12) wk at 5°C in the dark (*see* **Note 6**).
14. Bulblets are excised from the explant, planted in soil, and grown in a gauze house (at 17°C, most bulblets will sprout within 2–4 wk).

Fig. 1. Preparation of a lily explant.

3.2. Narcissus

1. Remove roots and dry tunic.
2. Submerge bulbs completely in hot water of 54°C for 1 h (hot water treatment).
3. Dry bulbs on filter paper at room temperature for 1 d.
4. Remove outer scale and $^1/_3$ of the upper part of the bulbs. Halve the bulbs longitudinally.
5. Sterilize bulbs for 30 min in sodium hypochlorite.
6. Rinse three times with sterile water.
7. Cut "twin scales": two bulb scale segments 1 cm wide and 1.5–2 cm long, joined by a part of the basal plate (**Fig. 2**).
8. Culture explants in upright position with the basal plate on medium, in the dark at 15–20°C.
9. After about 14 wk, the bulb scale segments are excised from the explant. Shoot clumps attached to the basal plate are propagated by splitting them into smaller clumps. Removal of large shoots stimulates formation of new shoots. This cycle can be repeated about every 5–6 wk.
10. The propagation rate decreases every cycle, but is restored by an intermediate bulb formation phase after every three to four propagation cycles: single shoots are placed on bulbing medium for about 6 wk (20°C, dark), so the shoot base swells to form a bulblet. From the bulblets, "minitwin scales" are cut (ca. 5 mm wide and 5–10 mm high). Depending on the size of the bulblet, two to four minitwin scales can be cut (*see* **Note 7**).
11. Culture minitwin scales on basal medium for 10–12 wk. Outgrowth of new shoots is stimulated by removal of large shoots and leafy structures after about 6 wk.

Fig. 2. A twin-scale explant of narcissus with regenerating shoots.

12. Shoot clusters regenerated from the minitwin scales are propagated again by splitting.
13. Culture single shoots on bulb growth medium at 20°C in the dark for about 6 wk.
14. Culture bulblets on rooting medium at 15°C in the dark for 7–8 wk.
15. Rooted bulblets are planted in soil (*see* **Note 8**).

4. Notes

1. Before storage, the bulbs are given a cold treatment (2°C for 6–8 wk) after harvest, and then packed in moistened peat and stored at –1°C.
2. The optimal NAA concentration for regeneration (number of bulblets) decreases with duration of storage ***(15)***.
3. To obtain regeneration after a hot water treatment, bulbs should be stored dry at 30°C immediately after harvest.
4. At this point, a hot water treatment (1 h at 43°C) can be given to decrease endogenous contamination. Let the bulbs dry for 1 d at room temperature before cutting explants. After a hot water treatment, culture tubes should not be closed air-tight, since released gases may reduce viability of the explant. Some genotypes (e.g., *Lilium speciosum* "Rubrum No.10") are more sensitive to this than others.

5. Regeneration is highest in explants cut from the base of a scale. Some genotypes have a very strong regeneration gradient (only explants cut from the base of a scale show regeneration). In others, the whole scale can be used. The size of an explant strongly affects the size of the regenerating bulblets: from a large explant, large bulblets regenerate ***(4)***.
6. The cold treatment can also be given to bulblets excised from the explant. Store the bulblets on moistened filter paper.
7. Chow et al. ***(16)*** also developed a propagation method for narcissus. In this method, shoot clumps are subjected every other cycle to "severe cutting" (down to the basal plate region) to restore the propagation rate.
8. Rooting is essential for optimal growth after planting. Without roots, leaves remain small (ca. 0.5 cm) and die a few weeks later. Rooted bulblets do not need a cold treatment to break dormancy ***(17)***.

References

1. Hol, G. M. G. M. and Van Der Linde, P. C. G. (1992) Reduction of contamination in bulb-explant cultures of *Narcissus* by a hot-water treatment of parent bulbs. *Plant Cell Tiss. Org. Cult.* **31,** 75–79.
2. Langens-Gerrits, M., Albers, M., and De Klerk, G. J. (1998) Hot-water treatment before tissue culture reduces initial contamination in *Lilium* and *Acer*. *Plant Cell Tiss. Org. Cult.* **52,** 75–77.
3. Van Der Linde, P. C. G. and Schipper, J. A. (1992) Micropropagation of iris with special reference to *Iris x hollandica* Tub., in *Biotechnology in Agriculture and Forestry,* vol. 20 (Bajaj, Y. P. S., ed.), Springer-Verlag, Berlin, pp. 173–197.
4. Langens-Gerrits, M., Miller, W. B., Lilien-Kipnis H., Kollöffel, C., Croes, A. F. and De Klerk G. J. (1997) Bulb growth in lily regenerated in vitro. *Acta Hortic.* **430,** 267–273.
5. Gerrits, M., Kim, K. S., and De Klerk, G. J. (1992) Hormonal control of dormancy in bulblets of *Lilium speciosum* cultured in vitro. *Acta Hortic.* **325,** 521–527
6. Langens-Gerrits, M., Hol, T., Miller W. B., Hardin, B., Croes, A. F., and De Klerk, G. J. (1997) Dormancy breaking in lily bulblets regenerated in vitro: effects on growth after planting. *Acta Hortic.* **430,** 429–436.
7. Aguettaz, P., Paffen, A., Delvallée, I., Van Der Linde, P., and De Klerk, G. J. (1990) The development of dormancy in bulblets of *Lilium speciosum* generated in vitro. I. The effects of culture conditions. *Plant Cell Tiss. Org. Cult.* **22,** 167–172
8. Delvallée, I., Paffen, A., and De Klerk, G. J. (1990). The development of dormancy in bulblets of *Lilium speciosum* generated in vitro. II. The effect of temperature. *Physiol. Plant.* **80,** 431–436
9. Gerrits, M. and De Klerk, G. J. (1992) Dry-matter partitioning between bulbs and leaves in plantlets of *Lilium speciosum* regenerated in vitro. *Acta Bot. Neerl.* **41,** 461–468.
10. Kim, K. S., Davelaar, E., and De Klerk, G. J. (1994) Abscisic acid controls dormancy development and bulb formation in lily plantlets regenerated in vitro. *Physiol. Plant.* **90,** 463–469

11. Le Nard, M., Ducommun, C., Weber, G., Dorion, N., and Bigot, G. (1987) Observations sur la multiplication in vitro de la tulipe (*Tulipa gesneriana* L.) à partir de hampes florales prélevées chez des bulbes en cours de conservation. *Agronomie* **7,** 321–329.
12. Hulscher, M., Krijgsheld, H. T., and Van Der Linde, P. C. G. (1992) Propagation of shoots and bulb growth of tulip in vitro. *Acta Hortic.* **325,** 441–446
13. Paek, K. Y. and Thorpe, T. A. (1990) Hyacinth, in *Handbook of Plant Cell Culture,* vol. 5, *Ornamental Species* (Ammirato, P. V., Evans, D. R., Sharp, W. R., and Bajaj, Y. P. S., eds.), McGraw-Hill, New York, pp. 479–508.
14. Murashige, T. and Skoog, F. (1962) A revised medium for rapid growth and bioassays with tobacco tissue cultures. *Physiol. Plant.* **15,** 473–497
15. Van Aartrijk, J. and Blom-Barnhoorn, G. J. (1981) Growth regulator requirements for adventitious regeneration from *Lilium* bulb-scale tissue in vitro, in relation to duration of bulb storage and cultivar. *Sci. Hort.* **14,** 261–268
16. Chow, Y. N., Selby, C., and Harvey, B. M. R. (1992) A simple method for maintaining high multiplication of *Narcissus* shoot cultures in vitro. *Plant Cell. Tiss. Org. Cult.* **30,** 227–230
17. Langens-Gerrits, M. and Nashimoto, S. (1997) Improved protocol for the propagation of *Narcissus* in vitro. *Acta Hortic.* **430,** 311–313.

15

Clonal Propagation of Woody Species

Indra S. Harry and Trevor A. Thorpe

1. Introduction

Tissue-culture technology is widely used for the vegetative propagation of selected plants in agriculture and horticulture and, to a lesser extent, in forestry. The objective is to produce large numbers of plants with uniform quality. Historically, commercial applications of this technology were restricted to herbaceous plants. However, for the last two decades, considerable success has been obtained with woody plants *(1,2)*. These include both gymnosperms and angiosperms, i.e., softwoods and hardwoods, and both trees and shrubs. Economically, these trees are extremely important for wood products, including lumber, pulp and paper, forestry plantations, and reforestation. Large-scale clonal systems can be an asset for selected high-performance trees, and reliable protocols are necessary for further genetic manipulation. However, a major problem in the propagation of woody plants is that most success is achieved with juvenile tissue and not from proven mature trees *(3,4)*.

Traditional methods for propagation include rooted cuttings, grafting and layering. For in vitro propagation, the methods generally used are first, the induction of adventitious buds or axillary bud breaking, both of which produce shoots that are subsequently elongated and rooted, and second, somatic embryogenesis mainly from juvenile tissue. Only propagation via the first two methods are discussed here. To obtain axillary shoots, shoot tips, lateral buds, and small nodal cuttings are used as explants *(2)*. Axillary shoot bud proliferation is induced from preformed meristems, and in general, true-to-type and genetically stable plants are produced *(3)*. Cytokinins with or without an auxin are used for the development of these meristems. Adventitious buds are either induced directly on explants or on callus derived from primary explants. Again, cytokinins are used primarily for direct organogenesis, whereas a combination

From: *Methods in Molecular Biology, Vol. 111: Plant Cell Culture Protocols*
Edited by: R. D. Hall © Humana Press Inc., Totowa, NJ

of cytokinins and auxins are used for the latter. In addition, the methods used should be relatively simple, have high multiplication rates, be reproducible, and allow differentiation of shoots without intervening callus. Also, shoots should be rootable, and plantlets should survive transfer to ex vitro conditions ***(5–7)***.

In general, for gymnosperms, whole embryos or excised seedling parts are used as explants. In contrast, a major advantage of manipulating angiosperm trees in vitro, is that a wide range of material can be used as explants. These include seeds, seedling and floral parts, leaves, and shoots and buds from both juvenile or mature trees. This extensive source of explants is important, since the viable period of seeds from many hardwood trees is quite short. Although the desirable characteristics in tree species are obvious at maturity, propagation using mature tissue is still problematic. Other difficulties encountered with woody species include systemic infections, episodic growth, the production of polyphenols, tannins and volatile substances, and hyperdricity (vitrification) of cultures ***(3)***.

To illustrate the tissue-culture procedures commonly used for these two groups of plants, *Thuja occidentalis* or eastern white cedar and *Populus tremuloides* or aspen are used. For both species, a multistage process is used ***(5)***. For cedar, this includes adventitious bud induction on the explant, bud elongation, axillary budding, and rooting and hardening of plantlets before transfer to greenhouse conditions ***(8,9)***. For poplar, this process involves axillary bud breaking and shoot initiation, shoot multiplication and elongation, followed by rooting and acclimatization ***(6,10,11)***.

2. Materials

1. Stocks of various salt formulations, both for induction and elongation, e.g., QP (***12***; *see* **Table 1**) for cedar and MS (***13***; *see* Appendix) for poplar, and phytohormones including N^6-benzyladenine (BA), zeatin, and indole-3-butyric acid (IBA) (*see* **Note 1**): BA and zeatin are dissolved in 1 *M* NaOH, and stored at 4°C. BA is added to medium before autoclaving. However, the pH of the zeatin stock is adjusted to 5.7–5.8 and is filter-sterilized, and the required amount is added to autoclaved medium before the plates are poured. IBA solutions are made up when needed, the pH is adjusted to 5.0, and are filter-sterilized before use. The media needed for *Thuja* are:
   a. 1/2 QP, 1 µ*M* BA, 3% sucrose.
   b. 1/2 QP, 10 µ*M* zeatin, 2% sucrose.
   c. 1/2 QP, 3% sucrose.
   d. 1/2 QP, 2% sucrose and 0.05% activated charcoal.
   Those for poplar are:
   a. MS, 2 µ*M* BA, 3% sucrose.
   b. MS, 1 µ*M* BA, 3% sucrose.

Table 1
Quoirin and Lepoivre Nutrient Medium *(12)* as Modified in Ref. *14*

	mg/L
Major salts	
KNO_3	1800
NH_4NO_3	400
$Ca(NO_3)_2 \cdot 4H_2O$	1200
KH_2PO_4	270
$MgSO_4 \cdot 4H_2O$	360
Minor salts	
$MnSO_4 \cdot 4H_2O$	1.0
H_3BO_3	6.2
$ZnSO_4 \cdot 7H_2O$	8.6
KI	0.08
$CuSO_4 \cdot 5H_2O$	0.025
$Na_2MoO_4 \cdot 2H_2O$	0.25
$CoCl_2 \cdot 6H_2O$	0.025
Vitamins *(15)*	
Nicotinic acid	5.0
Pyridoxine HCl	0.5
Thiamine HCl	5.0
Iron *(13)*	
$Na_2EDTA \cdot H_2O$	74.5
$FeSO_4 \cdot 7H_2O$	55.7
Asparagine	100
Myo-inositol	100

c. $^1/_2$ MS (half-strength major salts only), 3% sucrose.

For all culture media, the pH is adjusted to 5.7–5.8 before adding 0.8% (w/v) Difco Bacto-agar for autoclaving. Petri dishes (25 mL medium) and glass jars (100 mL medium) are prepared.

2. 9-cm Petri plates with 25 mL induction medium.
3. Autoclavable glass jars with 100 mL of elongation, propagation, and rooting media.
4. Relatively clean seeds and, preferably, greenhouse-grown plant material for bud explants.
5. Sterilants, including commercial bleach, e.g., Javex® (6% NaOCl),H_2O_2, $HgCl_2$, 70% ethanol and sterile water, and wetting agent, e.g., Tween 20®.
6. Parafilm® or Stretch 'n Seal® for sealing plates and jars.
7. Laminar flow hood for aseptic work.

8. Instruments including scalpels and #10 and #11 blades, long forceps, fine-tip dissection forceps, and a dissecting microscope.
9. Growth cabinets maintained at 24–26°C, and 16-h photoperiod at about 80 μmol/m^2/s from wide-spectrum fluorescent tubes (*see* **Note 2**).

3. Methods: Softwoods

3.1. Softwoods

3.1.1. Explant Sterilization and Induction

1. Imbibe cedar (*T. occidentalis*) seeds overnight under running tap water.
2. Disinfest seeds for 20 min in 30% Javex® bleach (6% NaOCl) with 3–4 drops of Tween 20®/100 mL. Rinse three times with sterile distilled water. Treat seeds for 5 min with 10% H_2O_2. Rinse several times with sterile water (*see* **Note 3**).
3. For *T. occidentalis,* 1/2QP (1/2-strength major salts only) with 1 μ*M* BA for 21 d was the most effective induction treatment (*see* **Notes 4** and **5**). Excise embryos from disinfested seeds, and plate on this medium about 12 explants/plate.
4. After phytohormone exposure, transfer explants to the same medium without BA for 3 wk (*see* **Note 6**).

3.1.2. Bud Development

1. Transfer explants to 1/2QP, 2% sucrose, and 0.05% conifer-derived activated charcoal for shoot elongation.
2. Subculture every 3–4 wk. As shoots elongate, transfer to glass jars with elongation medium and seal with Parafilm®.
3. Divide explants into smaller pieces for faster shoot elongation; separate individual shoots after two passages and when shoots are >5 mm in height. Maintain shoots on shoot elongation medium.

3.1.3. Multiplication via Axillary Buds

1. Use 8–10 mm adventitious shoots (*see* **Note 7**).
2. Prepare media in glass jars with phytohormone(s) to activate the axillary meristems. For cedar, culture shoots on 1/2QP, 2% sucrose with 10 μ*M* zeatin for 4 wk, followed by culture on 1/2QP, 2% sucrose and 0.05% activated charcoal for axillary shoot elongation.
3. Separate axillary shoots when 5–10 mm, and culture on the same medium.

3.1.4. Rooting and Acclimatization

1. Select shoots >10 mm in height for rooting (*see* **Note 8**).
2. Prepare rooting medium or substrate; substrate can be a soilless commercial mix, or a 1:1 mixture of peat and vermiculite moistened with diluted (1/2QP, 1% sucrose) medium and autoclaved. Stir mixture after autoclaving to aerate and redistribute peat/vermiculite.
3. Treat shoots for 3–4 h in an auxin solution, i.e., immerse shoot bases or ends in IBA (1 m*M*), pH 5.0; transfer treated shoots to substrate of choice. Seal jars with Parafilm®. Roots appear after 4–6 wk.

4. Harden rooted shoots at 20°C for 1–2 wk before transfer to greenhouse conditions, using an appropriate soil mix for conifers.

3.2. Hardwoods

3.2.1. Explant: Sterilization and Shoot Initiation

1. Select dormant twigs (e.g., *P. tremuloides*) with nodal or axillary buds, preferably from greenhouse material, since it will be less contaminated than field material.
2. Cut twigs into small sections (±7 mm) each with an intact bud.
3. Surface-sterilize by dipping in 70% ethanol for 1 min and rinsing three times with sterile distilled water; then, soak sections in 20% bleach (6% NaOCl) with 2–3 drops Tween® for 15 min. Rinse several times with sterile distilled water.
4. Separate buds, and remove the outermost scales under dissecting microscope; removal of these scales reduces the amount of phenolic exudates and enhances bud breaking (*see* **Note 9**).
5. Plate on MS medium with 2 μ*M* BA, 3% sucrose, and 0.8% agar (*see* **Note 10**).
6. Seal plates with Parafilm® or Stretch n' Seal® (*see* **Note 11**), and incubate at 24–25°C under 16 h light for 2–4 wk.
7. Subculture at 2- to 3-wk intervals or shoots will begin to turn brown. Always remove the lower end of the shoot (±2–4 mm), thus providing a freshly cut surface for nutrient uptake.

3.2.2. Shoot Elongation and Multiplication

1. To enhance elongation, transfer shoots at 2-wk intervals.
2. For multiplication, select shoots ±10 mm, and subculture these on MS medium with 1 μ*M* BA, 3% sucrose, and 0.8% agar in glass jars. Seal jars with Parafilm® and subculture at 3- to 4-wk intervals (*see* **Notes 12** and **13**).
3. Observe shoots carefully for culture contaminants (*see* **Note 14**).

3.2.3. Rooting and Acclimatization

1. After 8–12 wk on multiplication medium, transfer individual shoots to 1/2MS medium without phytohormones in glass jars (*see* **Note 14**).
2. Transfer at 3-wk intervals until root primordia develop.
3. Remove plantlets from agar medium when they have well-developed root systems, and transfer to plastic boxes or enclosed seedling trays with sterile vermiculite. Maintain these under high light and humidity. Mist plantlets with sterile water every 12 h to maintain high humidity. Harden plants for 3–4 wk before transferring to greenhouse conditions (*see* **Note 15**).

4. Notes

1. For preparing stock solutions and information on commonly used media *see* refs. ***(5–7)***.
2. For more information on the above equipment, implements, and so on, *see* refs. ***(7–20)***.
3. Some softwood seeds require much more elaborate pretreatments, including stratification before excision and explantation (*see* ***21***). Stratification procedures

carried out at temperatures of 3–5°C for periods of days or weeks are useful for breaking seed dormancy.

4. Salt formulations generally used for conifers include von Arnold and Eriksson's AE or LP *(22)*, Bornman's MCM *(23)*, and Schenk and Hildebrandt *(24)*. These can be used at full- or half-strength. In general, for induction and elongation, various salt formulations should be tested at both full- and half-strength and sometimes at lower concentrations. Select medium based on overall explant appearance, percent explant response, number of adventitious shoots, and elongation rates *(4)*. Thus, both quantitative and qualitative factors must be considered.
5. For each species, the optimum BA concentration, exposure time, and sucrose concentration have to be determined. Other cytokinins, like kinetin, zeatin, and 2iP used alone or in combination with BA can sometimes improve organogenesis, and the effects of these have to be determined *(4)*.
6. For species with larger seeds and embryos, e.g., radiata pine, it is necessary to germinate the whole seed in a moistened, sterile, soilless mix *(25)*, or conversely, the excised embryo is germinated on 1% sucrose–agar for 3–4 d, e.g., Canary Island pine *(26)*. Individual cotyledons can then be excised and used as explants.
7. Multiplication is a most important step in developing a micropropagation protocol, since this step allows for the production of large numbers of shoots, either via axillary budding (preferably) or by adventitious budding (which may involve bud formation in callus). For common approaches used, *see* refs. *(2–4)*. Without adequate rates of multiplication, the developed method will probably not be economical *(7)*. Activated charcoal is often needed. For some softwoods, only conifer-derived charcoal was effective *(27)*.
8. For some conifers, rooting and acclimatization may be carried out concurrently, but for most, two distinct sets of operations are needed. In such cases, rooting is often achieved or at least initiated under in vitro sterile conditions, and may require separate steps of root induction and root development (*see* **2,6,28**).
9. The production of toxic secretions, including polyphenols and tannins, is a major problem that often has to be overcome on introducing hardwoods into culture *(3)*. Several approaches have been developed to deal with this problem *(29)*, but in some cases, the only successful method is the very labor-intensive process of frequent transfer of the explants onto fresh medium *(3)*.
10. Although 2 µ*M* BA is recommended for this species, a range of 2–5 µ*M* can be used effectively. MS is used extensively for hardwoods, but other formulations like Gamborg's B5 *(30)* and McCown's Woody Plant Medium *(31)* are also effective.
11. Comparisons were made with Parafilm® (permeable) and Stretch 'n Seal® (impermeable), but no significant differences were observed.
12. The addition of 0.05% activated charcoal did not significantly improve shoot initiation or elongation in our experiments. Frequent transfer was the best option.
13. In comparisons with different types of culture jars, we found that poplar shoots grew slower in Magenta jars when compared to glass jars.
14. Low density of shoots in the culture vessel and regular transfers will help in the detection of contaminants and maintenance of aseptic cultures. However, when

contamination occurs, precautions must be taken to prevent the spread of the contaminant whether it is introduced or systemic to the plant material.

15. In-vitro grown plantlets cannot be transferred directly to ex vitro conditions. These plantlets have to be gradually acclimatized to higher light intensities and lower humidity under greenhouse or field conditions. They are particularly susceptible to ex vitro stresses because of reduced epicuticular wax, and have initially, a slow stomatal response to water stress and limited capacity for photosynthesis, since they were cultured on sucrose-containing media (*see* ***32***). In addition, for hardwoods, leaves must be fully developed, since poor leaf formation results in shoot deterioration ***(33)***.

References

1. Zimmerman, R. H. (1986) Regeneration in woody ornamentals and fruit trees, in *Cell Culture and Somatic Cell Genetics of Plants,* vol. 3, (Vasil, I. K., ed.), Academic, New York, pp. 243–258.
2. Thorpe, T. A., Harry, I. S., and Kumar P. P. (1991) Application of micropropagation to forestry, in *Micropropagation* (Debergh, P. C. and Zimmerman, R. H., eds.), Kluwer Academic, Dordrecht, pp. 311–336.
3. Harry, I. S. and Thorpe, T. A. (1990) Special problems and prospects in the propagation of woody species, in *Plant Aging: Basic and Applied Approaches,* (Rodriguez, R., Tamés, R. S., and Durzan, D. J., eds.), Plenum, New York, pp. 67–74.
4. Dunstan, D. I. and Thorpe, T. A. (1986) Regeneration in forest trees, in *Cell Culture and Somatic Cell Genetics in Plants,* vol. 3 (Vasil, I. K., ed.), Academic, New York, pp. 233–241.
5. Thorpe, T. A. and Patel, K. R. (1984) Clonal propagation: Adventitious buds, in *Cell Culture and Somatic Cell Genetics of Plants,* vol. 1 (Vasil, I. K., ed.), Academic, New York, pp. 49–60.
6. Ahuja, M. R. (1987) *In vitro* propagation of poplar and aspen, in *Cell and Tissue Culture in Forestry,* vol. 3 (Bonga, J. M. and Durzan, D. J., eds.), Martinus Nijhoff, Dordrecht, pp. 207–223.
7. Harry, I. S. and Thorpe, T. A. (1994) *In vitro* methods for forest trees, in *Plant Cell and Tissue Culture* (Vasil, I. K. and Thorpe, T. A., eds.), Kluwer Academic, Dordrecht, pp. 539–560.
8. Harry, I. S., Thompson, M. R., Lu, C.-Y., and Thorpe, T. A. (1987) *In vitro* plantlet formation from embryonic explants of eastern white cedar (*Thuja occidentalis* L). *Tree Physiol.* **3,** 273–283.
9. Nour, K. A. and Thorpe, T. A. (1993) *In vitro* shoot multiplication of eastern white cedar. *In Vitro Cell. Dev. Biol.* **29,** 65–71.
10. Lubrano, L. (1992) Micropropagation of Poplars (*Populus spp.*), in *Biotechnology in Agriculture and Forestry—High-Tech and Micropropagation II,* vol. 18 (Bajaj, Y. P. S., ed.), Springer-Verlag, Berlin, pp. 151–176.
11. Thorpe, T. A. (1995) Aspen Micropropagation. Part of a report Aspen Decay and Stain and Genetic Fingerprinting of Clones, Report A5012, Canadian Forest Service, and Land and Forest Services, Edmonton, Alberta.

12. Quoirin, M. and Lepoivre, P. (1977) Étude de milieux adaptés aux cultures *in vitro* de *Prunus. Acta Hortic.* **78,** 437–442.
13. Murashige, T. and Skoog, F. (1962) A revised medium for rapid growth and bioassays with tobacco tissue cultures. *Physiol. Plant.* **15,** 473–497.
14. Aitken-Christie, J. and Thorpe, T. A. (1984) Clonal propagation: Gymnosperms, in *Cell Culture and Somatic Cell Genetics of Plants,* vol. 1, *Laboratory Procedures and Their Applications* (Vasil, I. K., ed.), Academic, New York, pp. 82–95.
15. Gamborg, O. L., Murashige, T., Thorpe, T. A., and Vasil, I.K, (1976) Plant Tissue Culture Media. *In Vitro* **12,** 473–478.
16. Chandler, S. F. and Thorpe, T. A. (1985) Culture of plant cells: techniques and growth medium, in *Cell Biology,* vol. C1, *Techniques in Setting Up and Maintenance of Tissue and Cell Cultures,* C112 (Kurstak, E., ed.), Elsevier, Ireland, pp. 1–21.
17. Gamborg, O. L. and Phillips, G. C. (eds.) (1995) *Plant Cell, Tissue and Organ Culture—Fundamental Methods,* Section 1. Springer, Berlin, pp. 3–42.
18. Biondi, S. and Thorpe, T. A. (1981) Requirements for a tissue culture facility, in *Plant Tissue Culture Methods and Applications in Agriculture* (Thorpe, T. A., ed.), Academic, New York, pp. 1–20.
19. Bhojwani, S. S. and Razdan, M. K. (1983) *Developments in Crop Science,* vol. 5, *Plant Tissue Culture: Theory and Practice.* Elsevier, Amsterdam, p. 502.
20. Brown, D. C. W. and Thorpe, T. A. (1984) Organization of a plant tissue culture laboratory, in *Cell Culture and Somatic Cell Genetics of Plants,* vol. 1, *Laboratory Procedures and Their Applications* (Vasil, I. K., ed.), Academic, New York, pp. 1–12.
21. Thorpe, T. A. and Harry, I. S. (1991) Clonal propagation of conifers, in *Plant Tissue Culture Manual* C3 (Lindsey, K., ed.), Kluwer Academic, Dordrecht, pp. 1–16.
22. von Arnold, S. and Eriksson, T. (1981) *In vitro* studies of adventitious shoot formation in *Pinus contorta. Can. J. Botany* **59,** 870–874.
23. Bornman, C. H. (1983) Possibilities and constraints in the regeneration of trees from cotyledonary needles of *Picea abies in vitro. Physiol. Plant.* **80,** 534–540.
24. Schenk, R. U. and Hildebrandt, A. C. (1972) Medium and techniques for induction and growth of monocotyledonous and dicotyledonous plant cell cultures. *Can. J. Botany* **50,** 199–204.
25. Biondi, S. and Thorpe, T. A. (1982) Growth regulator effects, metabolic changes, and respiration during short initiations in cultured cotyledon explants of *Pinus radiata. Bot. Gaz.* **143,** 20–25.
26. Martinéz Pulido C, Harry, I. S., and Thorpe, T. A. (1990) *In vitro* regeneration of plantlets of Canary Island pine *(Pinus canariensis). Can. J. Forest Res.* **20,** 1200–1211.
27. Rumary, C. and Thorpe, T. A. (1984) Plantlet formation in black and white spruce. I. *In vitro* techniques. *Can. J. Forest Res.* **14,** 10–16.
28. Mohammed, G. H. and Vidaver, W. E. (1988) Root production and plantlet development in tissue-cultured conifers. *Plant Cell Tiss. Org. Cult.* **14,** 137–160.
29. Pierik, R. L. M. (1987) *In Vitro Culture of Higher Plants.* Martinus Nijhoff, Dordrecht.

30. Gamborg, O. L., Miller, R. A., and Ojima, K. (1968) Nutrient requirements of suspension cultures of soybean root cells. *Exp. Cell Res.* **50,** 151–158.
31. Lloyd, G. and McCown, B. (1980) Commercially feasible micropropagation of mountain laurel, *Kalmia latifolia,* by use of shoot-tip culture. *Int. Plant Propagators Society Combined Proceedings* **30,** 421–427.
32. Chalupa, V. (1987) European Hardwoods, in *Cell and Tissue Culture in Forestry,* vol. 3 (Bonga, I. M. and Durzan, D. J., eds.), Martinus Nijhoff, Dordrecht, pp. 224–246.
33. McCown, D. D. and McCown, B. H. (1987) North American hardwoods, in *Cell and Tissue Culture in Forestry,* vol. 3 (Bonga, J. M. and Durzan, D. J., eds.), Martinus Nijhoff, Dordrecht, pp. 247 260.

16

Spore-Derived Axenic Cultures of Ferns as a Method of Propagation

Matthew V. Ford and Michael F. Fay

1. Introduction

A method by which many fern species can be successfully grown from spores in axenic culture is described. Unlike the conventional method of sowing the spores on compost, this method allows spore populations free from contamination by spores of other species to be sown. The method can be used for the production of mature sporophytes or to provide a controllable system for biosystematic studies of, or experimentation with, fern gametophytes ***(1,2)***.

For many years, it has been known that spores of most ferns require exposure to light for germination and growth ***(1,3,4)***. However, dark germination has been recorded in *Pteridium aquilinum* ***(5,6)***, *Blechnum spicant* ***(7)***, and *Onoclea sensibilis* ***(8,9)***, among others. This ability to germinate in the dark may vary with specimen age, exposure to different temperatures, and plant hormone treatment ***(1,3,10)***. In general, the percentage of dark germination is low, and subsequent growth is abnormal or retarded.

Many workers have investigated the relative effectiveness of different spectral regions of light on spore germination, and have obtained results indicating the involvement of phytochrome ***(11–13)***. In general, exposure to red light induces germination ***(14,15)***. This is reversed by exposure to far-red light and blocked by blue light ***(16)***. Light quality also influences prothallial growth ***(17–20)***. Spectral filters are not necessary for germination of the majority of ferns using the method we describe here, since white light provides an adequate spectral balance. Light intensity has also been reported to have a marked influence on germination and subsequent prothallial growth ***(21)***. This phenomenon can be observed in nature in the various habitat preferences of ferns.

From: *Methods in Molecular Biology, Vol. 111: Plant Cell Culture Protocols*
Edited by: R. D. Hall © Humana Press Inc., Totowa, NJ

Spore sowing density can prove to be critical at its extremes *(22)*, with slower germination at lower densities and reduced germination at higher densities. This appears to be the result of volatile substances, possibly including ethylene, which are produced by germinating spores. Overcrowding may also lead to aberrations in prothallial growth. Competition for light and mineral resources may also cause problems. The production of large amounts of antheridogens, often found in high-density gametophyte populations, can also prove problematic, since they induce a predominance of maleness in the cultures *(23)*, hence preventing fertilization and sporophyte formation.

It can be seen from the brief resume given above that much work has been carried out on the physiological aspects of spore germination and gametophyte growth. The following method takes into account much of the available knowledge, but has been designed to be simple and effective.

2. Materials

1. Media: Murashige and Skoog (MS) medium (*24* and *see* Appendix) is obtained as a ready-mixed powder, without sucrose, growth regulators, or agar. The initial medium is made up at half the recommended strength ($^1/_2$MS), together with 15 g sucrose dissolved in 1 L of purified water (Milli-Q, distilled or equivalent). For subsequent transfers, the medium is supplemented with 1 g/L activated charcoal *(25)*. Adjust the pH to 5.6 using dilute HCl or NaOH as appropriate. Dissolve 9 g of agar in the medium using a microwave oven (2 min for each 250 mL of medium in a 1000-W oven), and stir thoroughly. This method was developed using Sigma purified agar (Product No. A7002). Dispense the medium into the desired vessels (we normally use $^1/_2$– and 1-lb honey jars), and sterilize in the autoclave for 15 min at 121°C and 1.05 kg/cm^2. For sowing, we use disposable sterile Petri dishes, and in this case, the medium is autoclaved in conical flasks and then dispensed into the Petri dishes in a laminar airflow bench. If the medium is not to be used immediately, store the vessels at 4°C in the dark.
2. Sterilization: Sodium hypochlorite (BDH, 10–14% available chlorine) is used as the sterilant, diluted to 3–10% v/v in deionized water (0.3–1.4% available chlorine), with the addition of a few drops of Tween 80/L. Spores are sterilized in filter paper packets made from a 4.25-cm circle of filter paper folded in from four sides, as shown in **Fig. 1**.
3. Compost: Some ferns, such as *Cyathea* and *Angiopteris,* are fairly sturdy when they are brought out of culture. These can be weaned in the first compost given below. Others (*Adiantum, Cheilanthes, Nephrolepis,* and so forth) are quite delicate when first out of culture, and thus require a much finer compost, such as the second one given below. In all cases, good drainage must be provided. For the more robust ferns, use: 2 parts leaf mold; 2 parts coir; 1 part bark with added fertilizer, e.g., 2 g/L Vitax UN2. For more delicate ferns, the same compost is used after sieving through a 5 mm sieve.

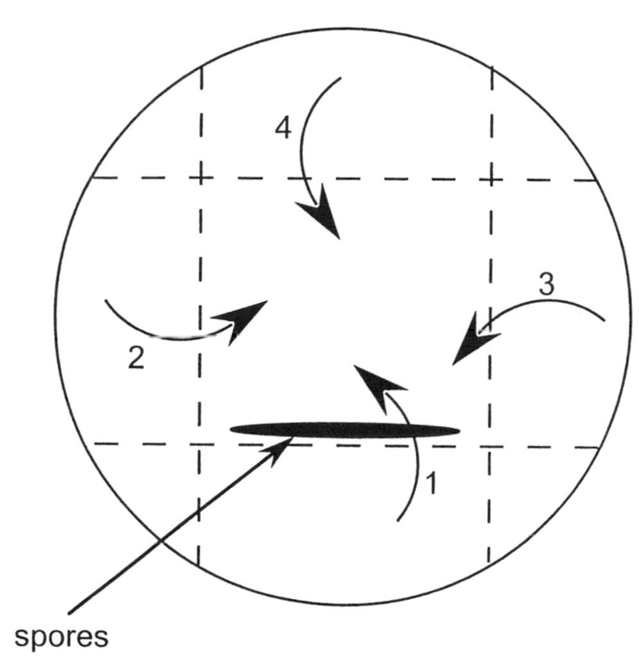

Fig. 1. Filter paper packet used for fern spore sterilization. The spores are placed on the sheet of filter paper (4.25-cm diameter), which is then folded as indicated to form a packet. This is then sealed with a staple.

3. Methods

3.1. Spore Collection and Storage

1. For each species a frond, or piece of frond, with mature sporangia is collected and kept warm and dry in a folded white card, or paper, awaiting dehiscence ***(3)***. Depending on the maturity of the sporangia collected, this usually takes place within 48 h. More than enough spores can be obtained from a single frond. It has been calculated that the approximate spore content of an individual frond can range from 750,000 to 750,000,000 depending on the species ***(10)***. Spore viability declines with age and depends on storage conditions, but this also varies between species. Ferns with chlorophyllous spores (those in *Osmundaceae, Gleicheniaceae, Grammitidaceae,* and *Hymenophyllaceae*) should be sown as soon as possible, because they lose their viability after only a few days ***(26,10)***. Nonchlorophyllous spores, in general, retain their viability for substantial periods when kept at room temperature, but there are exceptions (e.g., spores from the tree fern family, *Cyatheaceae,* lose viability rapidly after only a few weeks) ***(10)***. If storage of the spores is unavoidable, they are best kept in sealed packets in a desiccator, or in sealed tubes at 4°C.

3.2. Sterilization

1. Ensure that the spore sample is reasonably free from chaff and other debris. This can be done either manually with fine forceps, for the larger pieces, or by giving the white card a short, sharp, but gentle tap. This should separate chaff from spore.
2. Place some spores in the filter paper packet, and seal with a metal staple (remembering that too small or too great a quantity may result in density problems).
3. Soak the spore packets for 5–10 min in sterile deionized water, and carefully expel any air within them by gently squeezing with forceps.
4. Remove the packets from the water, put them in a plastic beaker, and place on a magnetic stirrer. Sodium hypochlorite (BDH, 10–14% available chlorine) is used as the sterilant, diluted to 10% v/v in deionized water (1–1.4% available chlorine), with the addition of a few drops of Tween 80/L. Add enough NaOCl solution to more than cover the packets, and set the speed on the stirrer to provide gentle agitation. Leave to agitate for 10 min.
5. Move the beaker into a laminar airflow bench. Standard aseptic techniques should now be followed.
6. Remove each packet from the solution, and rinse in separate tubes of sterile deionized water three times.

3.3. Sowing

1. Carefully take the packet from the water with large forceps, and squeeze gently to expel any excess water.
2. Remove the staple end of the packet with scissors.
3. With the Petri dish open, using two pairs of forceps, unfold the packet, and wipe the spores in a swirling motion over the surface of the medium.
4. If any obvious clumping of spores is observed, gently spread them out with the flat end of a spatula.
5. Seal the Petri dish with Parafilm or Nescofilm, and place in the growth room (temperature 22 ± 2°C; light intensity 400–4200 lx; photoperiod 16 h) or in an incubator. The light intensity used depends on the habitat type of the species being cultured.

3.4. Germination and Gametophyte Growth

1. Normally a period of 3–96 h in the dark after sowing is required for imbibition before the spores become light-receptive *(27)*. This preinduction phase of germination is not directly accommodated in this method because the photoperiod provided in the growth room has proven adequate in supplying the necessary environment. If, however, synchronous or uniform germination is desired, then a dark period before exposure to light is advisable.
2. The induction phase of germination will go largely unnoticed. It is followed by the postinduction phase, where dark processes are triggered, resulting in the protrusion of a rhizoid and the protonema *(28)*. The time from sowing to this stage varies greatly between species and can range from a few days to many months.

Fig. 2. A mass of gametophytes of *Actiniopteris semiflabellata* (left) forming a few sporophytes around the edge and a later stage (right) with well-developed sporophytes ready for separation from the gametophytes (approx x3).

Some spores will inevitably be killed by the surface sterilization process, but a good percentage of germination should still result.

3. Normally after a few weeks, the characteristic thalloid gametophyte form will be observed (*see* **Fig. 2**). From this stage, the rate of growth varies between species and with the light conditions to which the gametophyte has been subjected. If growth to the mature gametophyte appears particularly retarded, try moving some dishes to higher and lower light intensities.
4. During growth, the following problems may occur:
   a. The population is too dense.
   b. The medium begins to dry and split.
   c. The gametoyphytes begin to grow brown around their bases and discolor the medium.
   d. Mass sporophyte formation begins.

   The solution to problem (a) is to ease the gametophytes gently away from the medium and move them to fresh medium in either jars or dishes, depending on the size of the plants. A spatula with a flattened end bent to a 90° angle is useful for this operation. By gently pulling the latter across the surface of the media, the rhizoids (vital for the uptake of nutrients and moisture and anchorage to the substrate) are eased out of the media with minimal physical damage. Remove any media adhering to the plant during transfer.

   If problem (b) occurs, movement to fresh media is required as soon as possible. Gametophytes have been shown to be sensitive to drying ***(29)***. This can lead to a rapid death of the population if left too long.

If problem (c) happens, then either (i) the medium is not suitable for this species and it needs to be transferred to another medium (*see* **Note 1**), or (ii) the nutrients are being rapidly exhausted, and more frequent transfer to fresh medium is necessary.

Situation (d) quite often occurs where the level of moisture in the environment is sufficient to facilitate fertilization. This becomes a problem if the container is too small or the sporophytes arise too close to their neighbors. This can lead to a high degree of competition and stunted growth. It is therefore necessary to transfer the plants to a fresh medium, ensuring adequate space is provided for unrestricted growth.

If the gametophytes form clumps, then gently tease them apart and move to fresh medium.

3.5. Sporophyte Formation and Potting

1. Many ferns are homosporous, producing spores that germinate to form bisexual, haploid gametophytes. These have both antheridia (male) and archegonia (female) parts, and produce gametes that fuse and form the diploid sporophyte ***(30)***. Other ferns are heterosporous, producing micro- and megaspores that germinate to form microgametophytes (male) and megagametophytes (female), respectively. These spores must be sown together if the ultimate aim is to produce sporophytes (e.g., in *Marsilea* and *Platyzoma*).
2. Some species will produce sporophytes readily (**Fig. 2**), and others will do so only after several transfers to fresh medium. A few require supplementary moisture. This can be provided simply in one of two ways:
   a. Rinsing in sterile deionized water between transfers.
   b. Adding a small amount of sterile deionized water to the jar, swirling gently, and pouring off the excess (care has to be taken here not to add so much water that the structure of the medium is damaged). Do not attempt this if the gametophytes are not firmly anchored to the medium.

   In the early stages, the sporophyte is dependent on the gametophyte for moisture and nutrients, which are transferred to it via a foot.
3. Subsequently roots will be formed, and when these are large enough, the sporophytes may be pricked out onto fresh medium and cultured separately. It is often advisable to leave the parent gametophyte attached to the sporophyte, reducing the risk of physical damage. The gametophytes may then senesce and die, or continue growing, in which case they can then be removed and cultured separately, if so desired.
4. As the sporophyte grows, more frequent transfers to larger vessels may prove necessary. When a reasonable number of roots have been formed, the plant is ready for potting. This can be done using one of two methods:
   a. Direct potting of the fern in the compost described in the **Subheading 2.** Adhering media are carefully removed by rinsing in tepid deionized water before potting. Good drainage must be provided, and a covering layer of gravel helps to conserve water and prevents the growth of algae. The potted

ferns are grown in shaded frames in a glass house at temperatures between 20 and 30°C and high relative humidity (90–98%). They can then be gradually acclimatized to the recommended growth conditions for the species in question.

b. Transferring to sealed jars of sterile compost. Some ferns are slow to produce or produce only small amounts of root. This intermediate step can help to encourage root growth and reduce the potential damage to roots when the plant is finally potted as in (a).

4. Notes

1. Some ferns may prefer a different medium to that given here, e.g., *Angiopteris boivinii* will grow on ½MS, but grows better on Knudson's C medium ***(29)*** if frequent transfers are made. Useful media variations for experimentation are: Knudson's C and Knudson's C modified (with 1 mL microsalt solution/L) media; or Moore's medium (personal communication, Moore, London University):

NH_4NO_3	1.0 g/L
KH_2PO_4	0.2 g/L
$MgSO_4 \cdot 7H_2O$	0.2 g/L
$CaCl_2 \cdot 2H_2O$	0.1 g/L

   Dissolve each component separately. Then combine, and add 1 mL/L of solution (a) and 2 mL/L of solution (b).

   a. Microsalt stock solution:

$B_2O_3 \cdot H_2O$	6.20 g/L
$Na_2MoO_4 \cdot 2H_2O$	0.25 g/L
$ZnSO_4 \cdot 7H_2O$	8.60 g/L
KI	0.83 g/L
$CuSO_4 \cdot 5H_2O$	0.025 g/L
$CoCl_2 \cdot 6H_2O$	0.025 g/L

   (Dissolve separately and then combine; store in the dark at 0°C.)

   b. Ferric citrate stock solution: $C_6H_5O_7Fe \cdot 5H_2O$, 1 g/L. (Store in the dark at 0°C).

   After combining all components, adjust pH to 5.6, and add 9 g/L agar.
2. The time period and the strength of the sterilant can be varied. Increasing the time or concentration will improve the sterilization of those specimens that are repeatedly contaminated. Reducing the time or concentration will decrease the percentage of spore fatality in especially delicate samples.
3. For some species with short spore viability, manual dissection of the sporangia may be required if dehiscence has not occurred within 48 h. This is carried out in a draft-free environment using a clean blade and fine forceps.
4. If fungal contamination is severe, the culture must be thrown away. If mild, sections of the population may be saved by moving the uncontaminated parts to a sugar-free medium (such as Moore's) or to ½MS supplemented with a fungicide, e.g., 1% Benlate. Growth on the former may be slow and pale, but the plants should reach maturity. Transfer back to a sugar-based medium should be possible at a later stage, but the fungus may manifest itself again, after lying dormant in the absence of sugar.

5. Abnormal growth may occur at either the gametophyte or sporophyte stage.
   a. Elongation of the prothallus may occur. This is likely to be a density problem and should be treated as such.
   b. Sporophytes are sometimes formed without the fusion of gametes taking place, i.e., apogamously. This sometimes occurs in old cultures of gametophytes or in cultures of species that are particularly susceptible to this problem. Ethylene has been implicated as a promoter of this process ***(31)***. The apogamous (haploid) sporophyte may arise from the prothallial cushion, apical region, or on the end of an apical protuberance or podium ***(32)***.
   c. The sporophytes grow down into the media as well as upward. This is geotrophic confusion, possibly brought on by either ethylene or too much light below the culture. This can be averted by thinning out the population and blocking light from below.
6. The pH used in this method was chosen because it suits a range of ferns ***(33)***. However, for obligate calcicoles or calcifuges, the pH may have to be adjusted. If lower pH values are desired, then a higher concentration of agar is needed to maintain a sufficient gel strength (12 g/L or more). If pH values >7.0 are required, it is better to use $Ca(OH)_2$ rather than NaOH, since high concentrations of sodium can prove toxic.
7. Withering of fronds during transfer occurs in particularly sensitive ferns or if the transfer takes a long time. This is the result of desiccation caused by the stream of air in the flow bench. A screen, using one or more sterile dishes set up in the bench, helps to prevent this problem.

Acknowledgments

We would like to thank L. Goss and N. Rothwell for their advice on compost and glass house conditions.

References

1. Miller, J. H. (1968) Fern gametophytes as experimental material. *Bot. Rev.* **34,** 362–441.
2. Windham, M. D. and Haufler, C. H. (1986) Biosystematic uses of fern gametophytes derived from herbarium specimens. *Am. Fern J.* **76,** 114–128.
3. Dyer, A. F. (1979) The culture of fern gametophytes for experimental investigation, in *The Experimental Biology of Ferns* (Dyer, A. F., ed.), Academic, London, pp. 254–305.
4. Williams, S. (1938) Experimental morphology, in *Manual of Pteridology* (Verdoorn, Fr., ed.), Martinus Nijhoff, The Hague, pp. 105–140.
5. Ragavan, V. R. (1970) Germination of bracken fern spores. Regulation of protein and RNA synthesis during initiation and growth of the rhizoid. *Exp. Cell Res.* **63,** 341–352.
6. Weinberg, E. S. and Voeller, B. R. (1969) Induction of fern spore germination. *Proc. Natl. Acad. Sci. USA* **64,** 835–842.

7. Orth, R. (1937) Zur Keimungsphysiologie der Farnsporen in verschiedenen Spektralvezirken. *J. Wis. Bot.* **84,** 358–426.
8. Miller, J. H. and Greany, R. H. (1974) Determination of rhizoid orientation by light and darkness in germinating spores of *Onoclea sensibilis. Am. J. Bot.* **51,** 329–334.
9. Miller, J. H. and Miller, P. M. (1970) Unusual dark-growth and antheridial differentiation in some gametophytes of the fern *Onoclea. Am. J. Bot.* **57,** 1245–1248.
10. Page, C. N. (1979) Experimental aspects of fern ecology, in *The Experimental Biology of Ferns* (Dyer, A. F., ed.), Academic, London, pp. 552–589.
11. Ragavan, V. R. (1971) Phytochrome control of germination of the spores of *Asplenium nidus. Plant Physiol.* **48,** 100–102.
12. Ragavan, V. R. (1973) Blue light interference in the phytochrome-controlled germination of the spores of *Cheilanthes farinosa. Plant Physiol.* **51,** 306–311.
13. Sugai, M., Takeno, K., and Furuya, M. (1977) Diverse responses of spores in the light-dependent germination of *Lygodium japonicum. Pl. Sci. Lett.* **8,** 333–338.
14. Towill, L. R. and Ikuma, H. (1973) Photocontrol of the germination of *Onoclea* spores I. Action spectrum. *Plant Physiol.* **51,** 973–978.
15. Towill, L. R. and Ikuma, H. (1975) Photocontrol of the germination of *Onoclea* spores II. Analysis of germination processes by means of anaerobiosis. *Plant Physiol.* **55,** 150–154.
16. Furuya, M. (1985) Photocontrol of spore germination and elementary processes of development in fern gametophytes, in *Biology of Pteridophytes* (Dyer, A. F. and Page, C. N., eds.), *Proc. Royal Soc. Edin.* **86B,** 13–19.
17. Greany, R. H. and Miller, J. H. (1976) An interpretation of dose-response curves for light induced cell elongation in the fern protonemata. *Am. J. Bot.* **63,** 1031–1037.
18. Howland, G. P. and Edwards, M. E. (1979) Photomorphogenesis of fern gametophytes, in *The Experimental Biology of Ferns* (Dyer, A. F., ed.), Academic, London, pp. 394–434.
19. Miller, J. H. and Miller, P. M. (1967) Action spectra for light induced elongation in fern protonemata. *Physiologia Pl.* **20,** 128–138.
20. Miller, J. H. and Miller, P. M. (1974) Interaction of photomorphogenetic pigments in fern gametophytes: Phytochrome and a yellow-light absorbing pigment. *Pl. Cell Physiol. Tokyo* **8,** 765–769.
21. Ford, M. V. (1984) Growth responses of selected fern prothalli to various light regimes. Unpublished.
22. Smith, D. L. and Robinson, P. M. (1971) Growth factors produced by germinating spores of *Polypodium vulgare* (L.). *New Phytol.* **70,** 1043–1052.
23. Bell, P. R. (1979) The contribution of the ferns to an understanding of the life cycles of vascular plants. IV. Sexuality in gametophytic growth, in *The Experimental Biology of Ferns* (Dyer, A. F., ed.), Academic, London, pp. 64, 65.
24. Murashige, T. and Skoog, F. (1962) A revised medium for rapid growth and bioassays with tobacco tissue cultures. *Physiol. Plant.* **15,** 473–497.
25. Ford, M. V. (1992) Growing ferns from spores in sterile culture, in *Fern Horticulture, Past, Present and Future Perspectives* (Ide, J. M., Jermy, A. C., and Paul, A. M., eds.), Intercept, Andover, pp. 295–297.

26. Lloyd, R. M. and Klekowski, E. J., Jr. (1970) Spore germination and viability in the Pteridophyta: Evolutionary significance of chlorophyllous spores. *Biotropica* **2,** 129–137.
27. Jarvis, S. J. and Wilkins, M. B. (1973) Photoresponses of *Matteuccia struthiopteris* (L.) Todaro. *J. Exp. Bot.* **24,** 1149–1157.
28. Brandes, H. (1973) Spore Germination. *Ann. Rev. Plant Physiol.* **24,** 115–128.
29. Knudson, L. (1946) A new nutrient solution for the germination of orchid seed. *Am. Orchid Soc. Bull.* **15,** 214–217.
30. Hyde, H. A., Wade, A. E., and Harrison, S. G . (1978) *Welsh Ferns, Clubmosses, Quillworts, and Horsetails,* National Museum of Wales, Cardiff.
31. Elmore, H. W. and Whittier, D. P. (1973) The role of ethylene in the induction of apogamous buds in *Pteridium* gametophytes. *Planta* **111,** 85–90.
32. Walker, T. G. (1985) Some aspects of agamospory in ferns—the Braithwaite System, in *Biology of Pteridophytes* (Dyer, A. F. and Page, C. N., eds.). *Proc. Royal Soc. Edin.* **86B,** 59–66.
33. Otto, E. A., Crow, J. H., and Kirby, E. G. (1984) Effects of acidic growth conditions on spore germination and reproductive development in *Dryopteris marginalis* (L.). *Ann. Bot.* **53,** 439–442.

IV

Applications for Plant Protoplasts

17

Protoplast Isolation, Culture, and Plant Regeneration from *Passiflora*

Paul Anthony, Wagner Otoni, J. Brian Power, Kenneth C. Lowe, and Michael R. Davey

1. Introduction

The family *Passifloraceae* contains over 580 woody or herbaceous species *(1)*, the majority of species within the genus *Passiflora* being found in tropical South America. *Passiflora edulis* fv. flavicarpa is considered to be the most important species *(2)* because of its value in the fruit juice industry. Additionally, this species is resistant to the soil-borne pathogen *Fusarium oxysporum* and, consequently, is frequently used as a rootstock onto which is grafted *P. edulis* Sims. Interspecific sexual hybridization has been attempted in *Passiflora* breeding programs, using wild-type germplasms to transfer disease resistance and other potentially desirable traits into cultivated species. However, fertile hybrids have been difficult to obtain *(3)*. Somatic hybridization provides a means of circumventing such sexual incompatibilities. Indeed, novel fertile somatic hybrids have been produced between *P. edulis* fv. flavicarpa and *Passiflora incarnata (4)*. Since the latter species can survive winter temperatures of –16°C, such transfer of cold tolerance to the commercial crop would permit cultivation in more temperate climates *(5)*. Somatic hybrid plants have also been produced between *P. edulis* fv. flavicarpa and *Passiflora amethystina, Passiflora cincinnata, Passiflora giberti,* and *Passiflora alata*, respectively *(6)*.

A prerequisite to any somatic hybridization program is the successful isolation and culture of protoplasts (wall-less cells) of at least one of the parental species, leading to the efficient regeneration of normal, fertile plants. Protoplast isolation is influenced by several factors, including the choice of plant species and/or cultivar, together with the source tissue (e.g., leaves, cell suspensions, cotyledons, roots, pollen tetrads) and its physiological status. The

From: *Methods in Molecular Biology, Vol. 111: Plant Cell Culture Protocols*
Edited by: R. D. Hall © Humana Press Inc., Totowa, NJ

nature of the cell wall of the source tissue(s) is an important parameter, since it determines the composition and concentration of the enzyme mixture required for efficient wall degradation and protoplast release. High protoplast viability is crucial immediately after isolation, and also during subsequent culture. To date, plant regeneration from *Passiflora* protoplasts has been reported for *P. edulis* fv. flavicarpa ***(7,8)***, *P. amethystina,* and *P. cincinnata* ***(8)***.

The following protocol is an excellent starting point for those interested in dicot protoplast culture, since it concerns a well-tested and highly responsive system.

2. Materials

2.1. Glasshouse-Grown Seedlings of P. edulis and P. giberti

1. Seeds of *P. edulis* fv. flavicarpa and *P. giberti* (J. C. Oliveira and C. Ruggiero, Departmento de Fitotecnia, FCAVJ-Univeridade Estadual Paulista [UNESP], Jaboticabal, Sao Paulo, Brazil).
2. Levington M3 soilless compost (Fisons, Ipswich, UK) and John Innes no. 3 compost (J. Bentley Ltd., Barrow-on-Humber, UK).
3. "Vac-trays" (H. Smith Plastics Ltd., Wickford, Essex, UK).

2.2. Isolation of Protoplasts Directly from Seedling Leaves of P. edulis

1. "Domestos" bleach (Lever Industrial Ltd., Runcorn, UK), or any commercially available bleach solution containing approx 5% available chlorine.
2. Enzyme 1 solution ***(7)***: 2 g/L Macerozyme R10 (Yakult Honsha Co Ltd., Nishinomiya Hyogo, Japan), 10 g/L Cellulase R10 (Yakult Honsha Co. Ltd.), 1 g/L Driselase (Kyowa Hakko Co. Ltd., Tokyo, Japan), 1.1 g/L 2-[*N*-morpholino]ethanesulfonic acid (MES), 250 mg/L polyvinylpyrrolidone (PVP-10; Sigma), 250 mg/L cefotaxime ("Claforan"; Roussel Laboratories, Uxbridge, UK) with CPW salts ***(9)***, pH 5.8. Filter-sterilize.
3. Modified CPW salts solution (*9*): 27.2 mg/L KH_2PO_4, 101 mg/L KNO_3, 246 mg/L $MgSO_4 \cdot 7H_2O$, 0.16 mg/L KI, 0.025 mg/L $CuSO_4 \cdot 5H_20$, 1480 mg/L $CaCl_2 \cdot 2H_2O$, pH 5.8.
4. CPW13M solution: modified CPW salts solution with 130 g/L mannitol (**item 3**), pH 5.8. Autoclave (15 min, 121°C, steam pressure).
5. Seaplaque agarose (FMC Bioproducts, Rockwell, USA). Autoclave.

2.3. Culture of Leaf Protoplasts of P. edulis

1. KM8P medium: based on the formulation of Kao and Michayluk ***(10)*** with modifications ***(11)*** (**Table 1**) and supplemented with 250 mg/L cefotaxime. For semisolid KM8P medium, mix double-strength KM8P medium with an equal volume of molten (40–60°C) 1.6% (w/v) aqueous Seaplaque agarose, the latter prepared (as with all media) with reverse-osmosis water. Add filter-sterilized cefotaxime to the molten medium at 40°C.

Table 1
Formulation of Media Macronutrients, Micronutrients, Vitamins, and Other Supplements

Component	Concentration, mg/L			
	KM8P	KM8	KPR	K8
Macronutrients				
NH_4NO_3	600	600	600	600
KNO_3	1900	1900	1900	1900
$CaCl_2 \cdot 2H_2O$	600	600	600	600
$MgSO_4 \cdot 7H_2O$	300	300	300	300
KH_2PO_4	170	300	300	300
Sequestrene 330 Fe	28	28	28	28
Micronutrients				
KI	0.75	0.75	0.75	0.75
H_3BO_3	3.0	3.0	3.0	3.0
$MnSO_4 \cdot 4H_2O$	10.0	10.0	10.0	10.0
$ZnSO_4 \cdot 7H_2O$	2.0	2.0	2.0	2.0
$NaMoO_4 \cdot 2\ H_2O$	0.25	0.25	0.25	0.25
$CuSO_4 \cdot 5H_2O$	0.025	0.025	0.025	0.025
$CoCl_2 \cdot 6H_2O$	0.025	0.025	0.025	0.025
Vitamins				
Myo-inositol	100	100	100	100
Nicotinamide	1.0	1.0	1.0	1.0
Pyridoxine HCl	1.0	1.0	1.0	1.0
Thiamine HCl	1.0	1.0	1.0	1.0
D-Ca Pantothenate	1.0	1.0	0.5	0.5
Folic acid	0.4	0.4	0.2	0.2
Abscisic acid	0.02	0.02	0.01	0.01
Biotin	0.01	0.01	0.005	0.005
Choline chloride	1.0	1.0	0.5	0.5
Riboflavin	0.2	0.2	0.1	0.1
Ascorbic acid	2.0	2.0	1.0	1.0
Vitamin A	0.01	0.01	0.005	0.005
Vitamin D_3	0.01	0.01	0.005	0.005
Vitamin B_{12}	0.02	0.02	0.01	0.01
Na pyruvate	20	20	10	10
Citric acid	40	40	20	20
Malic acid	40	40	20	20
Fumaric acid	40	40	20	20
Other supplements				
Fructose	250	250	125	125
Ribose	250	250	125	125

(continued)

Table 1 *(continued)*

Component	Concentration, mg/L KM8P	KM8	KPR	K8
Other supplements				
Xylose	250	250	125	125
Mannose	250	250	125	125
Rhamnose	250	250	125	125
Cellobiose	250	250	125	125
Sorbitol	250	250	125	125
Mannitol	250	250	125	125
Vitamin-free casamino acids	250	250	125	125
Coconut milk	20 mL/L	20 mL/L	10 mL/L	10 mL/L
2,4-Dichlorophenoxyacetic acid	0.2	0.1	0.5	0.1
Zeatin	0.5	0.2	0.5	0.2
α-Naphthaleneacetic acid	1.0	1.0	1.0	1.0
Sucrose	250	20,000	250	20,000
Glucose	100,000	10,000	100,000	10,000
pH	5.8	5.8	5.8	5.8

2. Fluorescein diacetate (FDA): 3 mg/mL stock solution in acetone ***(12)***. Store in the dark at 4°C.
3. KM8 medium: based on the formulation of Kao and Michayluk ***(10)*** with modifications ***(11)*** (**Table 1**). Filter-sterilize.

2.4. Plant Regeneration from Protoplast-Derived Tissues of P. edulis

1. MSR1 medium: based on the formulation of Murashige and Skoog (**13**; *see* Appendix) with 5.0 mg/L α-naphthaleneacetic acid (NAA), 0.25 mg/L 6-benzylaminopurine (BAP), 50 mg/L cysteine, 50 mg/L glutamine, 50 mg/L glutamic acid, 0.5 mg/L biotin, 0.5 mg/L folic acid, 30 g/L sucrose, and 8 g/L agar, pH 5.8. Autoclave the medium, and add filter-sterilized solutions of the amino acids and vitamins when the medium has cooled to about 40°C.
2. MSR2 medium: MS-based medium with 1.0 mg/L BAP, 30 g/L sucrose, and 8 g/L agar, pH 5.8. Autoclave.
3. MSR3: half-strength MS-based medium containing 3.0 mg/L indole-3-butyric acid (IBA), 0.5 mg/L NAA, 30 g/L sucrose, and 8 g/L agar, pH 5.8. Autoclave.

2.5. Initiation of Cell Suspensions of P. giberti

1. "Domestos" bleach: as in **Subheading 2.2, item 1.**
2. MS1 medium: based on the formulation of Murashige and Skoog ***(13)***, supplemented with 7.2 mg/L 4-amino-3,5,6-trichloropicolinic acid (Picloram; Aldrich Chemical Co., Milwaukee, WI) and 8 g/L agar (Sigma, Poole, UK). Autoclave.
3. Nescofilm: Bando Chemical Ind. Ltd. (Kobe, Japan).

Table 2
Formulation of Modified AA2 Medium Macronutrients, Micronutrients, Vitamins, and Other Supplements

Component	Concentration, mg/L
Macronutrients	
$CaCl_2 \cdot 2H_2O$	440
KH_2PO_4	170
$MgSO_4 \cdot 7H_2O$	370
KCl	2940
Micronutrients	
KI	0.83
H_3BO_3	6.2
$MnSO_4 \cdot 4H_2O$	22.3
$NaMoO_4 \cdot 2\ H_2O$	0.25
$ZnSO_4 \cdot 7H_2O$	8.6
$CuSO_4 \cdot 5H_2O$	0.025
$CoCl_2 \cdot 6H_2O$	0.025
$FeSO_4 \cdot 7H_2O$	27.85
Na_2EDTA	37.25
Vitamins	
Myo-inositol	100
Nicotinic acid	0.5
Pyridoxine HCl	0.1
Thiamine HCl	0.5
Glycine	75
L-Glutamine	877
L-Aspartic acid	266
L-Arginine	228
Other supplements	
2,4-Dichlorophenoxyacetic acid	2.0
Gibberellic acid	0.1
Kinetin	0.2
Sucrose	20,000
pH	5.8

4. Modified AA2 medium: based on the formulation of Müller and Grafe ***(14)*** (**Table 2**). Filter-sterilize.

2.6. Isolation of Protoplasts from Cell Suspensions of P. giberti

1. Enzyme solution 101 ***(15)***: 10 g/L Cellulase RS (Yakult Honsha Co. Ltd.), 1 g/L Pectolyase Y23 (Seishim Pharmaceutical, Tokyo, Japan), 1.1g/L MES in CPW13M solution ***(9)***, pH 5.8. Filter-sterilize.
2. CPW13M solution: as in **Subheading 2.2, item 1.**

2.7. Culture of Cell Suspension-Derived Protoplasts of P. giberti

1. KPR medium *(16)*: (**Table 1**). Filter-sterilize.
2. FDA solution: as in **Subheading 2.3, item 2.**
3. K8 medium: based on the formulation of Kao and Michayluk *(10)* with modifications *(11)* (**Table 1**). Filter-sterilize.

2.8. Plant Regeneration from Protoplast-Derived Tissues of P. giberti

1. MSE1 medium: half-strength MS-based medium *(13)* containing 20 g/L sucrose, 4.8 mg/L Picloram, and semisolidified with 2 g/L Phytagel (Sigma), pH 5.8. Autoclave.
2. MSE2 medium: half-strength MS-based medium *(13)* containing 20 g/L sucrose, 0.5 mg/L gibberellic acid (GA_3), and semisolidified with 2 g/L Phytagel, pH 5.8. Autoclave.
3. MSE3 medium: half-strength MS-based medium *(13)* containing 20 g/L sucrose and semisolidified with 2 g/L Phytagel, pH 5.8. Autoclave.

3. Methods

3.1. Glasshouse-Grown Seedlings of P. edulis and P. giberti

1. Prepare a 1:1 (v:v) mixture of Levington M3 compost and John Innes No. 3 compost, and fill the individual compartments of plastic propagator trays.
2. Place 2–3 seeds (approx 1 cm deep) into each compartment of the plastic propagator trays and irrigate from above (*see* **Note 1**).
3. Maintain propagators at a maximum day temperature of 28 ± 2°C and a minimum night temperature of 18 ± 2°C in the glasshouse under natural daylight supplemented with a 16-h photoperiod provided by Cool White fluorescent tubes (180 μmol/m^2/s).

3.2. Isolation of Protoplasts Directly from Seedling Leaves of P. edulis

1. Surface-sterilize fully expanded, young leaves excised from 45- to 60-d-old plants (**Fig. 1A**), in 7% (v/v) "Domestos" bleach solution for 20 min. Wash thoroughly with sterile, reverse osmosis water (three changes).
2. Cut the leaves transversely into 1-mm strips (*see* **Note 2**) and incubate (30 min) approx 1 g f. wt portions of material in 20-mL aliquots of CPW13M solution (*see* **Subheading 2.4., item 4.**) contained in 9-cm Petri dishes. Seal the dishes with Nescofilm.
3. Remove the CPW13M solution, and replace with Enzyme 1 solution (*see* **Note 3**), incubating 1 g f. wt of tissue in 10-mL enzyme solution. Seal the dishes with Nescofilm, and incubate on a slow shaker (40 rpm) at 25 ± 2°C for 16 h in the dark.
4. Filter the enzyme–protoplast mixture through a nylon sieve of 64-μm pore size (*see* **Note 4**). Place the filtrate in 16-mL capacity screw-capped centrifuge tubes.

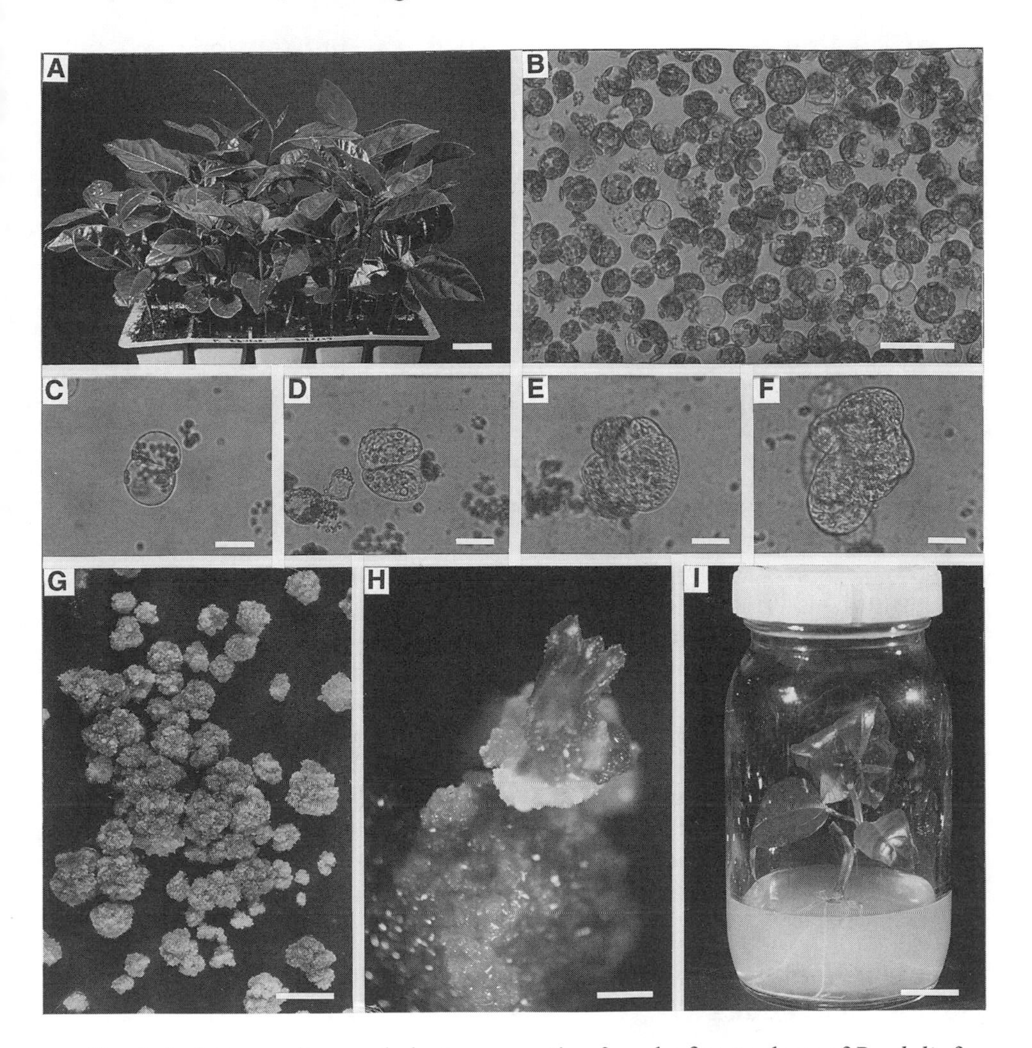

Fig. 1. Isolation, culture and plant regeneration from leaf protoplasts of *P. edulis* fv. flavicarpa. **(A)** Forty-five-d-old glasshouse-grown seedlings suitable for the isolation of leaf protoplasts (bar = 2.6 cm). **(B)** Freshly isolated leaf protoplasts (bar = 100 µm). **(C, D, E)** Protoplasts undergoing mitotic division after 8, 10, and 12 d of culture, respectively (bars = 75 µm). **(F)** A protoplast-derived colony after 15 d of culture (bar = 75 µm). **(G)** Protoplast-derived tissues (bar = 5 mm). **(H)** Protoplast-derived callus undergoing shoot regeneration (bar = 1.5 mm). **(I)** A rooted, protoplast-derived plant (bar = 1.5 cm).

5. Centrifuge (80*g*; 7 min). Discard the supernatants, and resuspend the protoplast pellets in CPW13M solution. Centrifuge and wash the protoplasts twice by resuspension and centrifugation in CPW13M solution.

6. Resuspend the protoplasts in a known volume (e.g., 10-mL) of CPW13M solution (**Fig. 1B**). Remove an aliquot of the suspension (e.g., 0.1-mL), and transfer to a haemocytometer. Count the protoplasts; yields of ± 12 × 10^6/g f. wt can be expected.
7. Assess protoplast viability using FDA (*see* **Note 5**). Add 100-µL of FDA stock solution to 16-mL of CPW13M solution. Mix one drop of this solution, by Pasteur pipet, with an equal volume of protoplast suspension. Observe the protoplasts under UV illumination (e.g., using a Nikon Diaphot TMD inverted microscope with high-pressure mercury vapor lamp HBO 100 W/2, a B1 FITC exciter filter IF 420–485 nm, dichromic mirror DM510, and eyepiece absorption filter 570). Viable protoplasts fluoresce yellow-green (*see* **Note 6**). Protoplast viability is usually ±90%.

3.3. Culture of Leaf Protoplasts of P. edulis

1. Centrifuge the protoplast suspension from **Subheading 3.2., step 6** at 80*g* for 7 min. Discard the supernatant.
2. Resuspend protoplasts in agarose-solidified KM8P medium (*see* **Note 7**) at a density of 1.5 × 10^5 /mL (*see* **Note 8**).
3. Dispense 40-µL droplets of KM8P medium containing suspended protoplasts in 5-cm Petri dishes (25 droplets/dish). Allow the droplets to solidify (*see* **Note 9**), and bathe the droplets in each dish in 2-mL vol of liquid KM8P medium containing cefotaxime. Seal the dishes with Nescofilm, and incubate at 25 ± 2°C in the dark.
4. Replace the KM8P bathing medium every 5 d, with KM8P medium mixed with KM8 medium, in the ratios of 3:1, 2:1, 1:1, and 0:1 (v:v).
5. Assess the protoplast plating efficiency (number of dividing [mitotic] protoplasts expressed as a percentage of the number of viable [FDA assessed] protoplasts originally plated) after 6–10 d of culture (**Fig. 1C,D**). The plating efficiency should be ±40%.

3.4. Plant Regeneration from Protoplast-Derived Tissues of P. edulis

1. Transfer protoplast-derived colonies (**Fig. 1E,F**) (*see* **Note 10**), 20 d after isolation of protoplasts, to agar-solidified MSR1 medium (50 colonies/20-mL aliquots of medium in 9-cm Petri dishes). Seal the dishes with Nescofilm and incubate at 25°C ± 2°C in the light under a 16-h photoperiod (25 µmol/m^2/s; Daylight fluorescent tubes).
2. After 25 d, transfer protoplast-derived tissues (**Fig. 1G**) to agar-solidified MSR2 medium (5 calli/45-mL aliquots of medium in 175-mL glass jars). Incubate as in **step 1**.
3. Subculture protoplast-derived tissues every 30 d to MSR2 shoot regeneration medium as in **step 2**. Shoots should appear progressively within a further 60 d of culture (**Fig. 1H**).
4. Excise developing shoots when 5–6 cm in height. Root the shoots by transfer to agar-solidified MSR3 medium for 7 d (**Fig. 1I**), followed by transfer to MS-

based medium, lacking growth regulators (1–3 shoots/50 mL medium for both stages in 175-mL jars). Maintain the shoots under growth conditions as in **step 1**.

3.5. Initiation of Cell Suspensions of P. giberti

1. Excise and sterilize leaves as in **Subheading 3.2., step 1**.
2. Place the leaves with their adaxial surfaces in contact with 20-mL aliquots of agar-solidified MS1 medium (*see* **step 2**) using 10 leaves/9-cm Petri dish. Seal the dishes with Nescofilm.
3. Incubate the cultures in the dark at 28 ± 2°C. After 28 d, transfer the cultures to the light under a 16-h photoperiod (25 μmol/m^2/s; Daylight fluorescent tubes) and incubate for a further 40 d.
4. Transfer 2–3 g f. wt of embryogenic callus (*see* **Note 11**) into 40-mL aliquots of liquid AA2 medium in 250-mL Erlenmeyer flasks. Incubate on a horizontal rotary shaker (90 rpm) in the dark at 28 ± 2°C.
5. Subculture 5–8 mL packed volume of cells (*see* **Note 12**) to 40-mL of fresh AA2 medium every 4 d, for the first 4 passages. Thereafter, subculture cell suspensions every 7 d (**Fig. 2A**).

3.6. Isolation of Protoplasts from Cell Suspensions of P. giberti

1. Harvest cells 4 d after subculture (Passage 5 onward; *see* **Subheading 3.5., step 5**) of suspensions using a nylon sieve of 100-μm pore size (*see* **Note 4**).
2. Transfer cells to 9-cm Petri dishes containing Enzyme solution 101 (1 g f. wt of tissue in 20-mL enzyme solution). Seal dishes with Nescofilm, and incubate on a slow shaker (40 rpm) for 16 h at 25 ± 2°C in the dark.
3. Filter the enzyme–protoplast mixture through a nylon sieve of 45-μm pore size (*see* **Note 4**). Place the filtrate in 16-mL capacity screw-capped centrifuge tubes.
4. Follow stages as in **Subheading 3.2., steps 5–7**.

3.7. Culture of Protoplasts of P. giberti

1. Resuspend protoplasts in liquid KPR medium (**Fig. 2B**) at a density of 1.0×10^5/mL, and dispense as 4-mL aliquots in 5-cm Petri dishes.
2. Seal the dishes with Nescofilm, and incubate at 25 ± 2°C in the dark.
3. Remove 0.5-mL aliquots of KPR medium at 5 and 10 d (*see* **Note 13**); replace with a similar volume of liquid K8 medium.
4. Assess the protoplast plating efficiency after 6–10 d of culture (**Fig. 2C,D**) as in **Subheading 3.3, step 4**. This should be ±10%.

3.8. Plant Regeneration from Protoplast-derived Tissues of P. giberti

1. Transfer protoplast-derived colonies (**Fig. 2E**) (*see* **Note 10**), 40 d after isolation, to MSE1 medium (5 colonies/10-mL medium in 5-cm Petri dishes). Seal the dishes with Nescofilm, and incubate at 27 ± 2°C in the light under a 12 h photoperiod (25 μmol/m^2/s; Daylight fluorescent tubes).

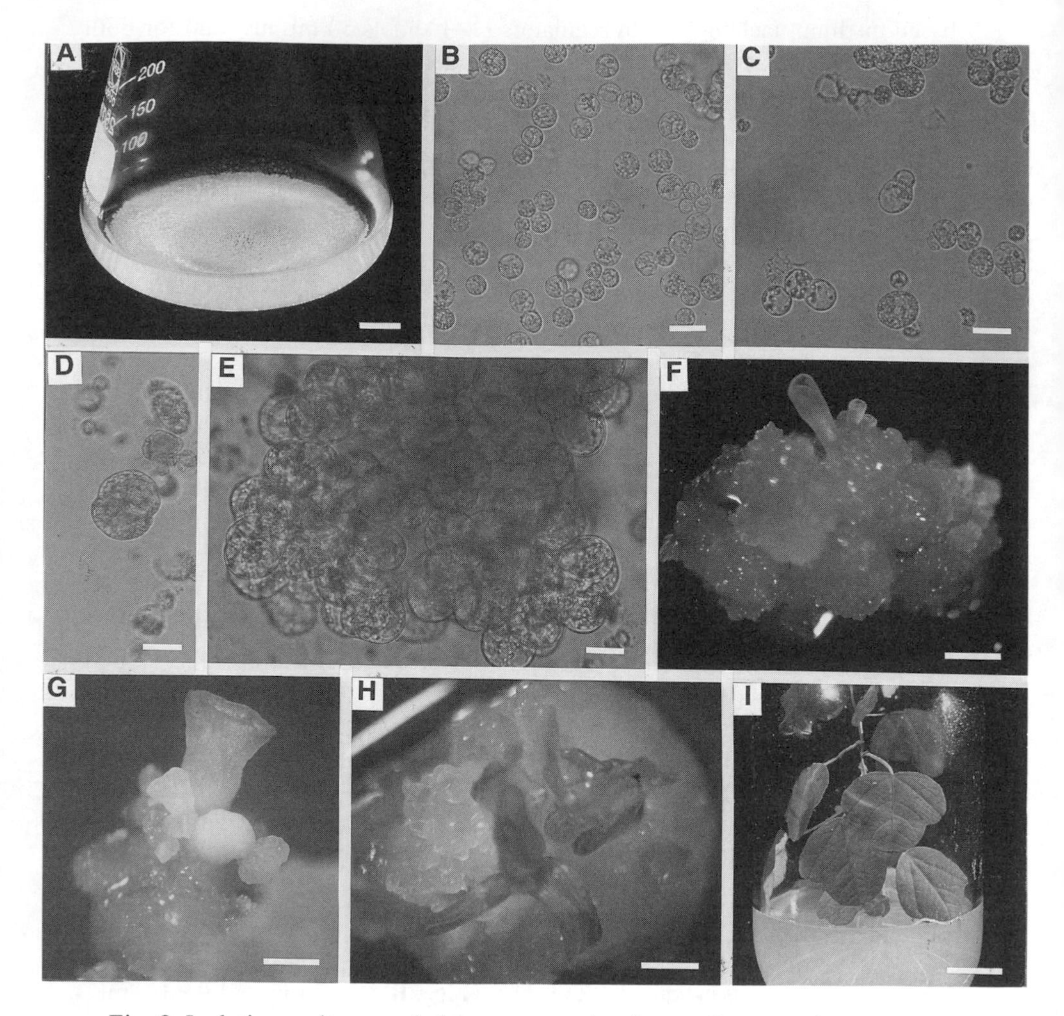

Fig. 2. Isolation, culture, and plant regeneration from cell suspension protoplasts of *P. giberti*.**(A)** Embryogenic cell suspension suitable for protoplast isolation (bar = 1 cm).**(B)** Freshly isolated cell suspension protoplasts (bar = 50 μm).**(C)** Protoplasts after 5 d in culture. Note the change from a spherical to a more oval shape as the protoplasts enter mitotic division (bar = 50 μm).**(D)** A protoplast that has undergone mitotic division to form two daughter cells (bar = 50 μm). **(E)** A protoplast-derived cell colony after 28 d of culture (bar = 50 μm). **(F, G)** Somatic embryos developing from protoplast-derived callus (bars = 5 mm). **(H)** Somatic embryo-derived shoots (bar = 1.5 mm). **(I)** A rooted protoplast-derived plant (bar = 1.5 cm).

2. After 20 d, transfer protoplast-derived colonies to MSE2 medium, as in **step 1**. Incubate at 27 ± 2°C in the light under a 16 h photoperiod (25 μmol/m^2/s; Daylight fluorescent tubes) to induce shoot regeneration by somatic embryogenesis (**Fig. 2F,G**).

3. Excise developing shoots when 1 cm in height (**Fig. 2H**). Root the shoots by transfer to MSE3 medium (1–3 shoots/50-mL medium in 175-mL jars) (**Fig. 2I**). Incubate the shoots as in **step 2**.

3.9. Characterization of Regenerated Plants

Morphologically, regenerated plants should be scored within 2–3 mo following transfer to the glasshouse, since the juvenile characteristics of leaf shape and pigmentation in vitro may be different from those of seed-derived plants raised directly in the glasshouse. Furthermore, this will enable the floral characteristics, fertility, and fruiting properties to be compared with those features of seed-derived (control) plants. The somatic chromosome complement can be assessed using root tip preparations and any shift in ploidy within populations of regenerated plants rapidly confirmed by flow cytometric analysis of isolated nuclei ***(17)***.

Further DNA molecular studies may be performed on those protoplast-derived plants that exhibit stable phenotypic, fertility, and cytological characters, which differ from those of seed-derived plants. These techniques include the use of AFLP ***(18)***, RAPD ***(19)***, RFLP ***(20)***, and simple sequence repeats (microsatellites) ***(21)***.

4. Notes

1. Store *Passiflora* seeds at 4°C. Sow *Passiflora* seeds in excess of the number of seedlings required, since germination within the genus *Passiflora* can be slow and erratic. A minimum of 10 seedlings will be required for each protoplast isolation. Leaves should be harvested before seedlings produce tendrils.
2. Repeatedly use a new scapel blade to ensure precise cutting, rather than tearing and bruising, of the leaf material.
3. The enzyme solutions should be prefiltered, using 0.2-µm pore size nitrocellulose 47-mm diameter membrane filters (Whatman) to remove insoluble impurities. This prevents premature blockage of the sterile microbial filters (0.2-µm pore size, 30-mm diameter; Minisart NML, Sartorius AG, Göttingen, Germany) during subsequent sterilization of the enzyme solutions.
4. Inexpensive, nylon sieves in a range of pore sizes may be obtained from Wilson Sieves, Common Lane, Hucknall, Nottingham, UK.
5. Prepare the FDA dilution in CPW13M solution, immediately prior to viability assessments, since even short-term storage will result in cleavage of FDA to fluorescein, especially if the solution is exposed to strong illumination.
6. FDA taken up into protoplasts is acted on enzymatically by esterases, to release fluorescein. This molecule is excited under UV illumination and fluoresces yellow-green. Only viable protoplasts with intact plasma membranes that retain fluorescein will fluoresce.
7. The agarose-solidified KM8P medium must be prepared as: (a) aqueous double-strength agarose and (b) liquid, double-strength medium components. The agar-

ose is sterilized by autoclaving (20 min, 121°C, saturated steam pressure); the liquid component is filter-sterilized. The double-strength liquid component is mixed with the molten agarose component at 50–60°C, immediately prior to use.
8. Ensure that the agarose-solidified KM8P medium is at 35–40°C, prior to mixing with, and resuspension of, the protoplast pellets.
9. Allow the KM8P droplets, containing protoplasts, to solidify for at least 1 h (at room temperature) before the addition of the liquid KM8P bathing medium. This permits the droplets to adhere to the bottom of the Petri dishes.
10. Transfer individual colonies using fine, jeweler's forceps; flame-sterilize, and cool the forceps immediately prior to use.
11. Embryogenic callus is yellow in color, dry, friable, and translucent in appearance.
12. Use a 10-mL graduated pipet, from which the end has been removed to increase the size of the orifice to transfer the cells. The packed cell volume can be determined by removing 10-mL aliquots of liquid medium, holding the pipet vertically, and allowing the cells to settle (10 min).
13. Angle the Petri dish, and allow the protoplasts to settle (5 min) to the bottom of the Petri dish before removal of the overlying liquid culture medium.

References

1. Oliviera, J. C. (1987) Melhoramento genético. in *Cultura do maracujazeiro*, vol. 1 (Ruggiero, C., ed.), Editora Legis Summa, Ribeirão Petro, pp. 218–246.
2. Vanderplank, J. (1991) *Passion Flowers,* Cassel Publishers, London.
3. Payan, F. R., and Martin, F. W. (1975) Barriers to the hybridization of *Passiflora* species. *Euphytica* **24,** 709–716.
4. Otoni, W. C., Blackhall, N. W., d'Utra Vaz, F. B., Casali, V. W., Power, J. B., and Davey, M. R. (1995) Somatic hybridization of the *Passiflora* species, *Passiflora edulis* fv. flavicarpa Degener. and *P. incarnata* L. *J. Exp. Bot.* **46,** 777–785.
5. Dozier, W. A., Jr., Rodriguez-Kabana, R., Caylor, A. W., Himelrick, D. G., McDaniel, N. R., and McGuire, J. A. (1991) Ethephon hastens maturity of passionfruit grown as an annual in a temperate zone. *Hort. Sci.* 26, 146–147.
6. Dornelas, M. C., Tavares, F. C. A., Oliviera, J. C., and Vieira, M. L. C. (1995) Plant regeneration from protoplast fusion in *Passiflora* spp. *Plant Cell Rep.* **15,** 106–110.
7. d' Utra Vaz, F. B., dos Santos, A. V. P., Manders, G., Cocking, E. C., Davey, M. R., and Power, J. B. (1993) Plant regeneration from leaf mesophyll protoplasts of the tropical woody plant, passionfruit (*Passiflora edulis* fv. flavicarpa Degener.): the importance of the antibiotic cefotaxime in the culture medium. *Plant Cell Rep.* **12,** 220–225.
8. Dornelas, M. C. and Vieira, M. L. C. (1993) Plant regeneration from protoplast cultures of *Passiflora edulis* var. *flavicarpa* Deg., *P. amethystina* Mikan. and *P. cincinnata* Mast. *Plant Cell Rep.* **13,** 103–106.
9. Frearson, E. M., Power, J. B., and Cocking, E. C. (1973) The isolation, culture and regeneration of *Petunia* protoplasts. *Dev. Biol.* **33,** 130–137.

10. Kao, K. N. and Michayluk, M. R. (1975) Nutritional requirements for growth of *Vicia hajastana* cells and protoplasts at a very low population density in liquid media. *Planta* **126,** 105–110.
11. Gilmour, D. M., Golds, T. J., and Davey, M. R. (1989) *Medicago* protoplasts: fusion, culture and plant regeneration, in *Biotechnology in Forestry and Agriculture*, vol. 8, *Plant Protoplasts and Genetic Engineering* I (Bajaj, Y. P. S., ed.), Springer-Verlag, Heidelberg, pp. 370–388.
12. Widholm, J. (1972) The use of FDA and phenosafranine for determining viability of cultured plant cells. *Stain Technol.* **47,** 186–194.
13. Murashige, T. and Skoog, F. (1962) A revised medium for rapid growth and bioassays with tobacco tissue cultures. *Physiol. Plant.* **56,** 473–497.
14. Müller, A. J. and Grafe, R. (1978) Isolation and characterisation of cell lines of *Nicotiana tabacum* lacking nitrate reductase. *Mol. Gen. Genet.* **161,** 67–76.
15. Jain, R. K., Khehra, G. S., Lee, S-H., Blackhall, N. W., Marchant, R., Davey, M. R., et al. (1995) An improved procedure for plant regeneration from indica and japonica rice protoplasts. *Plant Cell Rep.* **14,** 515–519.
16. Thompson, J. A., Abdullah, R., and Cocking, E. C. (1986) Protoplast culture of rice (*Oryza sativa* L.) using media solidified with agarose. *Plant Sci.* **47,** 179–183.
17. Hammatt, N., Blackhall, N. W., and Davey, M. R. (1991) Variation in the DNA content of *Glycine* species. *J. Exp. Bot.* **42,** 659–665.
18. Vos, P., Hogers, R., Bleeker, M., Reijans, M., van de Lee, T., Hornes, M., et al. (1995) AFLP: a new technique for DNA fingerprinting. *Nucleic Acids Res.* **23,** 4407–4414.
19. Williams, J. G. K., Kubelik, A. R., Livark, K. J., Rafalski, J. A., and Fingey, S. V. (1990) DNA polymorphisms amplified by arbitrary primers are useful as genetic markers. *Nucleic Acids Res.* **18,** 6531–6535.
20. Vaccino, P., Accerbi, M., and Carbellini, M. (1993) Cultivar identification in *Triticum aestivum* using highly polymorphic RFLP probes. *Theor. Appl. Genet.* **86,** 833–836.
21. Powell, W., Machray, G. C., and Provan, J. (1996) Polymorphism revealed by simple sequence repeats. *Trends Plant Sci.* **1,** 215–222.

18

Isolation, Culture, and Plant Regeneration of Suspension-Derived Protoplasts of *Lolium*

Marianne Folling and Annette Olesen

1. Introduction

An efficient protoplast regeneration system makes somatic hybridization and direct gene transfer attractive tools in the breeding of perennial ryegrass, one of the most important forage grass species in temperate regions. Despite many efforts during the last decade, regeneration from protoplasts of *Lolium perenne* L. has generally been low and further complicated by the frequent occurrence of albino plantlets *(1–3)*. Recently, however, the use of nurse culture *(4,5)* and conditioned medium *(5,6)* has considerably improved plating efficiencies and reproducibility.

Protoplasts of grasses are usually obtained by treating cell colonies from embryogenic suspension cultures with enzymes that digest cell-wall components. Since protoplasts cannot normally perform better than the donor suspension, the regeneration behavior of these donor suspensions is very important for success with the protoplast culture system. For ryegrass suspension cultures, regeneration capacity usually decreases and albino frequency increases with culture age *(7)*. This often leads to total loss of ability to regenerate green plantlets within 25 wk after culture initiation *(1,2)*. The genotype and subculturing regime strongly affect regeneration behavior of suspension cultures *(7)*, but responsive genotypes can be found within most varieties *(8)*. Based on correlation studies, we found regeneration ability in suspension and anther culture to be controlled to some extent by the same genes *(7)*.

Old suspension cultures can adapt to in vitro culture conditions, and protoplasts isolated from such cultures may divide in fairly simple media.

From: *Methods in Molecular Biology, Vol. 111: Plant Cell Culture Protocols*
Edited by: R. D. Hall © Humana Press Inc., Totowa, NJ

However, protoplasts isolated from younger morphogenic suspension cultures are usually fragile and require more complex media ***(6,9)***. Cocultivation of protoplasts with nurse cells or conditioned medium is often necessary for obtaining division in such protoplasts, but will usually also improve plating efficiency and the proportion of green plants from protoplasts that are able to divide without inclusion of nurse or conditioned medium ***(5)***. In our laboratory, the use of nurse culture, in combination with the use of protoplasts from highly regenerable and well established suspensions has proven very successful (*see* **Fig. 1**). By using of nurse cells, we obtained average plating efficiencies up to 12% from the best culture, 40–50% regeneration frequency, about 80% green plants, and a total yield of almost 60,000 green protoplast-derived plants (without cloning) per mL suspensions cells. Without nurse cells or conditioned medium, the same culture gave about 1200 green plants/mL suspension cells, and with conditioned medium, the yield of plants was approx 26,000/mL ***(5)***.

The main effect of nurse culture is improvement of initial cell division and microcolony formation. After this stage, nurse cells are removed, since prolonged cocultivation may result in competition for nutrients and accumulation of toxic waste products excreted from the nurse cells. Attempts have been made to characterize and identify the conditioning factor(s) ***(10–13)***. However, several compounds may be involved ***(9,14,15)***, and the indication that some are unstable ***(5)*** makes their identification difficult. This instability may explain why the nurse system generally is more efficient than conditioning media ***(5)***. Interactions between protoplasts and sources of conditioning or nurse cells may complicate their use ***(9,15)*** and should be taken into consideration. We found that nurse cells and conditioned medium obtained from the protoplast donor suspension had the best effect on the plating efficiency, microcolony growth, and frequency of green plants ***(5)***.

Here, we describe a protocol for the isolation, culture and plant regeneration of *Lolium* suspension-derived protoplasts involving nurse cells or conditioned medium. It includes the following steps:

1. Preparation of suspension cultures prior to protoplast culture (5 d).
2. Protoplast isolation and purification.
3. Protoplast culture with nurse or conditioned medium (2 wk).
4. Culture of microcolonies embedded in agarose drops in liquid protoplast colony plating medium (2 wk).
5. Macrocolony culture on solid protoplast colony plating medium. This step includes one subculture of individual colonies (2–6 wk depending on colony growth rate + 4 wk).
6. Plant regeneration (3–4 wk).

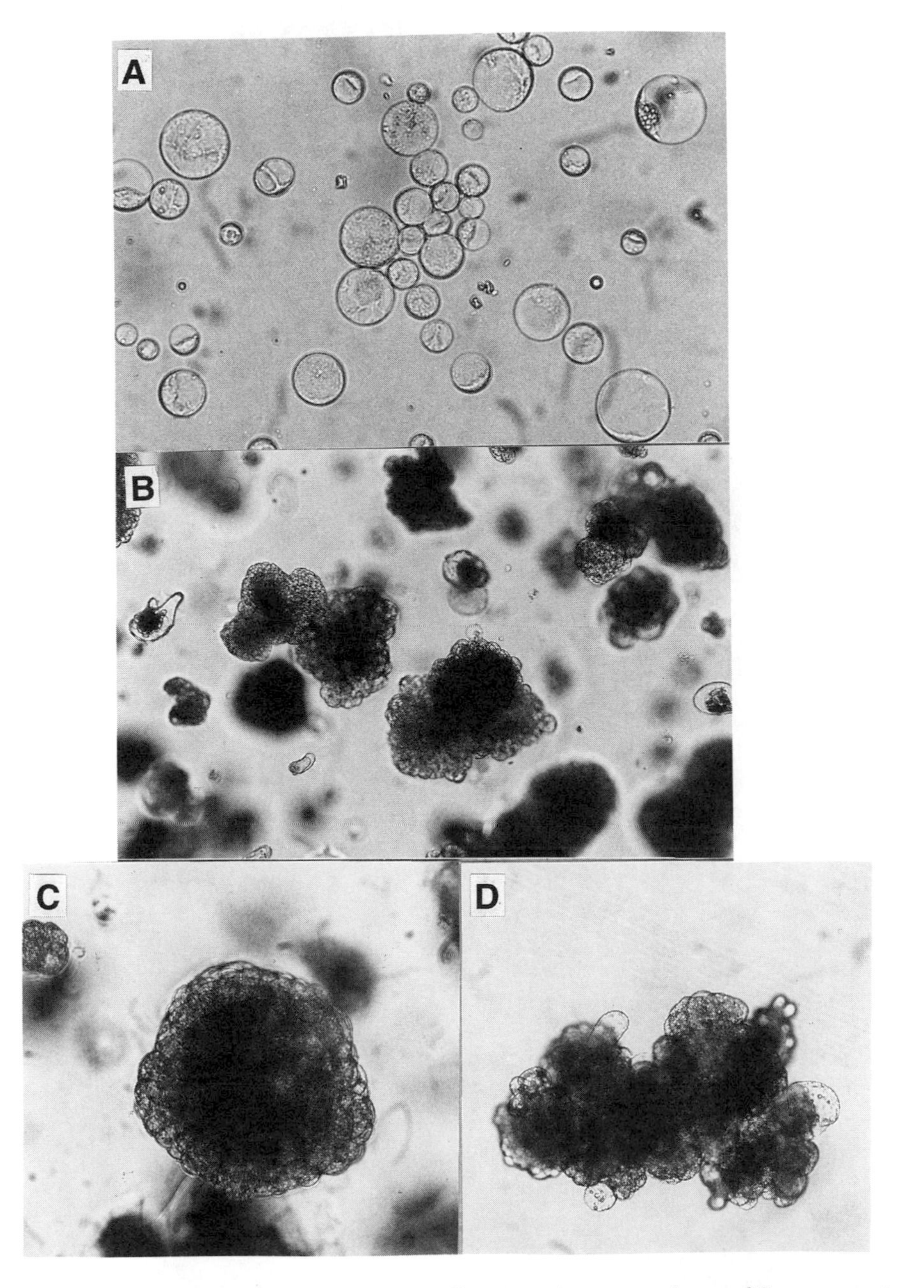

Fig. 1. **A–D.** Plant regeneration from cell suspension protoplasts of *L. perenne* L. **(A)** Freshly isolated protoplasts. **(B)** Colonies formed in agarose drops after 2 wk with nurse culture and 1 wk in liquid protoplast colony plating medium. **(C)** Colony consisting of small dense cells with a regular shape. This type of colony will often regenerate into plantlets. **(D)** Colony with loose irregular growth. This type of colony will usually not be able to regenerate plants.

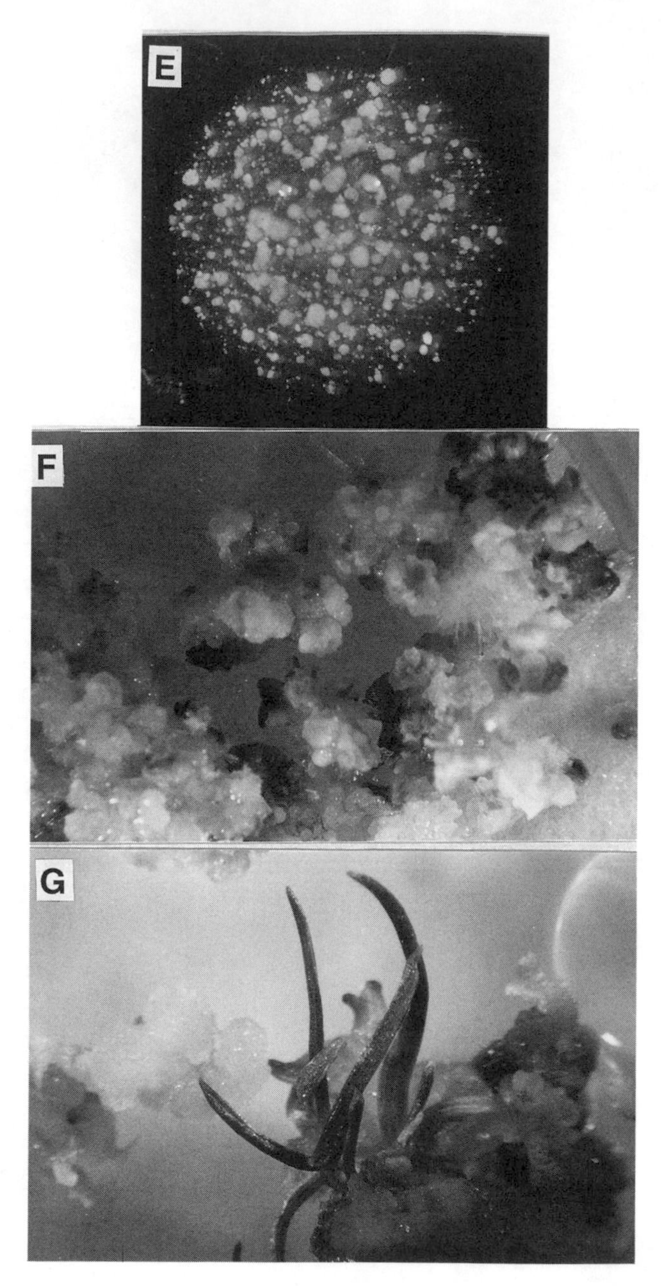

Fig. 1. **E–G. (E)** Agarose drop (40 μL) containing protoplast-derived colonies 4 wk after protoplast isolation. **(F)** Regenerating calli 2½ mo after protoplast isolation. **(G)** Shoots of *L. perenne* 3 mo after protoplast isolation.

2. Materials

All media are made with distilled deionized water and based on Murashige and Skoog's medium (MS) (***16***; and *see* Appendix). This is commercially available from, e.g., Flow Laboratories Inc. Hormone stocks described below can be stored at 4°C for up to 2 mo.

1. 2,4-Dichlorophenoxyacetic acid (2,4-D), 0.1 mg/mL stock: dissolve 100 mg 2,4-D in a small volume of 96% ethanol, and adjust volume to 1000 mL with water.
2. 6-benzylamino purine (BAP), 0.1 mg/mL stock: dissolve 100 mg BAP in a small volume of 0.5 *N* HCl, and adjust volume to 1000 mL with water.
3. Kinetin, 0.1 mg/mL stock: dissolve 100 mg kinetin in a small volume of 0.5 *N* HCl and adjust volume to 1000 mL with water.

2.1. Suspension Culture Medium

Suspension maintenance medium (SM) consists of MS salts and vitamins supplemented with 3% (w/v) sucrose and 3 mg/L 2,4-D. Adjust pH to 6.2 and autoclave.

2.2. Protoplast Isolation Solutions

1. Enzyme solution: 1% (w/v) Cellulase Onozuka RS (Yakult Pharmaceutical Industry Co. Ltd.), 1% (w/v) Meicelase (Meiji Seika Kaisha Ltd.), 0.3% (w/v) Macerozyme R-10 (Yakult Pharmaceutical Industry Co. Ltd.), and 0.1% (w/v) Pectolyase Y-23 (Seishin Pharmaceutical Industry Ltd.) dissolved in MS medium supplemented with 10 m*M* $CaCl_2$, 3% (w/v) sucrose, 8% (w/v) mannitol, and 2 mg/L 2,4-D. Stir solution at low rotation for 1–2 h, and pellet undissolved particles by centrifugation at 125*g* for 20 min at 4°C. Adjust pH to 6.0 and filter-sterilize supernatant (0.22-µm membrane filter). Store in 10-mL aliquots at –20°C for up to 1 yr.
2. Protoplast washing medium: PC medium 1X (*see* **Subheading 2.3., step 1**).

2.3. Protoplast Culture Media

1. Liquid protoplast culture medium (PC 1X): MS salts and vitamins supplemented with 10 m*M* $CaCl_2$, 11% (w/v) glucose, 0.1 mg/L 2,4-D, and 0.1 mg/L 6-benzylamino purine (BAP). Adjust pH to 6.0 and filter-sterilize (0.22-µm membrane filter). Medium can be kept at room temperature for up to 1–2 mo.
2. Solid protoplast culture medium (agarose plating medium): Prepare medium described above, but with a twofold concentration of all constituents (PC 2X) and filter-sterilize. Prepare an agarose solution with 2.4% (w/v) Sea-Plaque agarose (FMC) in distilled water and autoclave. While still hot and fluid, mix agarose solution and PC 2X medium in the ratio 1:1. Cool the medium to approx 32°C in water bath (*see* **Note 1**).
3. Conditioned medium: Harvest spent suspension culture medium from the protoplast-donor suspension 5 d after last subculture (*see* **Note 2**). Remove cells

by filtering the medium through a 22.5-µm nylon sieve. Adjust osmolarity to 800 mOsM/kg with a saturated glucose solution, and adjust pH to 6.0 (*see* **Note 3**). Filter-sterilize the medium (0.22-µm membrane filter), use immediately, or store at 4°C until next day.

2.4. Colony Plating and Plant Regeneration Media

1. Protoplast colony plating medium (PCP 1X): MS salts and vitamins supplemented with 6.84% (w/v) glucose, 0.1 mg/L 2,4-D, and 0.1 mg/L BAP. Adjust pH to 6.0 and filter-sterilize (0.22-µm membrane filter). For solid plating media, prepare medium at a twofold concentration (PCP 2X) and filter-sterilize (0.22-µm membrane filter). Prepare a 0.6% (w/v) Gelrite (Kelco) solution in distilled water and autoclave. While still at 90–100°C, mix Gelrite solution with PCP 2X in the ratio 1:1 (final concentration; PCP 1X with 0.3% [w/v] Gelrite), and pour into 9-cm plastic Petri dishes (approx 50 mL/dish).
2. Plant regeneration medium (PR): MS salts and vitamins supplemented with 3% (w/v) sucrose, 0.2 mg/L kinetin, and 0.8% (w/v) agar (Difco). Adjust pH to 6.2 and autoclave. Pour medium into 9-cm Petri dishes (approx 50 mL/dish).

3. Methods

3.1. Maintenance of Suspension Cultures

Suspensions may be initiated from embryos or meristems *(1,7)*. Success with suspension culture of *Lolium* depends primarily on the donor genotype, an optimal suspension maintenance scheme, and personal experience (*see* **Note 4**).

1. Suspension cultures are cultivated in 250-mL Erlenmeyer flasks, and maintained by weekly subculturing. Do this by tilting the flask to allow the cells to sediment, pour off 90% of the used medium, and add fresh SM medium to a total volume of 80–100 mL. Cultivate at 25°C in dim light (7 µmol/m^2/s) on a gyratory shaker at 150 rpm. Seal flasks with two layers of polyethylene (PE) film (*see* **Note 5**).
2. Suspensions are renewed at 21- or 28-d intervals by transferring using a wide-bore pipette (5-mm opening) 1–2 mL of small compact cell colonies (*see* **Note 6**) together with approx 10 mL spent suspension medium to new 250-mL flasks containing 50–70 mL of fresh SM medium (*see* **Note 7**).

3.2. Preparation of Suspension Cultures Prior to Protoplast Culture (see Note 8)

1. Cultures for protoplast isolation: 5 d prior to protoplast isolation and 7 d after last renewal by pipet subculture (*see* **Note 9**), transfer approx 1 mL of small compact cell colonies with a wide-bore pipet to a 250-mL Erlenmeyer flask.
2. Cultures for nurse culture: 5 d prior to protoplast isolation and 7 d after last subculture transfer approx 1.5–2 mL of small compact suspension colonies from the protoplast donor suspension culture (*see* **Note 10**) to a 250-mL Erlenmeyer flask.

3. To each flask, add 80 mL of fresh SM medium, and seal the flasks with two layers of PE film. Cultivate at 25°C in dim light (7 μmol/m^2/s) on a gyratory shaker at 150 rpm for 5 d.

3.3. Protoplast Isolation and Purification

1. Transfer approx 1 mL of small compact suspension colonies with a wide-bore pipet to a 9-cm Petri dish, and remove excess SM media. Add 10 mL of enzyme solution, and seal the dish with PE film.
2. Incubate dishes for 5 h in the dark at 25°C on a gyratory shaker at 30 rpm (*see* **Note 11**).
3. Filter the protoplast/enzyme solution through 91- and 22.5-μm nylon sieves (*see* **Note 12**). Tilt the sieves with the tip of a pipet, allowing protoplasts and medium to pass through the sieves, and wash with PC 1X medium (up to 10 mL). Carefully transfer the protoplast solution to a 30-mL container (Bibby Sterilin Ltd.) by use of a wide-bore pipet. Add additional PC 1X medium to a final volume of 30 mL. Centrifuge at 50*g* for 5 min at room temperature. Discard supernatant.
4. Resuspend protoplasts immediately in 20 mL of PC 1X medium (*see* **Note 13**), and centrifugate again.
5. Resuspend protoplast pellet in 10 mL of PC 1X medium, and store protoplasts overnight at 4°C or proceed immediately with next step.
6. Centrifuge protoplast suspension, and resuspend the protoplast pellet in 10–20 mL of PC 1X medium. Measure the protoplast density with a hemocytometer (Fuch-Rosenthal), and calculate total protoplast yield.

3.4. Protoplast Culture

1. Dispense aliquots of the protoplast suspension into centrifuge tubes, and centrifuge at 50*g* for 5 min at room temperature. Discard the supernatant, and resuspend the protoplast pellet in agarose plating medium to a final density of 2×10^5 protoplasts/mL (*see* **Note 14**).
2. Place, using a Pasteur pipet, one agarose drop (approx 24 drops/mL) in the middle of each well of a 24-well multidish (Nunclon® 15.5 mm, NUNC), and allow agarose to solidify in darkness for 1 h at room temperature. The protoplasts can either be cultivated in conditioned medium or cocultivated with nurse cells.
3. Use of conditioned medium: Dispense 0.5-mL aliquots of undiluted conditioned medium into each well. Fill the inner space between the wells with sterile water to reduce evaporation, seal with PE film, and incubate dishes in plastic bags at 25°C in darkness.
4. Use of nurse culture: Dispense 0.5-mL aliquots of liquid PC 1X medium to each well. Add approx 30 mg of small, dense suspension colonies (without suspension culture medium) to each well, fill the inner space between the wells with sterile water, seal, and incubate the dishes as described in **step 3**.

3.5. Microcolony Culture

1. After 14 d of culture when the microcolonies consist of 6–8 cells (*see* **Note 15**), determine plating efficiency (number of microcolonies/total number of cultured protoplasts) by counting microcolonies using an inverted microscope. For protoplasts cultured with nurse cells, wash agarose drops twice with liquid PCP 1X medium (*see* **Note 16**). These washes ensure removal of all nurse cells. No washing is needed for protoplast agarose drops cultured with conditioned medium.
2. Transfer agarose drops to new multidishes, and add 0.5 mL liquid PCP 1X medium. Culture for further 2 wk.

3.6. Culture of Micro- and Macrocolonies

1. Transfer one or two agarose drops with protoplast-derived colonies to solid PCP 1X medium (*see* **Note 17**). Chop agarose-drops lightly with a blunt scalpel (*see* **Note 18**), and spread the agarose pieces on the surface. Seal the dishes with PE film and incubate at 22°C in continuous white fluorescent light (70 µmol/m^2/s).
2. After 2 wk of culture, transfer the most rapidly growing colonies (1.5–2 mm) individually to fresh PCP (solid) dishes. Incubate at same temperature and under the same light conditions as described above. Leave the remaining small colonies on original PCP dishes to continue their growth by spreading them over the entire surface. Repeat transferring colonies from the original dishes until all colonies have been moved individually to fresh PCP medium (4–6 wk).

3.7. Plant Regeneration

1. After 4 wk of culture of individual colonies on PCP medium, transfer calluses to PR medium. Seal dishes with PE film and incubate at 20°C in continuous white fluorescent light (70 µmol/m^2/s) for 4 wk.
2. Count the number of calluses regenerating at least one plantlet, and transfer plantlets to vessels containing PR medium. Cultivate under high light intensities (70 µmol/m^2/s) with a 16/8 h day/night light regime.
3. Transfer rooted plants to peat/soil, and continue growth in greenhouse.

4. Notes

1. Alternatively, stored solid agarose solution may be reheated in a water bath or microwave oven and mixed 1:1 with stored PC 2X. To avoid damage to protoplasts, medium should never be warmer than 32°C when used for plating.
2. Do not harvest spent medium from cultures more than 3 wk after latest pipet subculturing (*see* **Subheading 3.1., step 2**). Spent medium harvested later than this or conditioned medium from other cultures than the protoplast-donor suspension may have a detrimental effect on protoplast division. The best conditioned medium is obtained from the protoplast-donor suspension taken 12 d (7 + 5 d) after pipet subculturing.
3. The osmolarity of SM medium must be raised to a value similar to PC medium to avoid bursting of the protoplasts. The pH of spent SM medium is usually reduced

to below 5.6 as a consequence of suspension growth, although it may occasionally be higher than 6.0.

4. Suspensions should be tested for regeneration capacity every 4th week. Do this by pipeting 1–2 mL of small dense suspension colonies to solid PCP 1X. Cultivate for 4 wk with light and temperature conditions as described in **Subheading 3.6., step 1**, and proceed with steps described in **Subheading 3.7.**
5. Type of sealing is very important for suspension growth. Foil sealing will increase dramatically the rate of suspension growth because of better aeration. However, this will often reduce regeneration longevity of the suspensions. We use household clingfilm (PE, 10 μm).
6. Tilt the flask and allow the cells to sediment. Then slowly rotate the flask in one hand so that the small dense cell colonies concentrate in the top layer of the sedimented cells for a few seconds. Transfer small compact cell colonies to new flasks.
7. The handling of suspension cultures is a critical step for protoplast culture. Preferably only one person should handle the cultures and thereby obtain experience for a correct subculturing scheme for each culture. For most cultures, pipet subculturing every 4th week is adequate, whereas others benefit from being subcultured by pipet every 2–3 wk. This is the case if suspensions turn darkish yellow before the 4th week after pipet subculturing. If the culture easily turns brown (especially older cell colonies), pipet small light-colored cell colonies to new flasks (avoid all brown cell colonies), and add 30–50 mL of fresh suspension medium. Repeat this with periodic intervals (each 2–3 wk) until browning has stopped. For fast-growing suspensions (usually relevant for nonembryogenic suspensions only), it may be necessary to discard some cells every week. Do this by spinning the flask, and immediately pour off some medium with cells before they sediment. For slow-growing suspensions, be careful not to dilute to much.
8. The dilution of cell colonies 5 d prior to protoplast isolation will usually improve yield and viability of protoplasts as well as quality of nurse cells.
9. Use of suspension cultures more than 14 d after latest pipet subculturing may reduce protoplast yield and viability.
10. The best nursing effect is obtained with cell colonies from the protoplast donor suspension. However, embryogenic suspension cultures often grow slowly, and for large experiments, it may be necessary to take nurse cells from a fast-growing nonembryogenic suspension so that adequate amounts of nurse cells can be obtained. In this case, screen several *Lolium* suspensions for their ability to support the protoplast culture.
11. After 5 h of enzyme treatment, some undigested cell colonies will remain, the amount depending on the suspension culture. However, prolonged digestion will reduce protoplast viability. For time-consuming transformation or fusion experiments, it may be advantageous to isolate protoplasts overnight in half-strength enzyme. Dishes with cells for overnight digestion are placed at very low rotation (almost no movement of cells).

12. This is easiest done by placing the 91-μm sieve inside the 22.5-μm sieve in a Petri dish. We use homemade sieves that touch the bottom of the dish, since this improves flow through the meshes and is more gentle to the protoplasts. These are constructed by securing pieces of nylon filters of appropriate meshes to circular glass rings with diameters of 4 cm (91-μm sieves) and 5.5 cm (22.5-μm sieves). Keep in mind that protoplasts are very fragile, and take great care in steps, such as pipeting and the resuspending of protoplast pellets.
13. Protoplasts may form aggregates that are difficult to dispense if resuspending is delayed.
14. Some suspension cultures are known to give rise to protoplast cultures with high plating efficiency (more than 10%) when cocultivated with a nurse culture. Reducing plating density to 1×10^5 protoplasts/mL for such cultures may increase plating efficiency further ***(5)***.
15. For very slow-developing protoplast cultures, 2 wk with a nurse culture may be too short a period for reaching the six- to eight-cell stage, and coculturing for a further 1 to 2 wk is then beneficial.
16. This is most easily done by using a small blunt scalpel to lift agarose drops from the wells and transfer them to a 9-cm Petri dish with liquid PCP 1X medium.
17. Culture on PCP medium matures the colonies and improves their ability to regenerate into plantlets. The number of agarose drops per plate depends on the plating efficiency. When plating efficiency is low (below 1%), proliferation of microcolonies may be stimulated by increasing the number of agarose drops per plate, since microcolonies at this stage still depend on the nursing effect of other microcolonies.
18. Protoplast-derived microcolonies have a very compact structure, and chopping the agarose drops with a blunt scalpel does not damage or break up individual colonies.

Acknowledgments

The authors wish to thank DLF-TRIFOLIUM Ltd. and the Danish Academy of Technical Science (ATV) for financial supports.

References

1. Dalton, S. J. (1988) Plant regeneration from cell suspension protoplasts of *Festuca arundinacea* Schreb. (tall fescue) and *Lolium perenne* L. (perennial ryegrass). *J. Plant Physiol.* **132,** 170–175.
2. Creemers-Molenaar, J., Van der Valk, P., Loeffen, J. P. M., and Zaal, M. A. C. M. (1989) Plant regeneration from suspension cultures and protoplasts of *Lolium perenne* L. *Plant Sci.* **63,** 167–176.
3. Zaghmout, O. M.-F. and Torello, W. A. (1992) Plant regeneration from callus and protoplasts of perennial ryegrass (*Lolium perenne* L.). *J. Plant Physiol.* **140,** 101–105.
4. Wang, Z. Y., Nagel, J., Potrykus, I., and Spangenberg, G. (1993) Plants from suspension-derived protoplasts in *Lolium* species. *Plant Sci.* **94,** 179–193.

5. Folling, M., Madsen, S., and Olesen, A. (1995) Effect of nurse culture and conditioned medium on colony formation and plant regeneration from ryegrass protoplasts. *Plant Sci.* **108,** 229–239.
6. Creemers-Molenaar, J., van Eeuwijk, F. A., and Krens, F. A. (1992) Culture optimization for perennial ryegrass protoplasts. *J. Plant Physiol.* **139,** 303–308.
7. Olesen, A., Storgaard, M., Madsen, S., and Andersen, S. B. (1995) Somatic *in vitro* culture response of *Lolium perenne* L.: genetic effects and correlations with anther culture response. *Euphytica* **86,** 199–209.
8. Olesen, A., Storgaard, M., and Madsen, S. (1996) Suspension culture performance in commercial varieties of perennial ryegrass. *Euphytica* **88,** 151–157.
9. Funatsuki, H., Lörz, H., and Lazzeri, P. A. (1992) Use of feeder cells to improve barley protoplast culture and regeneration. *Plant Sci.* **85,** 179–187.
10. Stuart, R. and Street, H. E. (1971) Studies on the growth in culture of plant cells. *J. Exp. Bot.* **22,** 96–106.
11. Somers, D. A., Birnberg, P. R., Petersen, W. L., and Brenner, M. L. (1987) The effect of conditioned medium on colony formation from "Black Mexican Sweet" corn protoplasts. *Plant Sci.* **53,** 249–256.
12. Birnberg, P. R., Somers, D. A., and Brenner, M. L. (1988) Characterization of conditioning factors that increase colony formation from "Black Mexican Sweet Corn" protoplasts. *J. Plant Physiol.* **132,** 316–321.
13. Schröder, R., Gärtner F., Steinbrenner, B., Knoop, B., and Beiderbeck, R. (1989) Viability factors in plant suspension cultures—some properties. *J. Plant Physiol.* **135,** 422–427.
14. Bellincampi, D. and Morpurgo, G. (1989) Evidence for the presence of a second conditioning factor in plant cell cultures. *Plant Sci.* **65,** 125–130.
15. Jørgensen, R. B., Andersen, B., and Andersen, J. M. (1992) Effects and characterization of the conditioning medium that increase colony formation from barley (*Hordeum vulgare* L.) protoplasts. *J. Plant Physiol.* **140,** 328–333.
16. Murashige, T. and Skoog, F. (1962) A revised medium for rapid growth and bioassays with tobacco tissue cultures. *Physiol. Plant.* **15,** 473–497.

19

Protoplast Fusion for Symmetric Somatic Hybrid Production in Brassicaceae

Jan Fahleson and Kristina Glimelius

1. Introduction

This chapter will focus on somatic hybridization in Brassicaceae. The results from a vast number of investigations have clearly shown that somatic hybridization in this family can overcome sexual barriers, and that the nuclear, mitochondrial, and chloroplast genomes from different sexually incompatible species can be combined. Reproducible protocols for protoplast culture and regeneration to plants have been established for a large number of crop species within this family, such as oilseed rape, turnip rape, cabbage, and Indian mustard ***(1)***. This has laid the foundation for the production of somatic hybrids ***(2,3)***. A substantial number of somatic hybrids have been produced between species within the genus *Brassica*, combining both diploid and allopolyploid *Brassica* species into hybrid plants. For example, the resynthesis of rapeseed ***(4–7)***, but also a combination of all three *Brassica* genomes (A, B, and C) have been performed ***(8,9)***. To widen the gene pool of rapeseed, several intergeneric somatic hybrids combining *Brassica napus* with species from the genera *Eruca* ***(10)***, *Sinapis* ***(11,12)***, *Raphanus* ***(13)***, *Moricandia* ***(14,15)***, and *Diplotaxis* ***(16,17)*** have been made. Furthermore, it has even been possible to combine species from different tribes, by obtaining somatic hybrids, e.g., between rapeseed and *Arabidopsis thaliana* ***(18–21)***, rapeseed and *Thlaspi perfoliatum* ***(22)***, as well as rapeseed and *Lesquerella fendleri* ***(23)***.

In general, it can be stated that the process of somatic hybridization has widened the gene pool of crops within the Brassicaceae family. In addition to this, successful results regarding somatic hybridization with the purpose of improving *Brassica oleracea* ***(24,25)*** and *Brassica juncea* have been made ***(26–30)***. It must be recognized, however, that even if sexual barriers are

From: *Methods in Molecular Biology, Vol. 111: Plant Cell Culture Protocols*
Edited by: R. D. Hall © Humana Press Inc., Totowa, NJ

circumvented by this method, interspecific incompatibility barriers can be present at other stages of plant development, inhibiting the production of fertile hybrid plants *(31)*. Furthermore, except in a few cases *(32,33)* the chloroplast genomes seem unable to recombine, whereas in the mitochondrial genomes, a considerable degree of recombination has been detected *(34–36)*.

The method for protoplast fusion described in this chapter is the mass-fusion method. High numbers of protoplasts from the parental species are mixed together and induced to fuse via a polyethylene glycol (PEG)/$CaCl_2$ treatment. The mass-fusion method is in itself random, since homofusions (fusions of protoplasts within the same species) as well as heterofusions (fusions of protoplasts between the two parental species) are obtained. In addition, the number of protoplasts participating in a fusion event cannot be controlled by the mass-fusion method.

The mass-fusion method has to be combined with an efficient system for selection of hybrid cells. Different approaches have been adopted, such as the use of metabolic mutants *(11)* or selection with iodacetamide (IOA) in combination with irradiation/media selection *(37,38)*. The selection system used in our laboratory is flow cytometry and cell sorting. This is a general system and relies on differential fluorescence from the parental protoplasts, obtained through staining of the protoplasts and/or chlorophyll autofluorescence from one of the parental protoplasts *(39)*. The hybrid cells will produce a double-fluorescent signal and can thus be selected.

Irrespective of the selection method used, a procedure for identification of hybrid individuals has to be developed. The identification procedure can be based on biochemical methods *(40)*, morphological features *(14)* or genetic markers *(21,22,31)*. Genetic markers based on DNA sequences, such as restriction-length polymorphisms (RFLPs) or species-specific repetitive DNA sequences, are generally robust markers and are not influenced by changes in cell metabolism or environment.

This chapter covers the first steps in the production of somatic hybrids within Brassicaceae, and describes isolation of protoplasts, protoplast fusion, preparation for flow cytometry and cell sorting, cultivation and regeneration. As a procedure to identify hybrids, we have also included a section covering isolation of species-specific DNA sequences. This last section includes some standard procedures that are not described in detail.

2. Materials

2.1. Plant Material and Culture Conditions

1. Etiolated hypocotyls, grown for 4–5 d in vitro, in darkness at 25°C.
2. Green leaves from in vitro grown plants, approximately 3-wk-old, grown at a light intensity of 12 W/m^2 and a day/night regime of 16/8 h at 25°C. *(See* **Note 1**.)

2.2. Equipment

All materials must be sterilized prior to use.

1. Protoplast filter (100-μm nylon mesh).
2. Forceps.
3. Scalpel.
4. Spatula.
5. Pipets, 10-, 5-, and 1-mL.
6. Pasteur pipets.
7. Petri dishes, 3.5-, 5-, and 9-cm.
8. Glass jars (baby food jar-type).
9. Flasks, 100-, 250-, and 500-mL.
10. Cylinders, 50-, 100-, and 500-mL.
11. Graded 15-mL glass centrifuge tubes.
12. Multiboxes, 48-wells (Costar®).
13. 15-mL Falcon tubes with rounded bottom.
14. Hemocytometer (Bürker chamber).
15. Flow hood.
16. Microscope, inverted, equipped with a UV lamp.
17. Fluorescent activated cell sorter (e.g., a Becton Dickinson instrument or equivalent).
18. X-ray film.
19. Cassettes for X-ray film.
20. Shaker.
21. Membranes for colony hybridization.
22. Toothpicks.
23. Oligolabeling kit for ^{32}P labeling of DNA.

2.3. Media and Chemicals

1. Calcium hypochlorite (7.5% w/v).
2. Tween 20.
3. Ethanol, 70 and 96%.
4. Cell-wall-degrading enzymes (cellulysin and macerase, Calbiochem): 1% cellulysin, 0.1% macerase (w/v) dissolved in the culture medium K_3 (*see* **Table 1** and **Note 2**).
5. Mannitol, 0.4 *M*, dissolved in the culture medium 8p (*see* **Table 1**).
6. Agarose, 0.13% and 0.4% in K_3 medium.
7. SeaPlaque agarose 1.47% (FMC®) in deionized water, autoclaved.
8. CPW 16 (***41***; *see* **Table 2**). Sterilized by filtration.
9. TVL (*see* **Table 2**), autoclaved
10. W5 (***42***; *see* **Table 2**), autoclaved.
11. KB (*see* **Table 1**), autoclaved.
12. 5(6)-Carboxyfluorescein diacetate, 0.22 *M*, dissolved in W5 (CFDA, Molecular Probes; *see* **Note 3**).
13. PEG solution (*see* **Table 2** and **Note 4**).

Table 1
Media Compositions for Cultivation of Protoplasts

Media	Components	Amount	
MS pH 5.7–5.8	A	20	mL/L
	B	14	mL/L
	C	10	mL/L
	D	10	mL/L
	E	10	mL/L
	MS II	10	mL/L
	Fe-EDTA	5	mL/L
	Glycine	1	mL/L
	Nicotinic acid	1	mL/L
	Vitamin B_6	1	mL/L
	Vitamin B_1	1	mL/L
	Edamin (casein hydrolysate)	1	g/L
	Inositol	100	mg/L
	Sucrose	10	g/L
	Gellan gum	3	g/L
K_3 pH 5.8–6.0	A	26	mL/L
	B	2	mL/L
	C	7	mL/L
	E	7	mL/L
	K	10	mL/L
	K_3	5	mL/L
	F	10	mL/L
	Fe-EDTA	5	mL/L
	Zn-EDTA	1	mL/L
	Vitamin B_1	10	mg/L
	Vitamin B_6	2	mL/L
	Nicotinic acid	2	mL/L
	Xylose	250	mg/L
	Inositol	100	mg/L
8p pH 5.6	A	20	mL/L
	B	3.75	mL/L
	C	8.1	mL/L
	D	10	mL/L
	E	13.6	mL/L
	Kao's micro	10	mL/L
	Fe-EDTA	5	mL/L

(continued)

Table 1 *(continued)*

Media	Components	Amount	
	Nicotinic acid	2	mL/L
	Vitamin B_6	2	mL/L
	Vitamin B_1	10	mg/L
	Ascorbic acid	2	mg/L
	Na pyruvate	20	mg/L
	Citric acid	40	mg/L
	Malic acid	40	mg/L
	Fumaric acid	40	mg/L
	Fructose	250	mg/L
	Ribose	250	mg/L
	Xylose	250	mg/L
	Mannose	250	mg/L
	Rhamnose	250	mg/L
	Cellobiose	250	mg/L
	Sorbitol	250	mg/L
	Mannitol	250	mg/L
	Inositol	100	mg/L
	Casamino acids	250	mg/L
	Coconut water	20	mL/L
	Sucrose	250	mg/L
	Glucose	72.1	g/L

14. $CaCl_2$:PEG 2:1 (*see* **Table 2**): The $CaCl_2$ solution is autoclaved.
15. $CaCl_2$:PEG 5:1 (*see* **Table 2**).
16. Stock solutions of media chemicals (*see* **Table 3**): Kept at +4°C to prevent fungal and/or bacterial growth.
17. MS medium, autoclaved (***43***; *see* **Table 1**).
18. K_3 medium, filter-sterilized (***42***; with modifications according to ***44,45***; *see* **Table 1**).
19. 8p medium, filter-sterilized (***46***; with modifications according to ***47,48***; *see* **Table 1**).
20. Restriction enzyme *Sau*3A.
21. Restriction enzyme buffer 10X (provided with the restriction enzyme).
22. Isopropylthio-β-D-galactoside (IPTG): 2 g of IPTG to 8 mL distilled H_2O. Adjust the volume to 10 mL, and sterilize by filtration through a 0.22-μm disposable filter. Store in 1-mL aliquots at –20°C.
23. 5-Bromo-4-chloro-3-indolyl-β-D-galactoside (X-gal). Dissolve X-gal in dimethyl-formamide (DMF), 20 mg of X-gal/mL DMF. Use polypropylene tubes, put in a box to prevent from light. Store at –20°C. (*See* **Note 5**.)
24. Phenol:chloroform mixed 1:1 (Tris-saturated phenol).

Table 2
Solutions for Protoplast Isolation and Fusion

Solution	Components	Amount	
CPW 16	KH_2PO_4	27.2	mg/L
pH 5.5–5.8	KNO_3	101	mg/L
	$CaCl_2 \cdot 2H_2O$	1480	mg/L
	$MgSO_4 \cdot 7H_2O$	246	mg/L
	KI	0.16	mg/L
	$CuSO_4 \cdot 5H_2O$	0.025	mg/L
	Sucrose	160	g/L
TVL	Sorbitol	54.654	g/L
pH 5.6–5.8	$CaCl_2 \cdot 2H_2O$	7.35	g/L
W5	$CaCl_2 \cdot 2H_2O$	18.4	g/L
	NaCl	9	g/L
	Glucose	1	g/L
	KCl	0.8	g/L
KB	KCl	11.18	g/L
pH 7.2	$CaCl_2 \cdot 2H_2O$	1.47	g/L
	Tris	13.1	g/L
PEG	PEG 1500	400	g/L
pH 7.0	Glucose	54	g/L
	$CaCl_2 \cdot 2H_2O$	7.35	g/L
$CaCl_2$	$CaCl_2 \cdot 2H_2O$	14.7	g/L
pH 7.0	Sorbitol	18.2	g/L

25. 0.5 *M*, disodium etylendiamintetraacetate (EDTA). Dissolve 186.1 g/L with intensive stirring, heating and adjusting pH to 8.0 is also required.
26. NaAc, 3 *M* (sodium acetate · $3H_2O$): Add 408.1 g to 800 mL H_2O. Adjust pH to 5.2 with glacial acetic acid, and dilute to 1 L. Autoclave.
27. TE buffer, 10 m*M* Tris-HCl (pH 7.5), 1 m*M* EDTA (pH 8.0): To 988 mL of H_2O, add 10 mL 1 *M* Tris-HCl + 2 mL of 0.5 *M* EDTA.
28. Adenosine triphosphate (ATP), 0.1 *M*: Dissolve 60 mg in 0.8 mL H_2O. Adjust pH to 7.0 with NaOH, 0.1 *N*. Adjust volume to 1 mL, and store in aliquots at –20°C.
29. Dithiothreitol (DTT), 0.1 *M*: Dissolve 3.09 g in 20 mL of 0.01 *M* sodium acetate (pH 5.2). Sterilize by filtration. Store in 1-mL aliquots at –20°C.
30. PEG 50%: PEG 8000 dissolved in deionized water.
31. T4 DNA ligase.
32. Ligation buffer (10X, provided with T4 DNA ligase).

Table 3
Stock Solutions of Media Chemicals

Component	Amount		Component	Amount	
A. KNO_3	95	g/L	Kao's microelements ***(22)***		
B. NH_4NO_3	120	g/L	KI	75	mg/L
C. $MgSO_4 \cdot 7H_2O$	3	g/L	$MnSO_4 \cdot 4H_2O$	1000	mg/L
D. KH_2PO_4	17	g/L	H_3BO_3	300	mg/L
E. $CaCl_2 \cdot 2H_2O$	44	g/L	$ZnSO_4 \cdot 7H_2O$	200	mg/L
F. $MnSO_4 \cdot H_2O$	1.69	g/L	$Na_2MoO_4 \cdot 2H_2O$	25	mg/L
$Na_2MoO_4 \cdot 2H_2O$	0.025	g/L	$CuSO_4 \cdot 5H_2O$	2.5	mg/L
$CuSO_4 \cdot 5H_2O$	0.0025	g/L	$CoCl_2 \cdot 6H_2O$	2.5	mg/L
$CoCl_2 \cdot 6H_2O$	0.0025	g/L			
H_3BO_3	0.620	g/L	MS II (microelements)		
G. KI	10	g/L	KI	83	mg/L
H. $(NH_4)_2SO_4$	150	g/L	$MnSO_4 \cdot 4H_2O$	2230	mg/L
K. $NaH_2PO_4 \cdot H_2O$	15	g/L	H_3BO_3	620	mg/L
K_3. $(NH_4)_2SO_4$	26.8	g/L	$ZnSO_4 \cdot 7H_2O$	1234	mg/L
L. $MgCl_2 \cdot 6H_2O$	35	g/L	$Na_2MoO_4 \cdot 2H_2O$	25	mg/L
$ZnSO_4 \cdot 7H_2O$	0.86	mg/mL	$CuSO_4 \cdot 5H_2O$	2.5	mg/L
Zn-EDTA	15	mg/mL	$CoCl_2 \cdot 6H_2O$	2.5	mg/L
Fe-EDTA					
Fe $SO_4 \cdot 7H_2O$	5.57	mg/mL			
Na_2EDTA	7.45	mg/mL			
Glycine	2	mg/mL			
Nicotinic acid	0.5	mg/mL			
Thiamine (B_1)	0.1	mg/mL			
Pyridoxine (B_6)	0.5	mg/mL			
2,4 Dichlorophenoxy acetic acid (2,4-D)	0.1	mg/mL			
1-Naphtylacetic acid (NAA)	0.1	mg/mL			
6-Benzylamino purine (BAP)	0.1	mg/mL			
Zeatin	0.1	mg/mL			

33. Competent *Escherichia coli* XL1, prepared according to standard methods.
34. Luria-Bertani medium (LB medium): To 950 mL of deionized H_2O add 10 g Bacto-tryptone, 5 g Bacto-yeast extract, and 20 g NaCl. Dissolve under stirring, adjust pH to 7.0 with 5 *N* NaOH, and dilute to 1 L. Autoclave.
35. Ampicillin stock solution, 100 mg/mL. Dilute to a final concentration of 50–100 µg/mL when used in media.

3. Methods

3.1. Sterilization and Cultivation of Plant Material

1. Sterilize approx 500 seeds of one parent for 1 h with continuous shaking in 50–100 mL of a 7.5% calcium hypochlorite solution to which a few drops of Tween 20 have been added. (*See* **Note 6**.)
2. Rinse seeds for 1 min in 70% ethanol, and thereafter wash three times for 2 min each in sterile water.
3. Germinate seeds in 9-cm Petri dishes containing 20 mL MS medium supplemented with 3% (w/v) sucrose. In case of etiolated hypocotyls, seeds are placed in a row at one end of a 9-cm Petri dish (about 20 seeds/dish) and then placed slightly tilted in the dark for 4–5 d at 25°C.
4. Green leaves of the other parent are obtained from 3-wk-old plants, grown in glass jars at a light intensity of 12 W/m^2, and a day/night regime of 16/8 h at 25°C.

3.2. Pretreatment of Plant Material

1. With a forceps and scalpel, carefully cut the etiolated hypocotyls into 0.5- to 1-mm segments. The leaves are gently cut into small pieces. For a typical experiment, 25 Petri dishes (about 500 seeds) and leaves from 2–3 in vitro plants are needed.
2. Place each material in a separate TVL solution to plasmolyze for 1 h.
3. Withdraw the solution with a pipet and replace it with enzyme solution (1% cellulysin, 0.1% macerase in K_3 medium).
4. Place the dishes in darkness, with the leaf dishes stationary and the hypocotyl dishes on a shaker, about 15 rpm, overnight.

3.3. Isolation of Protoplasts

1. Check the protoplast suspension under an inverted microscope to confirm the absence of contamination by bacteria or fungi.
2. Filter the protoplast suspensions through a 100-μm nylon mesh.
3. Add approximately one-third the volume of CPW 16.
4. Transfer the suspensions to centrifuge tubes and centrifuge for 5 min at 100*g*.
5. Remove the floating protoplasts with a Pasteur pipet and add approximately five times the volume of W5. (*See* **Note 7**.)
6. Centrifuge as before.
7. Remove the supernatant, and resuspend the hypocotyl protoplast pellets in a small volume of W5 (about 1.5 mL/centrifuge tube). The leaf protoplasts are resuspended in a small volume of TVL:KB 1:1.
8. Stain the hypocotyl protoplasts with 250–500 μL of 0.22 *M* 5(6)-carboxy-fluorescein diacetate (CFDA) for 10 min.
9. Add W5 (8.5 mL/centrifuge tube), and centrifuge the hypocotyl protoplasts as before.
10. Resuspend the hypocotyl protoplasts in a small volume of TVL:KB 1:1, and calculate the protoplast density in the haemocytometer.
11. Dilute the hypocotyl protoplasts to 9 x 10^5 protoplasts/mL and the leaf protoplasts to 9×10^5 protoplasts/mL.

3.4. Fusion of Protoplasts

1. Mix the protoplasts 1:1 with a Pasteur pipet and place approx 20 separate drops (about 0.4 mL/droplet) in a 9-cm Petri dish.
2. Turn the flowhood off, and let the protoplasts settle for 5 min.
3. Add 2 droplets of the PEG solution to each droplet of protoplasts. Leave for 3–5 min. (*See* **Note 8**.)
4. Tilt the Petri dish, and remove the PEG solution with a pipet. With another pipet, add 2–3 mL of the $CaCl_2$:PEG 2:1 solution for 5 min to wash the protoplasts (which now should be slightly attached to the bottom of the Petri dish).
5. Remove the $CaCl_2$:PEG 2:1 solution, and replace it with $CaCl_2$:PEG 5:1 solution for 5 min. (*See* **Note 9**.)
6. Wash the protoplasts with 4 mL 8p medium (without hormones). After this point, selection of hybrid cells may be performed in several ways. The following section describes how cells are prepared for flow cytometry and cell sorting. However, alternative selection methods are indicated elsewhere (*see* **Note 10**).

3.5. Preparation of Protoplasts for Flow Cytometry and Cell Sorting

1. Store protoplasts for at least 2 h at +8°C. Remove 8p medium, and cover the protoplasts with 7–8 mL of 8p medium containing 0.4 *M* mannitol and mixed with W5 (1:1 v/v). Leave at +8°C for at least 2 h. (*See* **Note 11**.)
2. Using a Pasteur pipet, carefully detach the protoplasts from the bottom of the Petri dish, and resuspend them in W5. Add about the same amount of W5 as in **step 5, Subheading 3.3.** (*See* **Note 12**.)
3. Centrifuge in 15-mL Falcon tubes at 75*g* for 5 min. Discard supernatant.
4. Resuspend the protoplasts in a small volume of 8p medium with 0.4 *M* mannitol to give a protoplast concentration of about 1×10^6 protoplasts/mL.
5. Sort the protoplasts in the flow cytometer (this machine is usually operated by a specialist and is adjusted to sort out protoplasts exhibiting a clear signal of double fluorescence). (*See* **Note 13**.) The protoplasts are sorted into separate wells of a multivial dish, Costar®, 48 wells, each well containing 100 µL of 8p medium supplemented with 4.5 µ*M* 2,4-dichlorophenoxyacetic acid (2,4-D), 0.5 µ*M* NAA, and 2.2 µ*M* BAP.

3.6. Culture of Hybrid Cells and Regeneration of Plants

1. The hybrid cells are cultured in a culture room with a light intensity of 12 W/m^2 and a day/night regime of 16/8 h at 25°C. The medium used is the same as mentioned in **Subheading 3.5.**
2. When the first divisions are observed (usually after 4–7 d), the cells are diluted with fresh culture medium, containing 0.55% Sea Plaque agarose , w/v (FMC) without hormones, to four times the original volume. (*See* **Note 14**.)
3. When the agarose has solidified, use a spatula, and move the beads to 2.5-cm Petri dishes with 3 mL 8p medium, containing the same hormone concentration as in the beads (1/4 of the original hormone concentration in the multivial dish).

4. At weekly intervals, replace the medium with fresh medium.
5. When small calli (1–2 mm in diameter) are observed, usually after 2–4 wk, they are removed from the beads and plated on 0.13% agarose (w/v) in K_3 medium with 3.4% sucrose (w/v) and the same hormone concentrations as in the beads.
6. After about 2 wk the calli are induced to differentiate into plants by transferring them to K_3 medium supplemented with 0.5% sucrose, 0.4% agarose (w/v), 0.6 μ*M* indole-3-acetic acid, 2.2 μ*M* BAP, and 2.3 μ*M* zeatin.
7. Transfer calli to fresh differentiation medium every 3 wk.
8. When shoots emerge from a callus and have reached a size of about 1 cm in height, transfer the shoot to a baby food jar with MS medium, 1% sucrose (w/v) without hormones.

3.7. Isolation of Species-Specific Repetitive Sequences

1. Isolate DNA from parental species according to standard methods. There are a variety of DNA isolation methods. One commonly used in our laboratory is described in ref. ***(35)***.
2. To a clean and autoclaved microfuge tube, add the following:
   0.2 μg DNA.
   0.5 μL *Sau*3A.
   1 μL restriction enzyme buffer (10X).
   Distilled water to a final volume of 10 μL.
3. For each species, set up five reactions, and digest DNA for 2, 4, 6, 8, and 10 min at room temperature. Stop digestion by adding 2 μL of 0.5 *M* EDTA.
4. Check the different digests on a 0.6% agarose gel. Determine the reaction that produces fragments in the range of 500–1000 bp.
5. Set up five reactions, and digest for the amount of time determined in **step 4**.
6. Pool the 5 samples, extract once with phenol:chloroform 1:1, and precipitate with 3X the volume of cold absolute ethanol together with 1/10 of the volume of 3 *M* NaAc. Put in –70°C freezer for 1–2 h.
7. Spin for 15 min at maximum speed in a microcentrifuge. Discard supernatant and wash pellet in 70% ethanol. Spin for 10 min, dry, and dissolve pellet in 25 μL of TE buffer.
8. Set up ligation reaction. Add the following components to a clean microfuge tube:
   25 μL digested DNA.
   6 μL ligation buffer (10X, supplied with the DNA ligase).
   1 μL dephosphorylated pUC 18 (Pharmacia).
   6 μL 0.1 *M* ATP.
   6 μL 0.1 *M* DTT.
   6 μL 50% PEG.
   1 μL T4 DNA ligase.
   9 μL distilled water.
9. Ligate overnight at 16°C.
10. Heat-inactivate the reaction at 65°C for 10 min.

11. Take 20 μL of the ligation mix, and transform competent *E. coli* XL1. Thaw competent cells slowly on ice. Add 80 μL of TE buffer and 200 μL of competent cells.
12. Place on ice for 30 min.
13. Activate membranes for 2 min at 42°C. Place on ice for 15 min.
14. Add 1 mL LB without ampicillin. Culture in 15-mL Falcon tubes with rounded bottom on a shaker (about 200 rpm) at 37°C for 1 h.
15. Prepare LB plates plus ampicillin (50–100 μg/mL). Spread X-gal and IPTG on the plate (40 μL of each).
16. Spread bacterial suspension over the plate, and culture overnight at 37°C.
17. Pick white colonies over to gridded membranes (Bio-Rad). The membranes are laid on an LB plate. Use a toothpick, and produce two identical membranes and a master plate. This means that each colony will be represented three times at the same position on each plate. Pick 500 colonies from each transformation (in total 1000 colonies since DNA from two species are used). Usually 100 colonies/membrane or plate is a suitable density.
18. Culture the colonies to a size of about 1 mm in diameter. Wash the filters according to manufacturer's instructions.
19. Label 25–50 ng of undigested parental DNA with ^{32}P using oligolabeling kit (Pharmacia). Set up two reactions, one for each DNA of the parental species.
20. Hybridize and wash filters according to manufacturer's instructions. Expose on X-ray film (7–8 h or overnight), identify those colonies that give a strong signal to one of the labeled DNAs, but not to the other. The selected colonies are restreaked on a fresh LB plate and then grown in a larger volume of liquid LB medium (100–500 mL) for plasmid isolation. (*See* **Note 15**.)

4. Notes

1. The best parents to start with are some of the cultivated *Brassica* species, e.g., *Brassica campestris, B. napus,* and *B. oleracea.* Usually there is a variation in the yield of hypocotyl protoplasts among different species and even varieties, whereas it is easier to obtain enough material from green leaves. In our courses for undergraduate students, we have used *B. napus* cv. "Hanna" as a source for hypocotyl protoplasts.
2. Dissolve the Enzyme powder under intensive stirring and store at +4°C overnight. Filter-sterilize.
3. Boil the suspension heavily, let cool, and sterilize by filtration.
4. Melt PEG by gentle heating. Dissolve glucose and $CaCl_2$ in distilled water in 50% of final volume. Mix with melted PEG. Cool, adjust pH, and dilute to final volume. Sterilize by filtration.
5. DMF is a highly toxic compound.
6. The percentage of contaminated seed should be very low. Normally, 3–4 Petri dishes (out of 25) have to be discarded owing to fungal growth. If fungal growth is discovered in a Petri dish, the whole dish has to be discarded.
7. In CPW 16, only the clean viable protoplasts will float. The debris will form a pellet to be discarded. The addition of W5 will pellet the protoplasts.

8. To illustrate what is happening when PEG is added, place a few drops of mixed protoplasts in a Petri dish under the inverted microscope, and add the PEG solution. It can be seen that the protoplasts shrink heavily, owing to the increase in the osmotic pressure within the cells.
9. The dilution of the PEG solution in two steps will help to recover the round shape of the protoplasts in a gentle way.
10. Selection of hybrid cells can be done in several ways. The procedure described here, used in our lab, involves the preparation of cells for flow cytometry and cell sorting ***(39,49)***. Other alternative selection methods could be micromanipulation ***(40)***, or use of IOA treatment in combination with irradiation ***(37–38)***. Irradiation or treatment with IOA has to be done at some point prior to fusion. See specific papers for details.
11. The storage of protoplasts in cold will help to recover the protoplasts from the rough PEG treatment. Also, the passage through the flow cytometer is somewhat rough and freshly fused protoplasts might break.
12. Be very gentle when releasing the protoplasts from the bottom of the Petri dish. Do not flush directly on the cells.
13. The double fluorescence from a hybrid cell is obtained because the CFDA staining of the hypocotyl protoplasts (green fluorescence) and the red autofluoresence from the chloroplasts in the leaf protoplasts. The typical filter equipment in the flow cytometer was as follows: fluorescence emission is first filtered with a 510-nm high-pass filter and split into red and green channels by a half-silvered mirror. The green channel has a selection filter centered at 530 nm, band width 250 nm. Corresponding data for the red channel is 588 nm, and band width 20 nm.
14. Make a 1.47% (w/v) of a SeaPlaque solution. Dissolve in deionized water by heating and then autoclaving. Mix with double-concentrated 8p medium (sterilized by filtration), cool to about 45–50°C, and add 300 μL of this solution to each well.
15. Before plasmid isolation, make glycerol stocks of the selected colonies.

References

1. Vamling, K. and Glimelius, K. (1990) Regeneration of plants from protoplasts of oilseed *Brassica* crops, in *Legumes and Oilseed Crops,* vol. 1 (Bajaj, Y. P. S., ed.), Springer-Verlag, Berlin, pp. 385–417.
2. Glimelius, K., Fahleson, J., Landgren, M., Sjödin, C., and Sundberg, E. (1991) Gene transfer via somatic hybridization in plants. *TiBtech* **9,** 24–30.
3. Waara, S. and Glimelius, K. (1995) The potential of somatic hybridization for crop breeding. *Euphytica* **85,** 217–233.
4. Schenck, H. R. and Röbbelen, G. (1982) Somatic hybrids by fusion of protoplasts from *Brassica oleracea* and *B. campestris. Z. Pflanzenz.*, **89,** 278–288.
5. Sundberg, E., Landgren, M., and Glimelius, K. (1987) Fertility and chromosome stability in *Brassica napus* resynthesized by protoplast fusion. *Theor. Appl. Genet.* **75,** 96–104.
6. Heath, D. W. and Earle, E. D. (1996) Resynthesis of rapeseed (*Brassica napus* L.): A comparison of sexual versus somatic hybridization. *Plant Breed.* **115,** 395–401.

7. Terada, R., Yamashita, Y., Nishibayashi, S., and Shimamoto, K. (1987) Somatic hybrids between *Brassica oleracea* and *B. campestris* selection by the use of iodoacetamide inactivation and regeneration ability. *Theor. Appl. Genet.* **73,** 379–384.
8. Sjödin, C. and Glimelius, K. (1989) *Brassica naponigra,* a somatic hybrid resistant to *Phoma lingam. Theor. Appl. Genet.* **77,** 651–656.
9. Sacristan, M. D., Gerdemann, K. M., and Schieder, O. (1989) Incorporation of hygromycin resistance in *Brassica nigra* and its transfer to *B. napus* through asymmetric protoplast fusion. *Theor. Appl. Genet.* **78,** 194–200.
10. Fahleson, J., Råhlén, L., and Glimelius K. (1988) Analysis of plants regenerated from protoplast fusions between *Brassica napus* and *Eruca sativa. Theor. Appl. Genet.* **76,** 507–512.
11. Toriyama, K., Kameya, T., and Hinata, K. (1987) Selection of a universal hybridizer in *Sinapis turgida* Del. and regeneration of plantlets from somatic hybrids with *Brassica* species. *Planta* **170,** 308–313.
12. Chevre, A. M., Eber, F., Margale, E., Kerlan, M. C., Primard, C., Vedel, F., and Delseny, M. (1994) Comparison of somatic and sexual *Brassica napus—Sinapis alba* hybrids and their progeny by cytogenetic studies and molecular characterization. *Genome* **37,** 367–374.
13. Lelivelt, C. and Krens, F. A. (1992) Transfer of resistance to the beet cyst nematode (*Heterodera schachtii* Schm.) into the *Brassica napus* L. gene pool through intergeneric somatic hybridization with *Raphanus sativus* L. *Theor. Appl. Genet.* **83,** 887–894.
14. Toriyama, K., Hinata, K., and Kamcya, T. (1987) Production of somatic hybrid plants "*Brassicomoricandia*" through protoplast fusion between *Moricandia arvensis* and *Brassica oleracea. Plant Sci.* **48,** 123–128.
15. O'Neill, C. M., Murata, T., Morgan, C. L., and Mathias, R. J. (1996) Expression of the C_3-C_4 intermediate character in somatic hybrids between *Brassica napus* and the C_3-C_4 species *Moricandia arvensis. Theor. Appl. Genet.* **93,** 1234–1241.
16. McLellan, M. S., Olesen, P., and Power, J. B. (1987) Towards the introduction of cytoplasmic male sterility (CMS) into *Brassica napus* through protoplast fusion, in *Progress in Plant Protoplast Research* (Puite, K. J., Dons, J. J. M., Huizing, H. J., Kool, A. J., Koornneef, M. and Krens, F. A., eds.), Wageningen, The Netherlands, pp. 187–188.
17. Klimaszewska, K. and Keller, W. (1988) Regeneration and characterization of somatic hybrids between *Brassica napus* and *Diplotaxis harra. Plant Sci.* **58,** 211–222.
18. Gleba, Y. Y. and Hoffmann, F. (1979) "Arabidobrassica": plant-genome engineering by protoplast fusion. *Naturwissenschaften* **66,** 547–554.
19. Gleba, Y. Y. and Hoffmann, F. (1980) "Arabidobrassica": a novel plant obtained by protoplast fusion. *Planta* **149,** 112–117.
20. Bauer-Weston, W. B., Keller, W., Webb, J., and Gleddie, S. (1993) Production and characterization of asymmetric somatic hybrids between *Arabidopsis thaliana* and *Brassica napus. Theor. Appl. Genet.* **86,** 150–158.

21. Forsberg, J., Landgren, M., and Glimelius, K. (1994) Fertile somatic hybrids between *Brassica napus* and *Arabidopsis thaliana*. *Plant Sci.* **95,** 213–223.
22. Fahleson, J., Eriksson, I., Landgren, M., Stymne, S., and Glimelius, K. (1994) Intertribal somatic hybrids between *Brassica napus* and *Thlaspi perfoliatum* with high content of the *T. perfoliatum*-specific nervonic acid. *Theor. Appl. Genet.* **87,** 795–804.
23. Skarzhinskaya, M., Landgren, M., and Glimelius, K. (1996) Production of intertribal somatic hybrids between *Brassica napus* L. and *Lesquerella fendleri* (Gray) Wats. *Theor. Appl. Genet.* **93,** 1242–1250.
24. Toriyama, K., Hinata, K., and Kameya, T. (1987) Production of somatic hybrid plants, "Brassicomoricandia," through protoplast fusion between *Moricandia arvensis* and *Brassica oleracea*. *Plant Sci.* **48,** 123–128.
25. Hagimori, M., Nagaoka, M., Kato, N., and Yoshikawa, H.(1992) Production and characterization of somatic hybrids between the Japanese radish and cauliflower. *Theor. Appl. Genet.* 82, 819–824.
26. Sikdar, S. R., Chatterjee, G., Das, S., and Sen, S. K. (1990) "Erussica," the intergeneric fertile somatic hybrid developed through protoplast fusion between *Eruca sativa* Lam. and *Brassica juncea* (L.) Czern. *Theor. Appl. Genet.* **79,** 561–567.
27. Chatterjee, G., Sikdar, S. R., Das, S., and Sen, S. K. (1988) Intergeneric somatic hybrid production through protoplast fusion between *Brassica juncea* and *Diplotaxis muralis*. *Theor. Appl. Genet.* **76,** 915–922.
28. Kirti, P. B., Narasimhulu, S. B., Prakash, S., and Chopra, V. L. (1992) Somatic hybridization between *Brassica juncea* and *Moricandia arvensis* by protoplast fusion. *Plant Cell Rep.* **11,** 318–321.
29. Kirti, P. B., Mohapatra, T., Khanna, H., Prakash, S., and Chopra, V. L. (1995) *Diplotaxis catholica* + *Brassica juncea* somatic hybrids: molecular and cytogenetic characterization. *Plant Cell Rep.* **14,** 593–597.
30. Kirti, P. B., Narasimhulu, S. B., Prakash, S., and Chopra, V. L. (1992) Production and characterization of intergeneric somatic hybrids of *Trachystoma ballii* and *Brassica juncea*. *Plant Cell Rep.* **11,** 90–92.
31. Fahleson, J., Eriksson, I., and Glimelius, K. (1994) Intertribal somatic hybrids between *Brassica napus* and *Barbarea vulgaris*. *Plant Cell Rep.* **13,** 411–416.
32. Medgyesy, P., Fejes, E., and Maliga, P. (1985) Interspecific chloroplast recombination in a *Nicotiana* somatic hybrid. *Proc. Natl. Acad. Sci. USA* **82,** 6960–6964.
33. Thanh, N. D. and Medgyesy, P. (1989) Limited choroplast gene transfer via recombination overcomes plastome-genome incompatibility between *Nicotiana tabacum* and *Solanum tuberosum*. *Plant Mol. Biol.* **12,** 87–93.
34. Pelletier, R. G. (1986) Plant organelle genetics through somatic hybridization. *Oxford Survey Plant Mol. Cell Biol.* **3,** 97–121.
35. Landgren, M. and Glimelius, K. (1990) Analysis of chloroplast and mitochondrial segregation in three different combinations of somatic hybrids produced within *Brassicaceae*. *Theor. Appl. Genet.* **80,** 776–784.
36. Landgren, M. and Glimelius, K (1994) A high frequency of intergenomic mitochondrial recombination and an overall biased segregation of *B. campestris* or

recombined *B. campestris* mitochondria were found in somatic hybrids within *Brassicaceae. Theor. Appl. Genet.* **87,** 854–862.
37. Yamashita, Y., Terada, R., Nishibayashi, K., and Shimamoto, K. (1989) Asymmetric somatic hybrids of *Brassica*: partial transfer of *B. campestris* genome into *B. oleracea* by cell fusion. *Theor. Appl. Genet.* **77,** 189–194.
38. Terada, R., Yamashita, Y., Nishibayashi, S., and Shimamoto, K. (1987) Somatic hybrids between *Brassica oleracea*: selection by the use of iodoacetamide inactivation and regeneration ability. *Theor. Appl. Genet.* **73,** 379–384.
39. Glimelius, K., Djupsjöbacka, M., and Fellner-Feldegg, H. (1986) Selection and enrichment of plant protoplast heterokaryons of *Brassicaceae* by flow sorting. *Plant Sci.* **45,** 133–141.
40. Sundberg, E. and Glimelius, K. (1986) A method for production of interspecific hybrids within *Brassicaceae* via somatic hybridization, using resynthesis of *Brassica napus* as a model. *Plant Sci.* **45,** 155–162.
41. Banks, M. S. and Evans, P. K. (1976) A comparison of the isolation and culture of mesophyll protoplasts from several *Nicotiana* species and their hybrids. *Plant Sci. Lett.* **7,** 409–416.
42. Menczel, L., Nagy, F., Kiss, Zs. R., and Maliga, P. (1981) Streptomycin resistant and sensitive hybrids of *Nicotiana tabacum* + *Nicotiana knightiana*: correlation of resistance to *N. tabacum* plastids. *Theor. Appl. Genet.* **59,** 191–195.
43. Murashige, T. and Skoog, F. (1962) A revised medium for rapid growth and bioassays with tobacco tissue cultures. *Physiol. Plant.* **15,** 473–497.
44. Kao, K. N., Constabel, F., Michayluk, M. R., and Gamborg, O. L. (1974) Plant protoplast fusion and growth of intergeneric hybrid cells. *Planta* **120,** 215–227.
45. Nagy, J. I. and Maliga, P. (1976) Callus induction and plant regeneration from mesophyll protoplasts of *Nicotiana sylvestris. Z. Pflanzenphysiol.* **78,** 453–455.
46. Kao, K. N. and Michayluk, M. R. (1975) Nutritional requirements for growth of *Vicia hajstana* cells and protoplasts at a very low population density in liquid media. *Planta* **126,** 105–110.
47. Glimelius, K. (1984) High growth rate and regeneration capacity of hypocotyl protoplasts in some Brassicaceae. *Physiol. Plant.* **61,** 38–44.
48. Gamborg, O. L., Miller, R. A., and Ojima, K. (1968) Nutrient requirements of suspension cultures of soybean root cells. *Exp. Cell Res.* **50,** 151–158.
49. Sundberg, E., Lagercrantz, U., and Glimelius, K. (1991) Effects of cell type used for fusion on chromosome elimination and chloroplast segregation in *Brassica oleracea* (+) *Brassica napus* hybrids. *Plant Sci.* **78,** 89–98.

20

Production of Cybrids in Rapeseed *(Brassica napus)*

Stephen Yarrow

1. Introduction

Protoplast fusion produces cells that contain a mixture of the DNA-containing organelles (nuclei, chloroplasts, and mitochondria) from both fusion parents. This chapter describes the production of rapeseed *(Brassica napus)* fusion-product cells where one of the nuclei is eliminated. In the absence of selection pressure, the remaining chloroplast/mitochondrial mixture segregates randomly as the cell divides, eventually creating different nuclear/chloroplast/mitochondrial combinations. These novel nucleo-cytoplasmic combinations are called cytoplasmic hybrids or "cybrids."

As defined here, a cybrid can include cases where the chloroplast/mitochondrial combination is novel, as well as alloplasmic substitutions, where the cytoplasm of one parent is substituted with another. The existence of new types of mitochondria or chloroplasts derived from the recombination of organelle DNA also constitutes a cybrid.

With few exceptions, novel chloroplast/mitochondrial combinations are unobtainable from conventional hybridization of plants, since the male gamete rarely contributes functional cytoplasm during fertilization. The identification of useful and important traits that are conferred by the organelles of certain plants has made the production of cybrids commercially important.

Cytoplasmic male sterility (CMS) is one such trait of great agronomic importance for crop species. By preventing self-fertilization, CMS can allow the economic production of single-cross F_1 hybrids. The production of cybrids as a means to incorporate or transfer CMS, conferred by the mitochondria, has been demonstrated in several crop species, including rapeseed *(B. napus)* ***(1)***, particularly noteworthy since this relatively recent technology is progressing

From: *Methods in Molecular Biology, Vol. 111: Plant Cell Culture Protocols*
Edited by: R. D. Hall © Humana Press Inc., Totowa, NJ

in this species to the commercial stage ***(2)***; broccoli *(Brassica oleracea)* ***(3)***; rice *(Oryza sativa)* ***(4)***; carrot *(Daucus carota L.)* ***(5)***; *Lolium perenne* ***(6)***; and, potato *(Solanum tuberosum)* ***(7)***. In some cases, cybrid technology has been used further to improve CMS lines, as demonstrated by Kirti et al. ***(8)***, whereby the chloroplasts associated with CMS-mitochondria previously transferred from *Raphanus sativus* to *Brassica juncea,* were causing chlorosis. These were substituted via "cybridization" with non-chlorosis causing chloroplasts from *Brassica juncea.*

Cybrid technology has been used to transfer other traits, including tolerance to the herbicide triazine in rapeseed *(B. napus)* ***(9,10)***, and resistance to the antibiotic streptomycin in potato ***(11)***. In addition, the technology has been used to study organelle interactions in citrus ***(12)***, tomato ***(13)***, and tobacco/petunia ***(14)***.

A variety of methods of producing and selecting for cybrids has been reported. For a detailed review on cybrid research, *see* Pelletier and Romagosa et al. ***(15)***, and Kumar and Cocking ***(16)***.

This chapter describes the production of rapeseed *(B. napus)* cybrids using a "donor-recipient" aided selection scheme first described for *Nicotiana* by Zelcer et al. ***(17)***. The nuclei of the "donor" parent are inactivated prior to fusion, so that effectively it only donates cytoplasm to the other "recipient" nuclear-bearing parent. Nuclear inactivation can be by γ or X-irradiation ***(10,18–20)***. The selection is complemented by the treatment of the "recipient" parent with an inhibitor, iodoacetic acid (IOA), that kills the cells by metabolic inactivation of the cytoplasm ***(10,19,20)***. Only fusion products survive due to complementation. This selection is only for fusion products and not for specific cybrids; these are formed after organelle segregation during cell division and plant regeneration (*see* **Fig. 1**).

The scheme described here has been successfully used to produce important cybrids in both rice ***(4)*** and rapeseed *(B. napus)* ***(10,21)***. The cybrid example described is the combination of CMS and cytoplasmic triazine tolerance (CTT) traits. In rapeseed, CTT is conferred by the chloroplast genome ***(22)*** and CMS, by the chondriome ***(9,10,23,24)***. For commercial production of single-cross-hybrid CTT canola, as proposed by Beversdorf et al. ***(1)***, both CMS and CTT traits must be present in the female parent. However, since these are generally only transmitted through the female line, it is not possible to combine these traits by conventional breeding approaches (*see* **Notes 1** and **2**). Protoplast fusion, however, allows the biparental transmission of both CMS and CTT traits ***(9,10,23)***. In addition, incorporating CTT into rapeseed hybrids can allow the crop to be grown in weed-infested areas. The weeds are eliminated with applications of triazine ***(25)***.

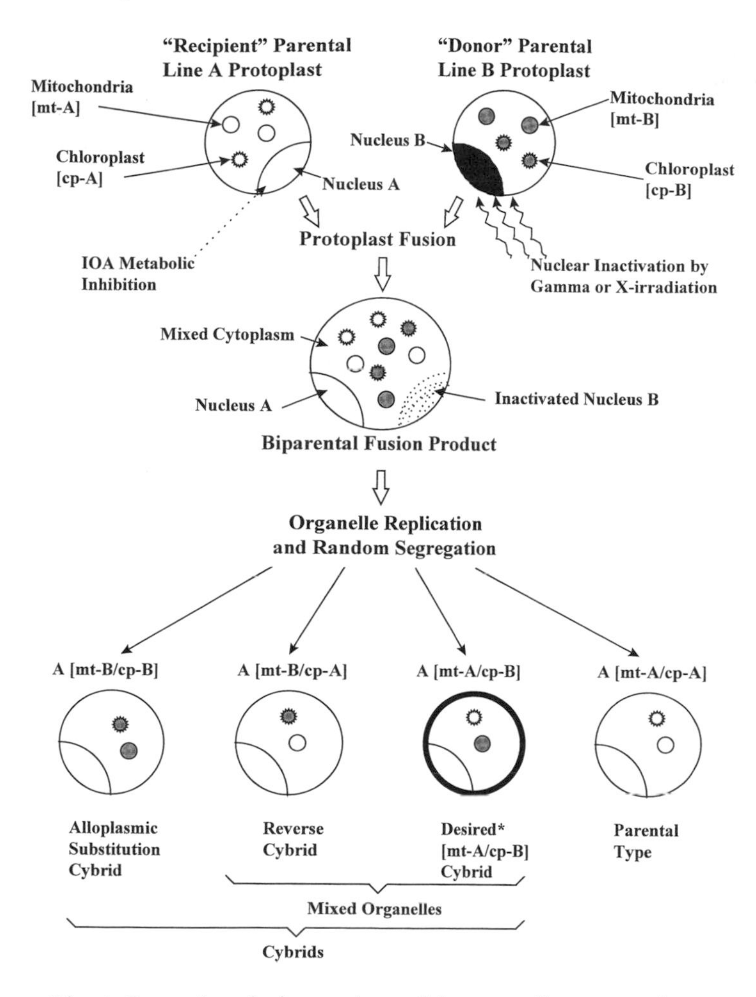

Fig. 1. Protoplast fusion and possible organelle segregation.

The procedures below describe how mesophyll protoplasts of the CTT rapeseed "donor" parent are fused with "recipient" CMS rapeseed hypocotyl protoplasts. Fusion is by a modified combination of polyethylene glycol (PEG) *(26,27)* and Ca^{+2} ions at high pH *(28)*. Protoplasts are cultured following a protocol that was devised for rapeseed *(29)*, with the addition of tobacco nurse cells *(9)*, that augment fusion product survival.

2. Materials

2.1. Protoplast Source Material

1. Mesophyll: CTT plants are grown from seed and established in "Metro-mix" potting compost, fertilized daily with "Peters" 20-10-20 fertilizer (both from W.R.

Grace and Co, Canada), maintained within growth chambers under the regimen of a 12-h photoperiod of 160 μmol/m²/s of photosynthetically active radiation (provided by 40 W "cool white" fluorescent tubes), at a constant temperature (23°C). Typically, four to six CTT plants are needed for each individual fusion operation.
2. Hypocotyl: CMS seedlings are grown from surface-sterilized seed.
3. 10% (v/v) Solution of 6% commercial preparation of hypochlorite.
4. Paper towels (wrapped in aluminum foil and autoclaved).
5. Forceps, sterilized by flaming.
6. MSS medium: basal salts of Murashige and Skoog's media (***30*** and *see* Appendix), containing 10 g/L sucrose, solidified with 4 g/L agarose (Sigma Type I), pH adjusted to 5.7, poured into 10-cm Petri plates (70 mL/dish).
6. Sieves, pore size <0.5 mm (autoclaved).

2.2. Protoplast Isolation

1. Soft nylon paint brush, sterilized by immersion in 70% ethanol for 5 min and dried on sterile paper towels.
2. Scalpel, with #11 blade, sterilized by flaming.
3. Sterile 250-mL Erlenmeyer flasks, capped with foil.
4. Sterile filtering funnels (double-layer cheesecloth lining a glass funnel).
5. Sterile Babcock bottles, capped with foil.
6. Sterile 230-mm cotton-plugged Pasteur pipets.
7. Porcelain tile (wrapped in foil and autoclaved).
8. Low-speed bench-type centrifuge.
9. Soak solution (for 1 L):

KNO_3	475 mg
$CaCl_2 \cdot 2H_2O$	110 mg
$MgSO_4 \cdot 7H_2O$	92.5 mg
KH_2PO_4	42.5 mg
Thiamine · HCl	0.125 mg
Glycine	0.5 mg
Nicotinic acid	1.25 mg
Pyridoxine · HCl	0.125 mg
Folic acid	0.125 mg
Biotin	0.0125 mg
Casein hydrolysate	25.0 mg
6-Benzylamino purine	0.5 mg
2,4-Dichlorophenoxyacetic acid	1.0 mg

pH 5.6, autoclaved. The soak solution can be stored at 4°C for several weeks.

10. Enzyme solution for 100 mL:

Macerozyme R-10*	0.1 g
Cellulase R-10*	1.0 g
2 (*N*-morpholine) ethanesulfonic acid (MES)*	0.1 g

Polyvinylpyrrolidone	1.0 g
Sucrose	12.0 g

pH 5.6, filter-sterilized through 0.2-μm Nalgene filter. The enzyme solution can be stored frozen for up to 6 mo. Mesophyll protoplast isolation requires a 1/10 strength enzyme solution, prepared by dilution with rinse solution *(see below)*. For the isolation of hypocotyl protoplasts, full-strength enzyme solution is used.

*Yakult Honsha Co., Ltd., Tokyo, Japan.

11. Rinse solution (for 1 L):

Sucrose	120 g
MES	0.5 g

pH 5.6, autoclaved. Stored at room temperature in sealed bottles for several months.

2.3. Protoplast Inactivation

1. *Iodoacetic acid* (IOA) solution: The IOA solution for protoplast inactivation is prepared from a 30 m*M* stock, diluted with rinse solution, pH adjusted to 5.6, and filter-sterilized with a 0.2-μm Nalgene filter. The final concentration of IOA necessary for protoplast inactivation is determined prior to fusion.
2. 30 m*M* stock: 0.279 g IOA in 50 mL H_2O. This stock solution can be stored at 4°C, in the dark for up to 3–4 mo, with no noticeable loss of activity.
3. X ray or γ-radiation source.

2.4. Protoplast Fusion

1. *PEG* solution (for 100 mL):

PEG 8000 (Sigma)	25.0 g
$CaCl_2 \cdot 2H_2O$	0.147 g
Sucrose	4.0 g

Autoclaved in foil wrapped vials. Can be stored at 4°C for up to 6–8 wk.

2. Ca^{2+} solution (for 100 mL):

$CaCl_2 \cdot 2H_2O$	0.735 g
Glycine	0.375 g
Sucrose	12.0 g

pH 10.5, filter-sterilized with 0.2-μm Nalgene filter. Solution must be prepared on the day of fusion.

2.5. Protoplast Culture

1. Preparation of media:
   a. Cell-layer (Cl) medium (*31* and *see* **Table 1**). Prepared on the day of fusion, in two forms:
      i. "Liquid," agarose-free—filter-sterilized with 0.2-μm Nalgene filter.
      ii. "Semisolidified," with double-strength agarose. Made up by mixing equal volumes of double strength "liquid" Cl (filter-sterilized with 0.2-μm Nalgene filter) with molten quadruple-strength solution of the appropri-

Table 1
Protoplast Media

Medium	Cl[a]	R[a]	DM[a]
Major elements, mg/L			
KNO_3	7600	1900	1900
$CaCl_2 \cdot 2H_2O$	1760	440	440
$MgSO_4 \cdot 7H_2O$	1480	370	370
KH_2PO_4	680	170	170
Iron and minor elements, mg/L			
$Na_2 \cdot EDTA$	18.5	18.5	18.5
$FeSO_4 \cdot 7H_2O$	13.9	18.5	18.5
H_3BO_3	3.1	3.1	3.1
$MnCl_2 \cdot 4H_2O$	9.9	9.9	9.9
$ZnSO_4 \cdot 7H_2O$	4.6	4.6	4.6
KI	0.42	0.42	0.42
$Na_2MoO_4 \cdot 2H_2O$	0.13	0.13	0.13
$CuSO_4 \cdot 5H_2O$	0.013	0.013	0.013
$CoSO_4 \cdot 7H_2O$	0.015	0.015	0.015
Organics, mg/L			
Thiamine · HCl	0.5	0.5	0.5
Glycine	2.0	2.0	2.0
Nicotinic acid	5.0	5.0	5.0
Pyridoxine · HCl	0.5	0.5	0.5
Folic acid	0.5	0.5	0.5
Biotin	0.05	0.05	0.05
Casein hydrolysate	50.0	100	100
Osmoticum, g/L			
Sucrose	85.6	34.25	34.25
Myo-inositol	5.7	0.1	0.1
D-Mannitol	5.7	18.25	18.25
Sorbitol	5.7	—	—
Xylitol	4.7	—	—
Hormones, mg/L			
NAA[b]	0.1	0.1	—
BAP[b]	0.4	—	—
2,4-D[b]	1.0	1.0	1.0
Kinetin	—	—	0.1
Agarose (g/L)			
Type VII (Sigma)	0.6	4.5	0.6

[a]From Shepard *(31)* , modified table reprinted by permission from The University of Minnesota Press.

[b]Abbreviations: NAA—1-napthaleneacetic acid; BAP—6-benzylaminopurine; 2,4-D—2,4-dichlorophenoxyacetic acid.

ate agarose (sterilized by autoclaving at 120°C for 20 min)—final agarose strength of 1.2 g/L. Maintained at 40°C.

b. Reservoir (R) and differentiation (DM) media (***31*** and *see* **Table 1**).

c. Proliferation medium (PM): MS medium (***30*** and *see* Appendix) containing 10 g/L sucrose, 1.0 mg/L 2,4-D†, 0.1 mg/L kinetin, 4.0 g/L agarose (Sigma Type I).

d. Regeneration medium (RM1): MS medium (***30*** and *see* Appendix) containing 2.0 g/L sucrose, 2.0 mg/L zeatin, 2.0 mg/L kinetin, 0.1 mg/L IAA†, 4.0 g/L agarose (Sigma Type I).

e. Regeneration medium (RM2): same as RM1 medium, except the sucrose is reduced to 1.0 g/L, and the zeatin and IAA† are each increased to 3.0 mg/L.

f. SP (shoot proliferation) medium: Gamborg's B5 medium (***32*** and *see* Appendix) containing 2.0 g/L sucrose, 0.03 mg/L GA_3† and 4.0 g/L agarose (Sigma Type I).

With the exception of the Cl medium, all media are made up by mixing equal volumes of:

a. Double-strength components, without agarose, filter-sterilized with 0.2-μm Nalgene filter with

b. Molten double-strength agarose, sterilized by autoclaving at 120°C for 20 min.

All media are adjusted to pH 5.6.

2. Preparation of R plates: Sterile deep Petri plates (100 × 25 mm) are filled with 70 mL of molten R medium. On setting, an X is cut into the medium, reaching right to the edges of the plate. Two diametrically opposite quadrants are then removed with a sterile spatula to complete the R plate preparation. Twenty or so R plates should be prepared in advance. Unused plates can be stored at 4°C in the dark, for up to 4 wk.
3. Preparation of RM1, RM2, and SP media plates: RM1, RM2, and SP media are dispensed into sterile deep Petri plates (100 × 25 mm), 70 mL/plate. Plates can be stored at 4°C in the dark for several weeks.

2.6. Triazine Tolerance Test

Triazine paste: A paste of the triazine herbicide Atrazine WP (80%; Ciba Giegy) is prepared by mixing 1.47 g of Atrazine in 100 mL water.

3. Methods

All procedures must be performed aseptically, within a laminar flow sterile air cabinet.

3.1. Production of Etiolated Hypocotyls

1. Seed of CMS-rapeseed are surface-sterilized by immersion in 10% (v/v) commercial hypochlorite solution (0.6% free chlorine) for 10 min and then rinsed for 5 min

---

†Abbreviations: 2,4-D- 2,4-dichlorophenoxyacetic acid; IAA- 3-indoleacetic acid; GA_3-gibberellic acid.

in sterile water. These procedures are best performed by placing the seed in presterilized sieves, for ease of handling and immersion into the different solutions.

2. For some batches of seed, this procedure may be ineffective. In these cases, seeds are surface-sterilized by immersion in a solution of 0.05% mercuric chloride (taking particular care in the handling of this substance) and 0.1% sodium dodecyl sulfate, for 10 min, and then rinsed six times in sterile water.
3. Seeds are transferred to paper towels to remove excess moisture and then onto MSS medium (approx 20 seeds/Petri plate).
4. Plates are incubated in the dark at constant temperature (25°C) for germination. Typically, 20–30 plates of CMS-rapeseed seed are required/fusion operation.

3.2. Isolation of Protoplasts

3.2.1. Mesophyll Protoplasts

1. Leaves from 3-wk-old CTT plants are surface-sterilized by immersing in 10% (v/v) commercial hypochlorite solution for 3 min, and rinsed once in sterile water and once again, for 30 s, in 70% ethanol.
2. The leaves are placed onto paper towels to remove excess moisture, care being taken not to dry the leaves completely to the point of wilting.
3. The lower epidermis of the leaves are abraded by gentle stroking with the paintbrush until bristle marks are just visible. Excessive brushing should be avoided to prevent damage to the subepidermal cells that could reduce protoplast yields.
4. Leaves are cut into 1-cm wide strips and transferred to the soak solution. Approximately 3.0 g of tissue (around 6–9 leaves) are immersed in each flask of soak (100 mL of soak/flask). Typically, two flasks of material are prepared for each individual fusion operation.
5. The flasks of leaf material are then incubated for 2–4 h in the dark, at 4°C. Longer incubations, of up to 12 h, are possible without undue harm.
6. Following cold incubation, the soak solution is carefully poured off and replaced with $^{1}/_{10}$ strength enzyme solution. Typically, leaf material is prepared early in the afternoon and the soak solution substituted with enzyme solution in the early evening.
7. Flasks are incubated for 15–18 h (i.e., overnight) at 25°C in the dark, on a rotary shaker, set to give a very gentle agitation. Successful enzyme digestion is evident when the leaf pieces are floating and have a translucent appearance. However, leaves that have sunk and/or appear undigested can still yield protoplasts, but the numbers are generally reduced.

   From this stage, all manipulations must be performed particularly carefully to avoid damaging the fragile protoplasts.
8. The contents of each flask is poured through a filtering funnel, into two babcock bottles. A retort stand is used to hold the funnel. Gentle agitation with a Pasteur pipet will assist the liberation of protoplasts.
9. The babcock bottles are then topped up with rinse solution and recapped with sterile foil.

10. Protoplasts will float to the top of the bottle following centrifugation at 500 rpm (350*g*) for 10 min (in a bench model centrifuge with babcock bottle-size swing-out buckets).
11. Protoplasts are "rinsed" free of enzyme solution by transferring them with a Pasteur pipet to a babcock bottle previously half filled with rinse solution, thereby combining into one the protoplasts from two babcocks.
12. The bottle is then filled completely with rinse solution, resuspending the protoplasts, recapped, and centrifuged once again, as described before.

3.2.2. Hypocotyl Protoplasts

1. Hypocotyls excised from etiolated CMS rapeseed seedlings (3–5 d after sowing) are chopped on a porcelain tile into 2–5 mm transverse segments. These are immediately placed into full-strength enzyme solution. Generally, 200–300 chopped hypocotyls are added to each flask filled with 100 mL of enzyme solution.
2. The hypocotyl material is incubated overnight (25°C in the dark, on a rotary shaker, set to give a very gentle agitation), and protoplasts isolated as described for mesophyll CTT material (*see* **Note 3**).

3.3. Prefusion Treatment of Protoplasts

3.3.1. Donor Parent

Nuclear inactivation of mesophyll CTT protoplasts can be by either γ-ray or X-ray irradiation. Whichever is used, it is necessary to establish the dosage and check the protoplast response in control cultures. Irradiation should produce protoplasts that do not divide, but remain "dormant" for several days, under normal culture conditions, before eventually dying. However, a low frequency of first divisions is acceptable, provided there is no colony formation. Irradiation is performed immediately after protoplast isolation in the babcock bottle (*see* **Note 4**).

3.3.2. Recipient Parent

The effective dosage level of IOA must be established prior to fusion experiments. The IOA concentration should be sufficient to kill the protoplasts, under normal culture conditions, within 24 h. For *B. napus* rapeseed, 3–4 m*M* IOA are typical.

1. Following isolation, CMS protoplasts are carefully transferred to a sterile centrifuge tube, of 10- to 25-mL capacity.
2. The tube is then filled with the diluted IOA solution, resuspending the protoplasts.
3. Following 10 min of incubation at room temperature, the tube is centrifuged at 350*g* for 10 min.
4. The IOA solution is removed by two consecutive centrifugations, each time transferring the floating protoplasts to separate babcock bottles filled with rinse solution. It is typical for about 25% of the protoplasts to be lost as a result of the centrifugation steps.

3.4. Protoplast Fusion

1. Protoplasts of both donor and recipient parents are carefully combined together in a sterile round-bottom tube (13 x 100 mm). Actual numbers of protoplasts from each parent will vary widely with each fusion owing to yield variations of different tissue samples and through varying degrees of cell death caused by centrifugation. Ratios of the CTT and CMS parents should be in the 1:1–1:4 range. Some adjustments may be necessary.
2. Prior to initiating the fusion process, approx $^1/_5$ of the mixed protoplast suspension is removed and transferred to a separate tube. These will constitute the "unfused mixed control" (UFMC) to be cultured separately *(see below)*.
3. The volume of the remaining protoplast suspension should be adjusted to 0.5–1.0 mL by the addition of extra rinse solution, or by centrifuging the tube and removing excess rinse solution from underneath the floating protoplasts.
4. To this is added an equal volume (0.5–1.0 mL) of PEG solution, followed by the gentle resuspension of the protoplasts by careful rotations of the tube for approx 30 s.
5. After 10 min of incubation at room temperature, 2 mL of Ca^{2+} solution are gradually added, further resuspending the protoplasts.
6. The tube is then filled with further additions of 2 mL of Ca^{2+} solution, with careful mixing each time, until the tube is full, and centrifuged at 400 rpm for 10 min.
7. Floating protoplasts are carefully removed and resuspended in rinse solution contained in a babcock bottle.
8. After a final centrifugation at 350*g* for 10 min, the fusion process is complete.

3.5. Protoplast Culture

Fusion protoplasts and the UFMC are plated in nurse culture plates prepared 24 h in advance.

3.5.1. Nurse Culture Preparation

1. *Nicotiana tabacum* (tobacco) plants are grown from seed under the same conditions as described **Subheading 2.1.** Conditions for the preparation of the leaf material and the isolation of protoplasts are the same as described in **Subheading 3.2.1.** Protoplasts are mitotically inactivated by irradiation as described in **Subheading 3.3.1.** (*see* **Note 5**).
2. Irradiated tobacco protoplasts are then transferred from the rinse solution to several milliliters of "liquid" CL medium and resuspended.
3. With a hemocytometer, the number of cells in the suspension is calculated from a small sample.
4. Using further volumes of "liquid" CL medium, the density of the protoplast suspension is adjusted to 2×10^5 protoplasts/mL.
5. Next, the protoplast suspension in "liquid" Cl medium is gently mixed with equal volume of molten "semisolid" Cl medium (containing double-strength agarose, at 40°C). Four milliliters of the adjusted protoplast suspension (in the "semi-

solid" Cl medium now with single strength agarose) is transferred to each empty quadrant in the R medium plates. Ten to 12 tobacco-filled nurse plates should be sufficient for each fusion and its corresponding UFMC.
6. Completed nurse plates are incubated at 25°C in the dark.

3.5.2. Culture of Fused and UFMC Protoplasts

1. Protoplasts from fusion and UFMC samples are separately resuspended in a few milliliters of "liquid" CL medium.
2. With the aid of a hemocytometer, the number of cells in the suspension is calculated and adjusted with further volumes of "liquid" CL medium to 10^5 protoplasts/mL.
3. One milliliter of the post-fusion suspension is then gently mixed into each nurse quadrant, giving a final *Brassica* protoplast density of 2×10^4 protoplasts/mL. The UFMC suspension is dispensed similarly.
4. Fusion and UFMC plates are then sealed with parafilm and incubated at 25°C in the dark.
5. One week to 10 d after plating, suspensions are transferred to regular Petri plates (contents of one quadrant/100 × 15 mm plate) and diluted with an equal volume of molten DM medium precooled to 40°C. The combined suspension and DM medium should be spread evenly over the bottom of the Petri plates to form a thin layer for maximum colony formation.

Subsequent culture conditions are changed to 25°C with a 16-h photoperiod of 150–200 μmol/m²/s of photosynthetically active radiation.

3.6. Callus Proliferation and Plant Regeneration

From 10–14 d after dilution, colonies in the fusion plates will have grown to approx 0.5–1.0 mm in size. Ideally, there should be no colony development in the UFMC plates, since the irradiation and IOA treatments should have inactivated or killed the unfused protoplasts (*see* **Note 5**, re. nurse controls). However, it is likely that some colonies will develop from the irradiated population, since the irradiation treatment is often not 100% effective. If the number of colonies in the UFMC plates approaches the colony count in the fusion plates, then the fusion must be discarded and the inactivation treatment checked and re-established.

1. Assuming there is no colony formation in the UFMC plates or at least considerably less than the fusion plates, then colonies of 0.5–1.0-mm size from the fusion plates are individually transferred to the surface of PM medium, 35–40 colonies/plate, for further proliferation.
2. Colonies that are 1.0- to 2.0-mm in diameter (usually after 10–14 d at which time they should be colorless or very pale green) are transferred to RM1 medium, 20–25 colonies/plate, to induce differentiation.

3. Within 2–3 wk, it is likely that some of the colonies will have regenerated small shoots. These colonies must then be transferred to SP medium to induce shoot proliferation and growout.
4. The remaining colonies are transferred to RM2 medium to further induce differentiation.
5. Any shoots developing on the RM2 medium should also be transferred to SP medium. It is unlikely that more than 1–5% of the colonies will respond on the differentiation medium, since the various treatments and fusion processes typically reduce the regeneration response considerably.
6. Shoots that have clearly differentiated meristems are transferred to peat pellets ("Jiffy 7," Jiffy Products [N.B.] Ltd., Shippegan, Canada; sterilized by autoclaving) to induce *root formation*.
7. Pellets are placed in sterile glass jars to maintain a high humidity. Peat pellets should be moistened with filter-sterilized fertilizer (*see* **Subheading 2.1., item 1**). To prevent premature flowering, the photoperiod is reduced to 10 h light/14 h dark.
8. When several roots are protruding from the sides of the peat pellets, the plantlets are transferred to potting compost for further plant development.
9. Regenerated plants are maintained at 23°C with a 16-h photoperiod of 160 μmol/m^2/s of photosynthetically active radiation to induce flowering.

3.7. Identification of Cybrids

1. On flowering, the flower morphology of the regenerants is examined for male sterility (absence of pollen production), which indicates the presence of CMS mitochondria.
2. Plants are then tested for their sensitivity to triazine herbicides. Two to three mature leaves still attached to each plant are painted with the atrazine paste. Those plants carrying CTT chloroplasts should remain undamaged. The other plants will suffer yellowing of the leaves and may be dead within 7–14 d.

Those plants that are male sterile and are tolerant to atrazine are the desired cybrids (*see* **Note 6**). Male fertile, atrazine-susceptible plants are also cybrids, but with the opposite organelle combination, so-called reverse-cybrids. In addition, male fertile, atrazine tolerant plants can also be considered cybrids, since by alloplasmic substitution, they possess the nucleus of the original CMS parent. However, if the original parents had identical nuclei, e.g., if they were both versions of the same rapeseed variety, then these plants cannot be considered as cybrids.

Male sterility sometimes appears in regenerants that have aneuploid chromosome complements, independent of the mitochondria. The genotype of the mitochondrial population can be distinguished by "restriction fragment-length polymorphism" (RFLP) analysis of the mitochondrial DNA. Plastomes can be similarly identified. The techniques involved are beyond the scope of this chapter, and the reader is referred to Kemble ***(33)***.

4. Notes

1. The described procedure can produce cybrids in 8–12 mo from sowing the seed of the protoplast source material to identifying the cybrids in the whole-plant regenerants. This time period predominantly depends on the rapidity of shoot differentiation. This in turn depends on the careful timing of each transfer onto the different culture media.
2. These procedures can be utilized with different parents to produce rapeseed cybrids with other desired organelle combinations. For example, similar techniques have transferred the CMS trait from spring-planted rapeseed to winter forms, producing cybrids that can be useful to single-cross-hybrid winter rapeseed production program *(34)*. In this case, conventional crossing also could have produced CMS winter lines, but this would have taken up to 3 yr. Fusion accomplished the same goal in just 1 yr, a significant saving of time.
3. In some circumstances, it may be necessary to use hypocotyl protoplasts as the "donor" parent when attempting to produce a particular rapeseed cybrid. Hypocotyl protoplasts will require higher dosages of irradiation for inactivation, since their nuclei are more resistant to radiation damage than their mesophyll counterparts.
4. When irradiating donor parent protoplasts prior to fusion, the exposure period should not be longer than 1–2 h to prevent undue cell-wall regeneration that can inhibit fusion. If the exposure period has to be longer, owing to the nature of the irradiation equipment, then the problem can be circumvented by directly irradiating the leaves instead, prior to preparations for protoplast isolation. Indeed, this may be more convenient, both for inactivating the donor parent and the nurse material.
5. The effectiveness of the mitotic inactivation may vary from time to time, both for the *Brassica* donor parent and for the tobacco nurse cultures. Therefore, nurse control cultures should be prepared, in addition to the UFMC controls. Instead of adding 1 mL of fusion or UFMC protoplasts, add 1 mL of protoplast-free "liquid" Cl medium to each nurse cell quadrant. If colonies appear in the nurse controls, then the fusion must be discarded, and the inactivation treatment checked and re-established.
6. **Figure 1** suggests that one-quarter of the regenerants will be the desired cybrids, which indeed there would be if there was random segregation of the organelles following cell divisions of the fusion product cells. However, the proportions of the different segregants is, in practice, usually not equal because of differences in protoplast tissue sources, differential responses in culture, and adverse effects of the cell inactivation processes.

References

1. Beversdorf, W. D., Erickson, L. R., and Grant, I (1985) Hybridization process utilizing a combination of cytoplasmic male sterility and herbicide tolerance. US Patent No. 4517763.
2. Renard, M., Delourme, R., Pelletier, G., Primard, C., Mesquida, J., Darrozes, G., and Morice, J. (1991) Un systeme de production de semences hybrides F1 utilisant

une sterilite male cytoplasmique. *Comptes Rendus de l'Academie d'Agriculture de France* **77,** 49–58.

3. Yarrow, S. A., Burnett, L., Wildeman, R. P., and Kemble, R. J. (1990). The transfer of "Polima" Cytoplasmic Male Sterility from oilseed rape *(Brassica napus)* to broccoli *(B. oleracea)* by protoplast fusion. *Plant Cell Rep.* **9,** 185–188.
4. Akagi, H., Taguchi, T., and Fujimura, T. (1995) Stable inheritance and expression of the CMS traits introduced by asymmetric protoplast fusion. *Theor. Appl. Genet.* **91,** 563–567.
5. Suenaga, L. (1991) Basic studies on transfer of cytoplasmic male sterility by means of cytoplasmic hybridization in carrot (*Daucus carota* L.). *J. Faculty Agriculture Hokkaido Univ. Japan* **65,** 62–118.
6. Creemers-Molenaar, J., Hall, R. D., and Krens, F. A. (1992) Asymmetric protoplast fusion aimed at intraspecific transfer of cytoplasmic male sterility (CMS) in *Lolium perenne* L Theor. *Appl. Genet.* **84,** 763–770.
7. Galun, E., Aviv, D., and Perl, A. (1991) Protoplast fusion for potato-cybrid formation. Report of the planning conference on application of molecular techniques to potato germplasm enhancement, International Potato Center, pp. 105–111.
8. Kirti, P. B. , Banga, S. S., Prakash, S., and Chopra, V. L. (1995) Transfer of Ogu cytoplasmic male sterility to *Brassica juncea* and improvement of the male sterile line through somatic cell fusion. *Theor. Appl. Genet.* **91,** 517–521
9. Yarrow, S. A., Wu, S. C., Barsby, T. L., Kemble, R. J., and Shepard, J. F. (1986) The introduction of CMS mitochondria to triazine tolerant *Brassica napus* L., var. "Regent," by micromanipulation of individual heterokaryons. *Plant Cell Rep.* **5,** 415–418.
10. Barsby, T. L., Chuong, P. V., Yarrow, S. A., Wu, S. C., Coumans, M., and Kemble, R. J. (1987) The combination of Polima cms and cytoplasmic triazine resistance in *Brassica napus. Theor. Appl. Genet.* **73,** 809–814.
11. Sidorov, V. A., Yevtushenko, D. P., Shakhovsky, A. M., Gleba, Y. Y., and Gleba, Yu-Yu (1994) Cybrid production based on mutagenic inactivation of protoplasts and rescuing of mutant plastids in fusion products: potato with a plastome from *S. bulbocastanum* and *S. pinnatisectum. Theor. Appl. Genet.* **88,** 525–529.
12. Grosser, J. W., Gmitter, F. G., Jr., Tusa, N., Recupero, G. R., and Cucinotta, P. (1996) Further evidence of a cybridization requirement for plant regeneration from citrus leaf protoplasts following somatic fusion. *Plant Cell Rep.* **15,** 672–676.
13. Bonnema, A. B., Melzer, J. M., and O'Connell, M. A. (1991) Tomato cybrids with mitochondrial DNA from *Lycopersicon pennellii. Theor. Appl. Genet.* **81,** 339–348.
14. Djurberg, I., Kofer, W., Hakansson, G., and Glimelius, K. (1991) Exchange of chloroplasts in a *Nicotiana tabacum* cybrid with *Petunia* chloroplasts in order to establish that a variegated and abnormal development of the cybrid is caused by a mitochondrial mutation. *Physiol. Plant.* **82,** A28.
15. Pelletier, G. and Romagosa, I. (1993) Somatic hybridization, in *Plant Breeding: Principles and Prospects.* Plant Breeding Series 1. (Hayward, M. D. and Bosemark, N. O., eds.), Chapman and Hall, London, pp. 93–106.

16. Kumar, A. and Cocking, E. C. (1987) Protoplast fusion: A novel approach to organelle genetics in higher plants. *Am. J. Bot.* **74,** 1289–1303.
17. Zelcer, A., Aviv, D., and Galun, E. (1978) Interspecific transfer of cytoplasmic male sterility by fusion between protoplasts of normal *Nicotiona sylvestris* and X-ray irradiated protoplasts of male sterile *N. tabacum. Z. Pflanzenphysiol.* **90,** 397–407.
18. Aviv, D., Arzee-Gonen, P., Bleichman, S., and Galun, E. (1984) Novel alloplasmic plants by "donor-recipient" protoplast fusion: cybrids having *N. tabacum* or *N. sylvestris* nuclear genomes and either or both plastomes and chondriomes from alien species. *Mol. Gen. Genet.* **196,** 244–253.
19. Sidorov, V. A., Menczel, L., Nagy, F., and Maliga, P. (1981) Chloroplast transfer in *Nicotiana* based on metabolic complementation between irradiated and iodoacetate treated protoplasts. *Planta* **152,** 341–345.
20. Ichikawa, H., Tanno-Suenaga, L., and Imamura, J. (1987) Selection of *Daucus* cybrids based on metabolic complementation between X-irradiated *D. capillifolius* and iodoacetamide-treated *D. carota* by somatic cell fusion. *Theor. Appl. Genet.* **74,** 746–752.
21. Sakai,T., Liu, H. J., Iwabuchi, M., Kohno-Murase, J., and Imamura, J. (1996) Introduction of a gene from fertility restored radish *(Raphanus sativus)* into *Brassica napus* by fusion of X-irradiated protoplasts from a radish restorer line and iodacetoamide-treated protoplasts from a cytoplasmic male-sterile cybrid of *B. napus. Theor. Appl. Genet.* **93,** 373–379.
22. Reith, M. and Straus, N. A. (1987) Nucleotide sequence of the chloroplast gene responsible for triazine resistance in canola. *Theor. Appl. Genet.* **73,** 357–363.
23. Pelletier, G., Primard, C., Vedel, F., Chetrit, P., Remy, R., Rouselle, P., and Renard, M. (1983) Intergeneric cytoplasmic hybridization in Cruciferae by protoplast fusion. *Mol. Gen. Genet.* **191,** 244–250.
24. Chetrit, P., Mathieu, C., Vedel, F., Pelletier, G., and Primard, C. (1985) Mitochondrial DNA polymorphism induced by protoplast fusion in *Cruciferae. Theor. Appl. Genet.* **69,** 361–366.
25. Beversdorf, W. D., Weiss-Lerman, J., Erickson, L. R., and Souza-Machado, Z. (1980) Transfer of cytoplasmic inherited triazine resistance from bird's rape to cultivated oilseed rape (*Brassica campestris* and *B. napus*). *Can. J. Genet. Cytol.* **22,** 167–172.
26. Kao, K. N. and Michayluk, M. R. (1974) A method for high frequency intergeneric fusion of plant protoplasts. *Planta* **115,** 355–367.
27. Wallin, A. K., Glimelius, K., and Eriksson, T. (1974) The induction of aggregation and fusion of *Daucus carota* protoplasts by polyethylene glycol. *Z. Pflanzenphysiol.* **74,** 64–80.
28. Keller, W. A. and Melchers, G. (1973) The effect of high pH and calcium on Tobacco leaf protoplast fusion. *Z. Naturforsch.* **28c,** 737–741.
29. Barsby, T. L., Yarrow, S. A., and Shepard, J. F. (1986) A rapid and efficient alternative procedure for the regeneration of plants from hypocotyl protoplasts of *Brassica napus. Plant Cell Rep.* **5,** 101–103.

30. Murashige, T. and Skoog, F. (1962) A revised medium for rapid growth and bioassays with tobacco tissue cultures. *Physiol. Plant.* **15,** 473–497.
31. Shepard, J. F. (1980) Mutant selection and plant regeneration from Potato mesophyll protoplasts, in *Genetic Improvement of Crops: Emergent Techniques* (Rubenstein, I., Gengenbach, B., Philips, R. L. , and Green, C. E., eds.), University Minnesota Press, pp. 185–219.
32. Gamborg, O. L., Miller, R. A., and Ojima, K. (1968) Nutrient requirements of suspension cultures of soybean root cells. *Exp. Cell Res.* **50,** 151–158.
33. Kemble, R. J. (1987) A rapid, single leaf, nucleic acid assay for determining the cytoplasmic organelle complement of rapeseed and related *Brassica* species. *Theor. Appl. Genet.* **73,** 364–370.
34. Barsby, T. L., Yarrow, S. A., Kemble, R. J., and Grant, I. (1987) The transfer of cytoplasmic male sterility to winter-type oilseed rape (*Brassica napus* L.) by protoplast fusion. *Plant Sci.* **53,** 243–248.

21

Microprotoplast-Mediated Chromosome Transfer (MMCT) for the Direct Production of Monosomic Addition Lines

Kamisetti S. Ramulu, Paul Dijkhuis, Jan Blaas, Frans A. Krens, and Harrie A. Verhoeven

1. Introduction

The transfer of single chromosomes carrying important genes between related, but sexually incongruent species, and the production of monosomic addition plants, can speed up gene introgression through homoelogous recombination or other mechanisms of gene transfer *(1,2)*. Monosomic addition lines form the most important material for the transfer of desirable alien genes from a wild donor species to a cultivated species. Because of sexual incongruity between the wild species and the cultivated species, the demands of breeders for the transfer of desirable traits, such as disease or stress resistance and apomixis, are insufficiently met by conventional breeding methods. DNA transformation using the isolated, cloned genes makes it feasible to transfer genes across sexual barriers or taxonomic boundaries. However, several of the agronomically important traits are encoded by polygenes, which are clustered within blocks on specific chromosomes or scattered throughout the genome, and therefore they are not yet amenable to this technique. In addition to sexual incongruity, and the possible gene clustering for a locus (e.g., controlling apomictic reproduction in maize-*Tripsacum* backcross progeny), the occurrence of male sterility, poor seed set, and the low frequency of desired traits hinder the transfer of economically important genes. Also, by backcrossing, it is often difficult to eliminate the recombined undesirable donor genes and prevent linkage drag "hitchhiking" genes *(1)*. On the other hand, distant hybrids that cannot be produced by conventional sexual crosses, can be obtained by symmetric somatic hybridization *(3)*. Thus far, the hybrids obtained after

From: *Methods in Molecular Biology, Vol. 111: Plant Cell Culture Protocols*
Edited by: R. D. Hall © Humana Press Inc., Totowa, NJ

asymmetric hybridization were genetically complex, with several donor chromosomes and unwanted genes, and were often sterile ***(4–7)***. Recently, we found that microprotoplast-mediated chromosome transfer (MMCT) is an efficient technique for transferring single chromosomes from one species to another ***(8,9)***. The establishment of procedures for the isolation of small subdiploid microprotoplasts and the fusion of these with recipient protoplasts enabled the development of MMCT technology ***(10,11)***. Using the MMCT technique, we have transferred single chromosomes of potato carrying the selectable marker gene for kanamycin resistance *(nptII)* and the reporter gene β- glucuronidase (*uid*A) to tomato *(Lycopersicon peruvianum)* and tobacco. Several monosomic addition hybrid plants could be regenerated within a short period of 3–4 mo after polyethylene glycol-induced mass fusion between donor potato microprotoplasts containing one or a few chromosomes and normal recipient *L. peruvianum* or tobacco protoplasts. These plants contained one potato chromosome carrying a single copy of *uid*A and one or two copies of the *neomycin phosphotransferase (nptII)* gene conferring kanamycin resistance, together with the complete set of chromosomes of the recipient tomato or tobacco ***(8,9)***. The monosomic addition plants were phenotypically normal, resembling the recipient lines. Genomic *in situ* hybridization (GISH) and RFLP analysis using chromosome-specific markers revealed that the monosomic addition plants contained potato chromosome 3 or 5, harboring the *nptII* and *uid*A genes ***(12)***. The alien genes *nptII* and *uid*A were stably expressed in both the tobacco and tomato backgrounds. RFLP analysis and GISH confirmed sexual transmission of the potato chromosome to the BC1 progeny plants. A few BC1 plants contained the *nptII* and *uid*A genes in the absence of the additional potato chromosome, indicating that the marker genes had been integrated into the tomato genome. The direct production of monosomic additions and the occurrence of donor DNA integration achieved through MMCT in sexually nonhybridizing, but related species, will be potentially useful for the transfer of economically important traits, introgressive breeding and for analyzing plant genome evolution ***(1,2,13,14)***. The intergenomic monosomic additions are useful tools for the construction of chromosome-specific DNA libraries ***(15)***, and for investigating the regulation of gene expression related to chromosome-gene domains ***(14,16)***. In view of these perspectives, we have described below in detail the essential requirements of the MMCT technique, as established in our investigations:

1. Induction of micronuclei at high frequencies in donor species.
2. Isolation of micronucleate protoplasts in donor species.
3. Isolation and enrichment of small subdiploid microprotoplasts in donor species.
4. Efficient fusion of the donor microprotoplasts with normal recipient protoplasts.
5. Selection of fusion products.
6. Plant regeneration from fusion products and the production of monosomic addition hybrids.

2. Materials

1. Donor species: We use a suspension cell culture of transformed potato (*Solanum tuberosum,* line 413, $2n = 3x = 36$) carrying the *nptII* and *uid*A genes ***(17,18)*** (*see* **Note 1**).
2. Recipient plant species, e.g., shoot cultures of the wild tomato species *L. peruvianum* (Hygromycin-resistant line PI 128650, $2n = 2x = 24$) ***(19)*** or tobacco (*Nicotiana tabacum,* cv. Petit Havana, streptomycin-resistant line SR 1, $2n = 4x = 48$) ***(20)***.
3. Spindle toxins: 32 μ*M* amiprophos-methyl (APM) (Tokunol M: *O*-methyl-*O*-*O*[4-methyl-6-nitrophenyl]-*N*-isopropyl-phosphoro thioamidate; Bayer Mobay Corporation, Agricultural Chemical Division, Kansas City, MO) is applied to the cell suspension culture from a stock solution (20 mg/mL) in water-free DMSO, and 7.5 μ*M* cremart (CR) (Butamiphos, *O*-ethyl-*O*-[3-methyl-6-nitrophenyl]*N*-sec-butylphosphorothioamidate; Sumitomo Chemical Company, Osaka, Japan) from a stock solution (10 mg/mL) in water-free dimethyl sulfoxide (DMSO) ***(21,22)***.
4. Inhibitors of DNA synthesis: 10 m*M* hydroxy urea (HU) is applied to the cell suspension culture as a concentrated, freshly prepared solution in the culture medium (40 mg/mL), and 15 μ*M* aphidicolin (APH) from a stock solution (20 mg/mL) in (DMSO) (both from Sigma Chemical Company, St. Louis, MO) ***(10,21)***.
5. Cell-wall-digesting enzyme mixture: cellulase Onozuka-R10 (1%), macerozyme Onozuka-R 10 (0.2%) (both Yakult Honsha Co., Tokyo, Japan), half-strength V-KM medium (*see* **Table 1**; ***23***) with 0.2 *M* glucose and 0.2 *M* mannitol, but no hormones ***(11)***. Adjust the osmolality to 500. m/sM/kg and the pH to 5.6. Centrifuge the mixture for 3 min at 3000 rpm and filter-sterilize. The mixture can be stored for several months at –20°C.
6. Cytochalasin-B (CB) (Sigma Chemical Company) is applied at 20 μ*M* from a stock solution (20 mg/mL) in DMSO.
7. Christ Omega ultracentrifuge with swing out rotor for tubes of about 6 mL, capable of producing at least $10^5 \times g$.
8. Percoll solution (Sigma).
9. Nylon sieves of decreasing pore size 48, 20, 15, 10, and 5 μm (Nybott; Swiss Silk Bolting Cloth Mfg. Co., Zurich, Switzerland).
10. W5 medium (*see* **Table 2**) is used for mixing of microprotoplasts with the recipient protoplasts for PEG mass fusion.
11. PEG stock solution for mass fusion is prepared as follows. PEG stock solution: 30% w/v PEG 4000, 0.3 *M* glucose, 66 m*M* $CaCl_2 \cdot 2H_2O$.
12. High-pH buffer is prepared from two stock solutions, stock A and stock B: stock A: 0.4 *M* glucose, 66 m*M* $CaCl_2 \cdot 2H_2O$, 10% DMSO; stock B: 0.3 *M* glycine, pH 10.5. High-pH buffer: Mix stock A and stock B in the ratio of 9:1 to prepare the high-pH buffer solution just before use.
13. Callus induction (CI) media: TM-2 for tomato ***(24,25)*** and H-460M for tobacco ***(26)*** (*see* **Tables 3** and **4**).
14. CI liquid media for selecting fusion products: TM-2 medium (**Table 3**) supplemented with 50 mg/L kanamycin (Kan 50) prepared from a stock solution (50 mg/L) in H_2O + 25 mg/L hygromycin (Hyg 25) prepared from a stock solution (25 mg/L) in H_2O is used for tomato ***(18)***. In the case of tobacco, H-460M medium

Table 1
Composition of V-KM Medium *(23)*

Component	Quantity (mg/L)	Component	Quantity (mg/L)
Macroelements		Vitamins *(cont)*	
KNO_3	740.0	Ascorbic acid	1.0
$MgSO_4 \cdot 7H_2O$	492.0	*p*-Aminobenzoic acid	0.01
KH_2PO_4	34.0	Nicotinic acid	0.5
$FeSO_4 \cdot 7H_2O$	14.0	Pyridoxine-HCl	0.5
Na_2 EDTA $\cdot 2H_2O$	19.0	Thiamine-HCl	5.0
$CaCl_2 \cdot 2H_2O$	368.0	Folic acid	0.2
Microelements		Biotin	0.005
H_3BO_3	1.5	Vitamin A	0.005
$MnSO_4 \cdot H_2O$	5.0	Vitamin D_3	0.005
$ZnSO_4 \cdot 7H_2O$	1.0	Vitamin B_{12}	0.01
$Na_2MoO_4 \cdot 2H_2O$	0.12	Sugars	
$CuSO_4 \cdot 5H_2O$	0.012	Mannitol	125.0
$CoCl_2 \cdot 6H_2O$	0.012	Sorbitol	125.0
KI	0.38	Sucrose	125.0
Organic acids		Fructose	125.0
Pyruvic acid	10.0	Ribose	125.0
Fumaric acid	20.0	Xylose	125.0
Citric acid	20.0	Mannose	125.0
Malic acid	20.0	Rhamnose	125.0
Vitamins		Cellobiose	125.0
d-Ca-panthotenate	0.5	Myo-inositol	50.0
Choline	0.5	pH	5.6

Table 2
Composition of W5 Medium

Component	Quantity, g/L
NaCl	9
$CaCl_2 \cdot 2H_2O$	18.3
KCl	0.37
Glucose	0.90
pH	5.8

Table 3
Composition of the Media Used for Tomato

Component	TM-2, mg/L, *(24,25)*	TM-3, mg/L, *(24,25)*	TM-4, mg/L, *(24,25)*
KH_2PO_4	170.0	170.0	—
$CaCl_2 \cdot 2H_2O$	440.0	440.0	150
KNO_3	1500.0	1500.0	1900.0
NH_4NO_3	—	—	320.0
$NH_4H_2PO_4$	—	—	230.0
$(NH_4)_2SO_4$	—	—	134.0
$MgSO_4 \cdot 7H_2O$	370.0	370.0	247.0
KI	0.83	0.83	0.83
H_3BO_3	6.20	6.20	6.20
$MnSO_4 \cdot 4H_2O$	22.30	22.30	22.30
$ZnSO_4 \cdot 7H_2O$	8.60	8.60	8.60
$Na_2MoO_4 \cdot 2H_2O$	0.25	0.25	0.25
$CuSO_4 \cdot 5H_2O$	0.025	0.025	0.025
$CoCl_2 \cdot 6H_2O$	0.025	0.025	0.025
$FeSO_4 \cdot 7H_2O$	13.90	13.90	13.90
Na_2EDTA	18.50	18.50	18.50
Nicotinic acid	2.50	5.0	5.0
Thiamine-HCl	10.0	5.0	5.0
Pyridoxine-HCl	1.0	0.50	0.50
Biotin	0.05	0.05	0.05
D-Ca-pantothenate	0.50	—	—
Choline chloride	0.10	0.10	0.10
Glycine	0.50	2.50	2.50
Casein hydrolysate	150.0	100.0	—
L-Cysteine	1.0	—	—
Malic acid	10.0	—	—
Ascorbic acid	0.50	—	—
Adenine sulfate	40.0	40.0	—
L-Glutamine	100.0	100.0	—
Myo-inositol	4600.0	100.0	100.0
Riboflavine	0.25	—	—
Sucrose	7% w/v	5% w/v	3% w/v
Mannitol	4560.0	—	—
Xylitol	3800.0	—	—
Sorbitol	4560.0	—	—
MES	97.60	97.60	97.60
Agar	—	7% w/v	7% w/v
NAA	1.0	—	—
Zeatin	1.0	—	2.0
2,4-D	—	0.2	—
BAP	—	0.5	—
GA3	—	—	0.2
pH	5.6	5.8	5.8

Table 4
Composition of the Media Used for Tobacco

Component	H-460M mg/L *(26)*	AG mg/L *(18)*	MSR-1 mg/L *(18)*
$NaH_2PO_4 \cdot H_2O$	170.0	—	—
$Ca(H_2PO_4)2 \cdot H_2O$	50.0	—	—
$CaCl_2 \cdot 2H_2O$	900.0	440.0	440.0
KNO_3	2500.0	1000.0	1900.0
NH_4NO_3	250.0	800.0	1650.0
$(NH_4)_2SO_4$	134.0	—	—
$MgSO_4 \cdot 7H_2O$	250.0	740.0	370.0
KH_2PO_4	—	136.0	170.0
$(NH_4)_2 \cdot$ succinate	—	100.0	—
H_3BO_3	1.24	0.62	6.2
$MnSO_4 \cdot H_2O$	3.38	1.7	17.0
$ZnSO_4 \cdot 7H_2O$	2.12	1.06	10.6
KI	0.17	0.08	0.8
$Na_2MoO_4 \cdot 2H_2O$	0.05	0.03	0.25
$CuSO_4 \cdot 5H_2O$	0.005	0.003	0.025
$CoCl_2, 6H_2O$	0.005	0.003	0.025
Na_2EDTA	37.30	37.30	37.30
$FeSO_4 \cdot 7H_2O$	27.80	27.80	27.80
Ca-panthotenate	1.0	1.0	1.0
Myo-inositol	100.0	100.0	100.0
Nicotinic acid	1.0	1.0	1.0
Pyridoxine	1.0	1.0	1.0
Thiamine-HCl	5.0	5.0	5.0
Biotin	0.01	0.01	0.01
Casein hydrolysate	125.0	—	—
Fructose	250.0	—	—
Xylose	250.0	—	—
Sorbitol	250.0	—	—
Mannitol	250.0	50 g	—
Glucose $\cdot H_2O$	80 g	—	—
Sucrose	250.0	30 g	30 g
NAA	1.0	0.1	—
BAP	0.2	1.0	0.25
2,4-D	0.1	—	—
Agar	—	8 g	8 g
pH	5.8	5.7	5.8

Table 5
Composition of Rooting Medium for Tomato: MS Medium Supplemented with 2% Sucrose

Component	Quantity mg/L	Quantity Component	mg/L
Macroelements		Vitamins	
$CaCl_2$	332.02	Glycine	2.00
KH_2PO_4	170.00	Myo-inositol	1000.00
KNO_3	1900.00	Nicotinic acid	0.50
$MgSO_4$	180.54	Pyridoxine HCl	0.50
NH_4NO_3	1650.00	Thiamine HCl	0.10 Vitamins
Microelements		Sugars	
$CoCl_2 \cdot 6H_2O$	0.025	Sucrose	20 g
$CuSO_4 \cdot 5H_2O$	0.025	pH	5.8
FeNaEDTA	36.70		
H_3BO_3	6.20		
KI	0.83		
$MnSO_4 \cdot H_2O$	16.90		
$Na_2MoO_4 \cdot 2H_2O$	0.25		
$ZnSO_4 \cdot 7H_2O$	8.60		

supplemented with 50 mg/L kanamycin (Kan 50) prepared from a stock solution (50 mg/L) in H_2O is used ***(18)***.

15. Solid media, i.e., callus growth media (CG) for selecting fusion products: TM-3 medium (**Table 3**) supplemented with 100 mg/L kanamycin (Kan 100) prepared from a stock solution (50 mg/mL) in H_2O + 50 mg/ L hygromycin (HYG 50) prepared from a stock solution (25 mg/mL) in H_2O is used for tomato ***(18)***. In the case of tobacco, AG medium (**Table 4**) supplemented with 100 mg/L kanamycin (Kan 100) prepared from a stock solution of (50 mg/mL) in H_2O is used ***(18)***.
16. Regeneration media: TM-4 medium is used for tomato (**Table 3**), and MSR-1 for tobacco (**Table 4**).
17. Rooting media: MS supplemented with 2% sucrose for potato microprotoplast (+) tomato protoplast fusions (**Table 5**), and MS supplemented with 2% sucrose and indole butyric acid (0.25 mg/L) for potato (+) tobacco fusions. The composition of rooting medium for shoots derived from potato (+) tobacco fusions is the same as above, i.e., MS medium, but is supplemented with indole butyric acid (0.25 mg/L).

3. Methods

3.1. Induction of Micronuclei

1. For species/genotypes, which respond to synchronization, e.g., *Nicotiana plumbaginifolia,* treat early log-phase suspension cells at day 1 after subculture with the inhibitors of DNA synthesis HU at 10 m*M* or APH at 15 μ*M* for 24 h.

2. Wash the suspension cells free of HU or APH with culture medium, and treat with APM at 32 μ*M* for 48 h or CR at 3.7 or 7.5 μ*M* for 48 h. **Figure 1A** shows the various steps and processes involved in micronucleation (*see* **Notes 2** and **3**).
3. For species/genotypes in which spontaneous cell synchrony occurs (e.g., potato line 413), treat early log-phase suspension cells at day 1 after subculture with CR at 3.7 or 7.5 μ*M* or with 32 μ*M* APM.
4. Incubate the APM- or CR-treated suspension cells in the cell-wall-digesting enzyme mixture, supplemented with CB (20 μ*M*) and either APH (32 μ*M*) or CR (7.5 μ*M*) for 16 h in 9-cm Petri dishes containing 1.5-mL packed cell volume and 15-mL enzyme solution on a gyratory shaker (30 rpm) at 28°C (*see* **Notes 4** and **5**).
5. Filter the suspension of protoplasts through 297- and 88-μm nylon meshes, and wash with an equal volume of half-strength V-KM medium, containing only macro- and microelements (**Table 1**) and 0.24 *M* NaCl (pH 5.6). Purify the protoplasts by repeated washing and centrifugation three to four times (*see* **Notes 6** and **7**). Resuspend the pellet in sucrose solution (0.43 *M*), and load onto an already prepared Percoll gradient *(see below)*.

3.2. Isolation of Microprotoplasts

1. Prepare in advance the Percoll gradient by adding 7.2% (w/v) mannitol to the Percoll solution (Sigma), filter-sterilize, and centrifuge for 1 h at $10^5 g$ to produce a continuous iso-osmotic gradient of Percoll. Afterwards, remove the top layer of about 5 mm.
2. Load the purified dense suspension of protoplasts (mono- and micronucleated) onto the continuous iso-osmotic gradient of Percoll (Sigma), and centrifuge at high speed, $10^5 g$ for 2 h in a Christ Omega Ultracentrifuge, using a swing-out rotor. Several bands containing a mixture of evacuolated protoplasts, microprotoplasts of various sizes, and cytoplasts form at various distances from the top of the centrifuge tube. **Figure 1B** gives a schematic representation of various steps for the isolation of microprotoplasts (*see* **Note 5**).

Fig. 1. *(opposite page)* **(A)** Various steps and processes involved in micronucleation. Induction of micronuclei in donor suspension cells after treatment with the inhibitors of DNA synthesis, HU (10 m*M*, 24 h) or APH (15 μ*M*, 24 h) followed by treatment with the spindle toxins APM (32 μ*M*, 48 h) or CR (3.7 or 7.5 μ*M*, 48 h). Afterwards, micronucleated cells are incubated in a cell-wall-digesting mixture, which also contains CB (20 μ*M*) and APH (32 μ*M*) or CR (7.5 μ*M*) for 16 h to isolate micronucleated protoplasts. **(B)** High-speed centrifugation ($10^5 \times g$) of a suspension of micronucleated protoplasts and mononucleate protoplasts for the isolation of individual microprotoplasts (MPPS) and protoplasts (PPS), and enrichment of MPPS by sequential filtration through nylon sieves of decreasing pore size to isolate smaller subdiploid microprotoplasts (reproduced with permission; ***18***).

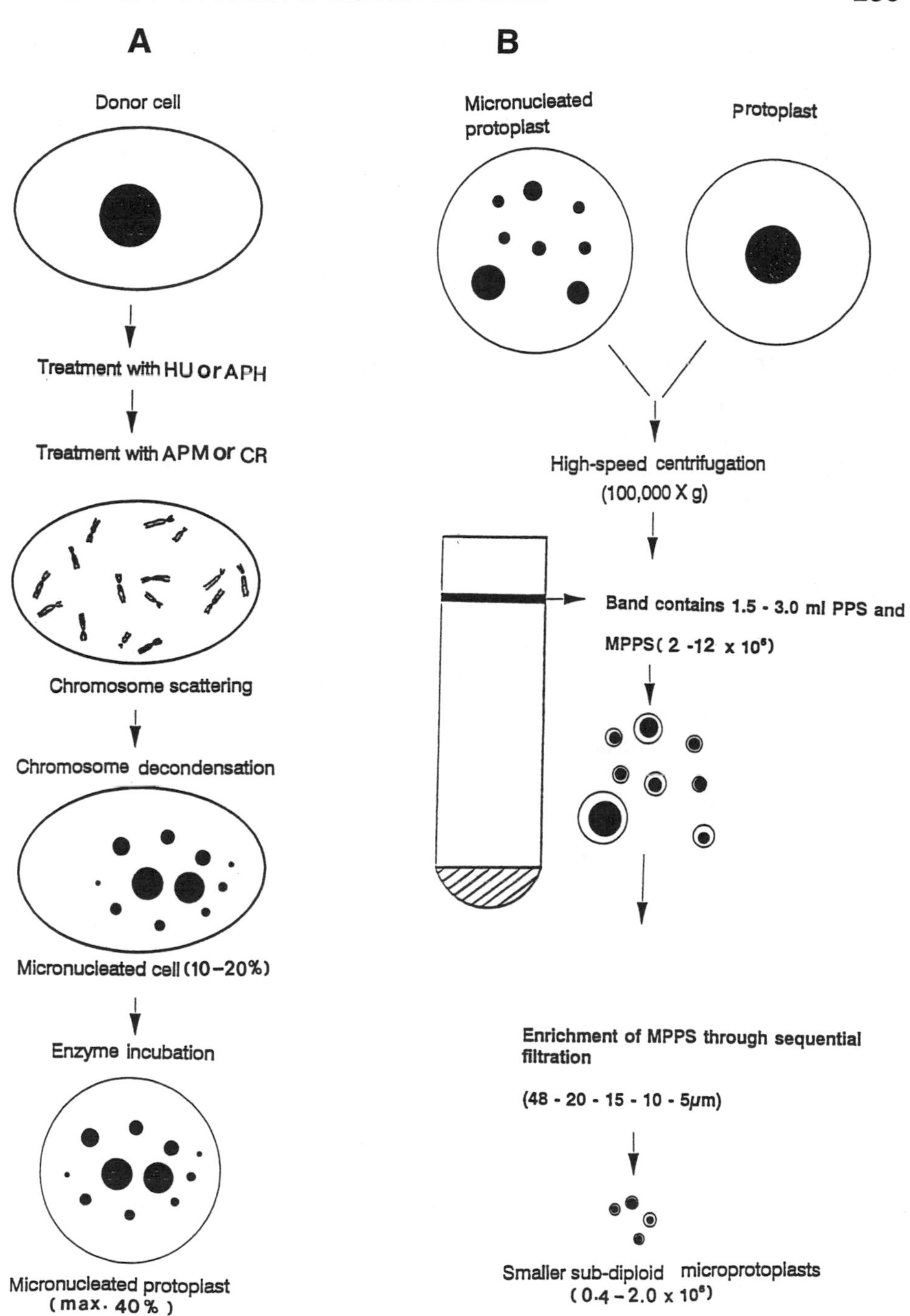
A
Donor cell
Treatment with HU or APH
Treatment with APM or CR
Chromosome scattering
Chromosome decondensation
Micronucleated cell (10–20%)
Enzyme incubation
Micronucleated protoplast
(max. 40%)
B
Micronucleated protoplast
Protoplast
High-speed centrifugation
(100,000 X g)
Band contains 1.5 - 3.0 ml PPS and
MPPS(2 -12 x 10⁶)
Enrichment of MPPS through sequential filtration
(48 - 20 - 15 - 10 - 5μm)
Smaller sub-diploid microprotoplasts
(0.4 – 2.0 x 10⁶)

Fig. 1

3.3. Isolation and Enrichment of Small Subdiploid Microprotoplasts

1. Collect the bands, and filter them sequentially through nylon sieves of decreasing pore size (48, 20, 15, 10, and 5 µm) (*see* **Notes 6** and **7**). Rinse with mannitol (0.4 *M*).
2. Collect the filtered fractions in mannitol (0.4 *M*), and centrifuge at 80*g* for 10 min. Recentrifuge the supernatant at 160*g* for 10 min.
3. Resuspend the pellets in mannitol, and analyze the samples by light and UV microscopy to estimate the number of FDA-stained microprotoplasts and by microdensitometry and flow cytometry to determine the nuclear DNA content of the microprotoplasts.

3.4. Fusion of Donor Microprotoplasts with Recipient Protoplasts

1. Mix the donor microprotoplasts and recipient protoplasts in a ratio of 2:1 (e.g., 3.0×10^5 microprotoplasts of potato: 1.5×10^5 protoplasts of tomato or tobacco). In 6-cm Falcon Petri dishes, place 5 separate 60-µL drops of of protoplasts suspended in W5 medium ***(18,24)*** to give a final plating density of 1×10^6 /mL for PEG-based mass fusion.
2. After 20 min, add PEG (PEG 4000) to give a final concentration of 8–10%.
3. After 7 min, carefully remove the PEG and W5 medium using a Pasteur pipet, and slowly add 20 µL high-pH buffer solution to each group of cells.
4. Twenty minutes later, when the microprotoplast–protoplast mixture has again settled, rinse the dish gently with 3 mL of H460 *M* medium for fusion samples of potato (+) tobacco and with TM2 medium for fusions of potato (+) tomato.
5. As a control for microprotoplast (+) protoplast fusions, carry out symmetric fusions between the donor cell suspension protoplasts and the recipient leaf protoplasts under the same conditions and at a similar plating density.

3.5. Selection of Fusion Products

1. Isolate the hybrid calli using selection pressure with antibiotics, e.g., kanamycin and hygromycin, as in the case of fusions of potato microprotoplasts (+) tomato protoplasts, or only kanamycin, as in the case of potato microprotoplasts (+) tobacco protoplasts: dilution and selection in CI liquid medium ***(18,24–26)*** on days 12 and 19, and selection on solid callus growth (CG) medium ***(18,24–26)*** on day 25 (*see* **Note 8**).
2. Transfer the resistant calli (when turned green) to regeneration medium containing no kanamycin or hygromycin.
3. Transfer the regenerated shoots to rooting medium, e.g., MS supplemented with 2% sucrose in the case of potato microprotoplasts (+) tomato protoplast fusions (**Table 5**), and MS supplemented with 2% sucrose and indole butyric acid (0.025 mg/L) in the case of potato–tobacco fusions.

3.6. Production of Monosomic Addition Hybrid Plants

1. Analyse the regenerated plants for the expression of donor marker genes, e.g., kanamycin resistance and GUS, as in potato (+) tomato fusions, or only kanamycin resistance as in potato (+) tobacco fusions to identify the hybrid plants ***(8,9)***.

2. Determine the chromosome composition of the hybrid plants by GISH to select the monosomic addition plants containing a single donor chromosome carrying the marker genes, and a complete chromosome set of the recipient parent, e.g., 1 potato chromosome and 24 tomato chromosomes, as in potato (+) tomato fusions ***(8,9,12)*** (*see* **Notes 9** and **10**).

4. Notes

1. *N. plumbaginifolia* Viviani (Doba line I-125-1 line resistant to kanamycin) has been used as a model plant species in our investigations to establish optimum conditions, together with the donor potato line 413 for the production of monosomic additions in tomato and tobacco backgrounds. Cremart at 7.5 μ*M* for *S. tuberosum* and APM at 32 μ*M* for *N. plumbaginifolia* gave the highest frequency of micronucleated cells.
2. Micronucleation in suspension cells/protoplasts occurs through modification of mitosis under the influence of the spindle toxins APM or CR. The metaphase-blocked cells/protoplasts with single or groups of chromosomes show no division of centromere or chromatid separation (no anaphase), but decondense and form micronuclei. Thus, the higher the frequency of metaphases, the greater the yield of micronuclei will be.
3. The yield of micronuclei depends not only on the frequency of micronucleated protoplasts, but also on the frequency of micronuclei/micronucleated protoplast. The sequential treatment with HU or APH, followed by APM or CR treatment, results in a synchronous progression of cells through S-phase to G2/M, enhancing the frequency of micronuclei.
4. Further, the presence of CB in the enzyme mixture can positively influence the micronucleation of protoplasts. In potato and *N. plumbaginifolia,* the frequency of micronuclei could be increased two- to threefold by incubating CR- or APM-treated suspension cells in an enzyme mixture containing CB and CR/APM. CB disrupts the microfilaments, whereas CR avoids the reformation of microfilaments in micronucleated protoplasts. Therefore, micronuclei formed prior to enzyme incubation can be maintained in a stable state without undergoing fusion and restitution. Additionally, the chromosomes in the remaining metaphases and prophases from the late dividing cells decondense and form micronuclei, thus increasing the percentage of micronucleated protoplasts.
5. The yield of microprotoplasts strongly depends on an efficient fragmentation of micronucleated protoplasts and the presence of CB and APM/CR during ultracentrifugation. To isolate microprotoplasts from micronucleated protoplasts, the protoplasts have to be fragmented, while maintaining the integrity of the plasma membrane. The presence of microfilaments opposes changes to the shape of protoplasts under the influence of *g* forces during ultracentrifugation, thus resulting in a lower yield of subprotoplasts. Therefore, the microfilaments have to be disrupted by incubation with CB, combined with APM/CR to avoid reformation of microtubules in micronucleated protoplasts. Also, high-speed centrifugation at $10^5 \times g$ is required for a longer duration (2 h) to obtain a better fragmentation and separation of microprotoplasts.

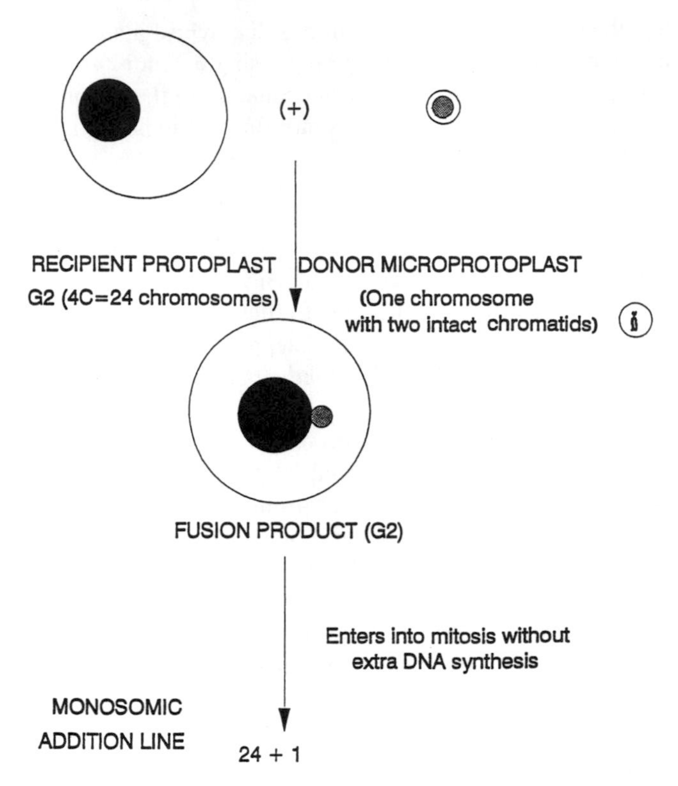

Fig. 2. Production of a monosomic addition line with a complete chromosome set (24 chromosomes) and one donor chromosome after fusion of a recipient protoplast with a donor microprotoplast (e.g., potato) (reproduced with permission; ***18***).

6. In addition to the yield of microprotoplasts, the frequency of smaller subdiploid microprotoplasts containing single chromosomes is critical for the successful production of monosomic addition hybrids after fusion. Therefore, sequential filtration with nylon sieves of decreasing pore size is essential to separate the small microprotoplasts from the larger ones.
7. A repeated washing of micronucleated protoplasts with a washing medium containing mannitol (0.4 *M*) is important to avoid clogging of sieves during the purification step, thus increasing the total yield of micronucleated protoplasts.
8. The population of small subdiploid microprotoplasts with a partial genome (containing one to four chromosomes) enriched after sequential filtration remains intact and viable (FDA-positive) for several days, but should not regenerate cell walls or undergo cell division (as shown for *N. plumbaginifolia* or potato microprotoplasts). The lack of cell division activity of the donor microprotoplasts with a partial genome can be an additional advantage in fusion experiments, because it avoids contamination of the donor partner while selecting for the fusion products.

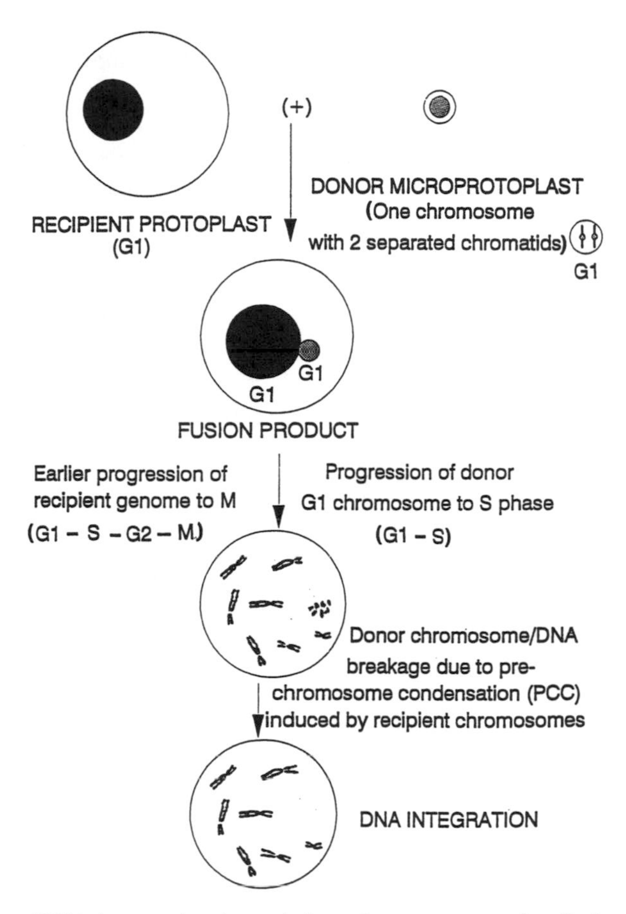

Fig. 3. Donor DNA integration in recipient chromosomes after fusion of a recipient protoplast with a donor microprotoplast (reproduced with permission; ***18***).

9. By microprotoplast fusion, the regeneration ability of the recipient line remains unchanged as compared to that after protoplast–protoplast fusion (symmetric fusion). Since the donor protoplast contains a single chromosome in interphase and a small amount of cytoplasm surrounded by an intact plasma membrane, its integration into the recipient genome is stable, thus producing monosomic addition hybrids with one donor chromosome and a complete chromosome set of the recepient species (**Fig. 2**).
10. From potato microprotoplast (+) tomato protoplast fusion, several monosomic addition plants ($2n = 24 + 1$) were regenerated within a short period of 3–4 mo. Also, donor DNA integration can occur in microprotoplast hybrid plants. A delayed replication of the addition chromosome and mitotic failure might lead to chromosome/DNA breakage and integration of donor DNA into the recipient genome (**Fig. 3**). The integration of donor DNA occurred in the progeny after

backcrossing of monosomic addition hybrids with the recipient species (i.e., in potato microprotoplasts [+] tomato protoplasts) ***(9,12)***, showing sexual transmission of the alien gene. The progeny with the integrated donor DNA can be identified by using the marker genes (e.g., the reporter uidA gene or the selectable *nptII* gene). In generatively produced addition lines, the integration of donor DNA/chromosome segments apparently occur by similar mechanisms: double-strand breaks of the donor chromosome during its disintegration at some stage in meiosis/mitosis, followed by repair-induced invasion and integration of donor DNA into the host chromosomes ***(1)***. Thus, the loss or disintegration of the additional donor chromosome can occur after the production of monosomic addition plants. Consequently, the frequency of the monosomic addition plants as well as the stability and maintainance of the added donor chromosome in subsequent somatic/generative cycles might depend on various factors, i.e., chromosome type, ploidy level of the recipient, genotypic background, and so forth ***(26)***. The frequency of donor DNA integration can be determined by using appropriate markers.

References

1. Sybenga, J. (1992) Cytogenetics in plant breeding. *Monographs on Theoretical and Applied Genetics,* vol. 17. Springer-Verlag, Heidelberg.
2. Fedak, G. (1992) Perspectives on wide crosses in barley. *Barley Genetics Newsletter,* vol. 6. Department of Agronomy, Colorado State University, Fort Collins, CP, pp. 683–689.
3. Jacobsen, E., de Jong, J. H., Kamstra, S. A., van den Berg, P. M. M. M., and Ramanna, M. S. (1994) The first and second backcross progeny of the intergeneric fusion hybrids of potato and tomato after crosses with potato. *Theor. Appl. Genet.* **88,** 181–186.
4. Famelaer, I., Gleba, Y. Y., Siderov, V. A., Kaleda, V. A., Parokonny, A. S., and Boryshuk, N. V. (1989) Intrageneric asymmetric hybrids between *Nicotiana plumbaginifolia* and *Nicotiana sylvestris* obtained by "gamma-fusion." *Plant Sci.* **61,** 105–117.
5. Wijbrandi, J., Zabel, P., and Koornneef, M. (1990) Restriction fragment length polymorphism analysis of somatic hybrids between *Lycopersicon esculentum* and irradiated *L. peruvianum:* evidence for limited donor genome elimination and extensive chromosome rearrangements. *Mol. Gen. Genet.* **222,** 270–277.
6. Puite, K. J. and Schaart, J. G. (1993) Nuclear genomic composition of asymmetric fusion products between irradiated transgenic *Solanum brevidens* and *S. tuberosum:* limited elimination of donor chromosomes and polyploidization of the recipient genome. *Theor. Appl. Genet.* **86,** 237–244.
7. Wolters, A. M. A., Jacobsen, E., O'Connel, M., Bonnema, G., Ramulu, K. S., and de Jong, J. H. (1994) Somatic hybridization as a tool for tomato breeding. *Euphytica* **79,** 265–277.
8. Ramulu, K. S., Dijkhuis, P., Rutgers, E., Blaas, J., Krens, F. A., and Verbeek, W. H. J. (1996) Intergeneric transfer of partial genome and direct production of monosomic addition plants by microprotoplast fusion. *Theor. Appl. Genet.* **92,** 316–325.

9. Ramulu, K. S., Dijkhuis, P., Rutgers, E., Blaas, J., Krens, F. A., and Dons, J. J. M. (1996) Microprotoplast mediated transfer of single chromosomes between sexually incompatible plants. *Genome* **39,** 921–933.
10. Verhoeven, H. A. and Ramulu, K. S. (1991) Isolation and chracterization of microprotoplasts from APM-treated suspension cells of *Nicotiana plumbaginifolia. Theor. Appl. Genet.* **82,** 346–352.
11. Ramulu, K. S., Dijkhuis, P., Famelaer, I., Cardi, T., and Verhoeven, H. A. (1993) Isolation of subdiploid microprotoplasts for partial genome transfer in plants: enhancement of micronucleation and enrichment of microprotoplasts with one or a few chromosomes. *Planta* **190,** 190–198.
12. Rutgers, E., Ramulu, K. S., Dijkhuis, P., Blaas, J., Krens, F. A., and Verhoeven, H. A.(1997) Identification and molecular analysis of transgenic potato chromosomes transferred to tomato through microprotoplast fusion. *Theor. Appl. Genet.* **94,** 1053–1059.
13. Sybenga, J. (1989) Genetic manipulation: generative vs. somatic, in *Biotechnology in Agriculture and Forestry 9. Plant Protoplasts and Genetic Engeneering II.* (Bajaj, Y. P. S., ed.), Springer-Verlag, Heidelberg, pp. 26–53.
14. Heslop-Harrison, J. S., Leitch, A. R., and Schwarzacher, T. (1993) The physical organization of interphase nuclei, in *The Chromosome.* (Heslop-Harrison, J. S. and Flavell, R. B. eds.), Bios, Oxford. pp. 221–232.
15. Schondelmaier, J., Martin, R., Jahoor, J., Houben, J., Graner, J., and Koop, H. U. (1993) Microdissection and microcloning of the barley (*Hordeum vulgare* L.) chromosome 1HS. *Theor. Appl. Genet.* **86,** 629–636.
16. Bennett, M. D.(1988) Parental genome separation in F1 hybrids between grass species, in *Proceedings of Kew Chromosome Conference III* (Brandham, P.B, ed.), Her Majesty's Stationary Office, London, pp. 195–208.
17. Gilissen, L. J., Ramulu, K. S., Flipse, E., Meinen, E., and Stiekema, W. J. (1991) Transformation of diploid potato genotypes through *Agrobacterium* vectors and expression of T-DNA markers in root clones, regenerated plants and suspension cells. *Acta Bot. Neerl.* **40,** 53–61.
18. Ramulu, K. S., Dijkhuis, P., Rutgers, E., Blaas, J., Verbeek, W. H. J., Verhoeven, H. A., and Colijn-Hooymans, C. M. (1995) Microprotoplast fusion technique: a new tool for gene transfer between sexually incongruent plant species. *Euphytica* **85,** 255–268.
19. Koornneef, M., Hanhart, C. G., and Martinelli, L. (1987) A genetic analysis of cell culture traits in tomato. *Theor. Appl. Genet.* **74,** 633–641.
20. Maliga, P., Sz.-Brenovits, A., Maarten, L., and Joo, F. (1975) Non mendelian streptomycin resistant tobacco mutant with altered chloroplasts and mitochondria. *Nature* **255,** 401–402.
21. Ramulu, K. S., Dijkhuis, P., Famelaer, I., Cardi, T., and Verhoeven, H. A. (1994) Cremart: a new chemical for efficient induction of micronuclei in cells and protoplasts for partial genome transfer. *Plant Cell Rep.* **13,** 687–691.
22. Verhoeven, H. A., Ramulu, K. S., Gilissen, L. J. W., Famelaer, I., Dijkhuis, P., and Blaas J. (1991) Partial genome transfer through micronuclei in plants. *Acta Bot. Neerl.* **40,** 97–113.

23. Bokelmann, G. S. and Roest, S. (1983) Plant regeneration from protoplasts of potato (*Solanum tuberosum* cv. Bintje). *Z. Pflanzenphysiol.* **109,** 259–265.
24. Shahin, E. A. (1985) Totipotency of tomato protoplasts. *Theor. Appl. Genet.* **69,** 235–240.
25. Derks, F. H. M., Hakkert, J. C., Verbeek, W. H. J., and Colijn-Hooymans, C. M. (1992) Genome composition of asymmetric hybrids in relation to the phylogenetic distance of the parents. Nucleus-chloroplast interaction. *Theor. Appl. Genet.* **84,** 930–940.
26. Nagy, J. I. and Maliga, P. (1976) Callus induction and plant regeneration from mesophyll-protoplasts of *Nicotiana sylvestris*. *Z. Pflanzenphysiol.* **78,** 453–455.
27. Sing, R.G (1993) *Plant Cytogenetics*. CRC, Boca Raton, FL.

22

Guard Cell Protoplasts

Isolation, Culture, and Regeneration of Plants

Graham Boorse and Gary Tallman

1. Introduction

The guard cells that flank stomata undergo environmentally induced, turgor-driven cellular movements that regulate stomatal dimensions. Under many environmental conditions, changes in stomatal dimensions regulate rates of transpiration and photosynthetic carbon fixation ***(1)***. Among the environmental signals that guard cells receive and transduce are light quality, light intensity, intercellular concentrations of leaf carbon dioxide, and apoplastic concentrations of abscisic acid (ABA) ***(2)***. How guard cells integrate the variety of signals present in the environment and activate the appropriate signal transduction mechanisms to adjust stomatal dimensions for prevailing environmental conditions is the subject of intense investigation ***(3)***.

Many fundamental studies of stomata have been performed with detached leaf epidermis, but for many types of experiments, epidermis is not an adequate material. Guard cells are among the smallest and least numerous of cell types in the leaf, and the presence of even a very small number of contaminating cells in epidermal preparations (e.g., epidermal cells that neighbor guard cells and mesophyll cells) can result in significant experimental artifacts ***(4)***. This is particularly true of experiments that involve guard cell biochemistry or molecular biology and/or experiments in which guard cell metabolism is measured. Furthermore, the presence of the relatively thick guard cell wall precludes certain types of studies (e.g., electrophysiology).

Good methods for making large ($\approx 1 \times 10^6$ cells), highly purified preparations (<0.01% contamination with other cell types) of guard cell protoplasts (GCP) were first reported about 15 years ago ***(4–7)***. The availability of these

From: *Methods in Molecular Biology, Vol. 111: Plant Cell Culture Protocols*
Edited by: R. D. Hall © Humana Press Inc., Totowa, NJ

methods ***(8)*** contributed significantly to a resurgence of interest in the basic cell biology of guard cells. GCP have been used to demonstrate that guard cells possess:

1. Both PSI and PSII activity ***(4)***.
2. A functional photosynthetic carbon reduction pathway ***(9)***.
3. A blue light-activated plasma membrane H^+-translocating ATPase ***(10–12)***.
4. Voltage-gated, inward rectifying plasma membrane K^+ channels ***(13,14)***.
5. Outward-rectifying plasma membrane anion channels (*see* ***15*** for a review of ion channel studies using GCP).

GCP have been used to investigate:

1. Regulation of K^+ channel activity in the guard cell plasma membrane ***(15,16)***;
2. Elevation of cytosolic free Ca^{2+} by movement of extracellular Ca^{2+} through ABA-activated, Ca^{2+}-permeable channels ***(15)***; and
3. Ca^{2+}-dependent and Ca^{2+}-independent ABA signal transduction pathways ***(15)***.

Recently, two partial-length cDNAs coding for different plasma membrane H^+-ATPase isoforms have been isolated from GCP of *Vicia faba* L. and have been used to localize expression of these genes ***(17)***.

Most studies using GCP have been short-term experiments of a few hours' duration. Until recently, no in vitro experimental systems existed that would enable the study of responses of GCP to environmental signals administered over longer periods. However, GCP have now been established and maintained in culture ***(18–21)***. As with any culture system, an advantage of culturing GCP is that culture conditions can be rigorously defined and carefully controlled. An additional advantage of culturing GCP is that the homogeneity of such preparations confers on them a uniformity of response that is seldom observed in cultures of mixed-cell types ***(20,21)***. In the latter, identification of the cell type responding to a change in culture condition may be difficult. If the responding cell type is identified, the basis for any particular response to a change in a culture condition (e.g., an increase or decrease in synthesis of a particular protein or transcription of a particular gene) may still not be identifiable, because the condition can evoke from each unique cell type a response that affects the response(s) of each and/or every other cell type.

Monocultures of GCP hold promise for studies of the signal transduction mechanisms underlying:

1. Cellular differentiation.
2. Chloroplast senescence.
3. Regulation of the cell cycle in plants, processes that are only activated by longer periods of exposure to environmental signals.

Each of these processes can be activated and directed in cultures of GCP by manipulating concentrations and/or ratios of plant growth regulators (auxin, cytokinins, or ABA ***[18,20,22]***) and/or temperature ***(20)***. For example, when GCP of *Nicotiana glauca* are cultured at temperatures <34°C, they dedifferentiate and divide to form a callus from which plants can be regenerated ***(22)***. In the process of dedifferentiation, their chloroplasts senesce ***(20)***. When they are cultured at temperatures >34°C in media containing 0.1 μ*M* ABA, they remain differentiated. GCP cultured at temperatures >34°C for 1 wk do not divide ***(20)***, but when they are transferred to temperatures <34°C, at least a small percentage redifferentiate and divide (unpublished). These data indicate that in monocultures of GCP, alteration of only one or two culture conditions (±ABA; temperature) is sufficient to:

1. Maintain cultured GCP in or trigger entry of cultured GCP into certain stages of the cell cycle.
2. Determine whether GCP remain differentiated or dedifferentiate.
3. Determine whether or not chloroplast senescence is activated.

Cultured GCP of *N. glauca* (tree tobacco) and *Beta vulgaris* (sugar beet) have been used to produce friable, embryogenic callus from which plants have been regenerated, demonstrating that GCP are totipotent ***(22–24)***. Regeneration of plants from cultured GCP of *B. vulgaris* is of major commercial importance, because callus derived from other cell types, tissues, and organs of this plant are recalcitrant to regeneration ***(24)***. Cultured GCP of *B. vulgaris* have already been used to produce transgenic plants with enhanced resistance to herbicides by a relatively rapid protocol ***(25)***.

GCP are isolated using a two-step procedure. To remove contaminating mesophyll and epidermal cells, detached epidermis is treated with a mixture of cellulase and pectinase dissolved in a hypotonic solution. Because the cell walls of contaminating epidermal and mesophyll cells are thinner and are of a different chemical composition than those of guard cells, they are digested more quickly than those of guard cells. As protoplasts of contaminating epidermal cells and mesophyll cells are released into the hypotonic medium, they swell and burst. After sufficient time has passed to destroy contaminating cells, the remaining cuticle containing guard cells is collected on a nylon net, rinsed, and transferred to a solution of cellulase and pectinase in a solution that is slightly hypertonic to guard cells. After a few hours of digestion, GCP are released into the medium. The cells are collected by filtration and centrifugation, washed, and suspended in culture media (**Fig. 1**).

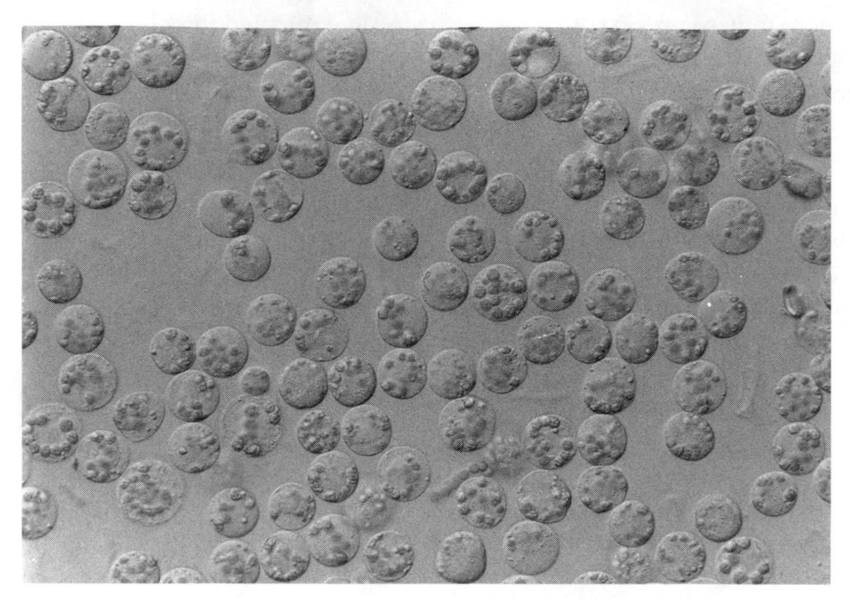

Fig. 1. Guard cell protoplasts isolated from leaves of *N. glauca* (Graham), tree tobacco. Differential interference contrast optics; large organelles in protoplasts are chloroplasts. Average diamter is ≈15 μm.

2. Materials

2.1. Plants

1. Seeds of Nicotiana glauca (Graham), tree tobacco (*see* **Note 1**).
2. Pots (0.16-, 2.0,- 10-L), potting soil, and sand. We have used Supersoil, Chino, CA, or Pro-Mix "HP" High-Porosity Growing Medium, Premier-Western US, Laguna Niguel, CA. Sand is 30 grade. Plants grow best in well-drained, porous media.
3. Environmental growth chamber: Conviron Model E7 (Conviron, Pembina, ND)
4. Fluorescent light bank: Lamps are model FT72T12.CW.1500, General Electric, Cleveland, OH, or equivalent.
5. Modified Hoagland's nutrient solution (**Table 1**).

2.2. Day Before Culture

1. 2-L Erlenmeryer flasks with cotton stoppers covered with cheesecloth *(4)*, paper towels, aluminum foil, and autoclave tape.
2. Pyrex casserole dish: 11 × 7 × 1.5 in. *(1)*.
3. Pyrex casserole dish: 11 × 7 × 1.5 in. *(1)* containing: fine-point forceps wrapped individually in foil *(2)*, glass plate: 4.5 × 4.5 × 0.13 in. *(1)*, 6-in. cotton swab *(1)*, single-edge razor blade *(1)*, Petri dish (glass, bottom only), 15 × 60 mm *(1)*.
4. Plastic beakers: 2 L *(1)*, 1 L *(2)*.

Table 1
Modified Hoagland's Nutrient Solution *(26)*

Constituent	g/L in stock solution	Final concentration in nutrient solution after dilution of stock, m*M*
1. $NH_4H_2PO_4$	21.0	1.0
2. KNO_3	109.0	6.0
3. $Ca(NO_3)_2 \cdot 4H_2O$	170.0	4.0
4. $MgSO_4 \cdot 7H_2O$	43.6	2.0
5. H_3BO_3	0.56	0.4
$MnSO_4 \cdot H_2O$	0.308	10.0
$ZnSO_4 \cdot 7H_2O$	0.042	0.8
$CuSO_4 \cdot 5H_2O$	0.018	0.4
MoO_3	0.011	0.4
NaCl	1.045	100.0
6. Iron solution		
First dissolve: Na_2 EDTA	6.0	0.09
Then add: $FeSO_4 \cdot 7H_2O$	4.5	0.09
7. KOH	4.0	71.3

To prepare nutrient solution from stock solutions, add 100 mL each of solutions 1–6 and 50 mL of solution 7 to a 20-L carboy. Add water to 18 L; shake to mix.

5. Plastic beaker: 2 L ***(1)*** containing: plastic, disposable Erlenmeyer flasks with screw caps—125 mL (*3*; loosen caps and cover with foil), plastic funnels with top diameter = 3.5 in., stem diameter = 0.5 in., (*2*; one funnel is lined with 220 × 220 µm mesh nylon net; the other is lined with a 30 × 30 µm mesh nylon net; nets are secured to rims of funnels with autoclave tape; wrap funnels in foil. Nylon nets are Nitex® from TETKO®, Inc., 111 Calumet St., Depew, NY 14043).

2.3. Isolation of Guard Cell Protoplasts

1. Plastic bag for holding leaf in moist paper towels.
2. Gyrotory shaking water bath, pH meter, balance.
3. Solutions—make fresh daily; filter-sterilize. All chemicals are reagent-grade. Powdered enzymes are stored refrigerated. Cellulase "Onozuka" RS from Yakult Pharmaceutical Ind. Co., Ltd., 1-1-19, Higashi-Shinbashi, Minato-Ku, Tokyo, 105 Japan. Pectolyase Y-23 from Seishin Pharmaceutical Co., Ltd., 4-13, Koami-cho, Nihonbashi, Chuo-ku, Tokyo 103, Japan. Media containing agar are sterilized by autoclaving at 121°C, 15 psi for 20 min.

   A: 0.5% polyvinylpyrrolidone 40 (PVP 40) and 0.05% ascorbic acid, pH 6.5; Dissolve 2.5 g of PVP 40 and 0.25 g of ascorbic acid in 450 mL of deionized water; adjust pH to 6.5 with NaOH, bring to final volume of 500 mL with deionized water, and mix.

B: In a small beaker combine 2.533 g sucrose, 0.0054 g $CaCl_2$, 0.4 g Cellulase Onozuka RS, 0.003 g Pectolyase Y-23, 0.185 g PVP 40, and 0.075 g bovine serum albumin (BSA). Add 35 mL of deionized water and stir until components are dissolved. The pH of the enzyme solution is adjusted initially to 3.4 with stirring for 7 min, and then raised to 5.5 before filter-sterilization (*see* **Note 2**). Final concentration of sucrose is ca. 0.2 *M*; final concentration of $CaCl_2$ is ca. 0.7 m*M*.

C: 0.2 *M* sucrose, 1 m*M* $CaCl_2$: Dissolve 13.692 g of sucrose and 0.0294 g of $CaCl_2$ in 190 mL of deionized water; adjust pH to 5.5 with HCl and/or NaOH. Bring final volume to 200 mL with deionized water and mix.

D: In a small beaker combine 2.140 g sucrose, 0.0037 g $CaCl_2$, 0.4 g Cellulase Onozuka RS, 0.003 g Pectolyase Y-23, 0.125 g PVP 40, and 0.075 g BSA. Add 23 mL of deionized water, and stir until components are dissolved. Adjust pH to 5.5 (*see* B and **Note 2**). Final concentration of sucrose is ca. 0.25 *M*; final concentration of $CaCl_2$ is ca. 1 m*M*.

4. Sterile 0.45 μm cellulose nitrate filters in disposable filter units: 115 mL ***(4)***, 250 mL ***(1)***, 500 mL ***(1)***. Enzyme solutions and culture media lacking agar are sterilized by filtration through 0.45-μm cellulose nitrate filters in disposable filter units (115 mL = model 125-0045; 250 mL = 126-0045; 500 mL = 450-0045, Nalgene Co., Rochester, NY). To prevent particulates from plugging filters, a 50-mm diameter prefilter (Gelman Sciences Type A/E Glass Fiber Filter, P/N 6/632, Gelman, Ann Arbor, MI) is used when enzyme solutions are sterilized.
5. Laminar flow cabinet.
6. Sterile latex gloves (*see* **Note 3**).
7. Disinfectant (*see* **Note 4**).
8. 5.25% sodium hypochlorite = Clorox bleach, Oakland, CA. Free chlorine concentration = 5.25%.
9. 95% Ethanol.
10. Centrifuge tubes: sterile, disposable, plastic conical with caps; 15 mL ***(3)***, 50 mL ***(3)***.
11. Syringe: plastic, 60 mL ***(1)***, disposable 0.45-μm membrane syringe filter (***1***; Corning disposable syringe filter: 0.45 μm; 25-mm diameter cellulose acetate membrane in acrylic holder, model # 21053–25; Corning, Inc., Corning, NY).
12. Glass pipets: sterile, disposable, glass, 10 mL ***(5)***, 1 mL ***(2)***.
13. Clinical centrifuge.
14. Hemocytometer, microscope, hand tally counter.
15. Incomplete medium I (**Table 2**).

2.4. Primary Cultures and Colony Formation

1. Plastic eight-well microchamber culture slides, Lab-Tek Chamber Slide™, Model 177402, Nunc, Inc., Naperville, IL.
2. Petri dish (***1***; 2.5 cm deep x 15 cm diameter), Parafilm (American National Can Co., Greenwich, CT).
3. Hormone stock solution (*see* **Subheading 3.3., step 27**).
4. Incubator (lighting optional).

Table 2
Media Used to Culture and Regenerate Plants from Guard Cell Protoplasts of *N. glauca* (Graham), Tree Tobacco

Constituent	Medium I, mg/L	Medium II, mg/L	Tobacco shoot medium, mg/L	Root medium, mg/L
Salts				
$Ca(NO_3)_2 \cdot 4H_2O$	180.0			
NH_4NO_3	82.5	825.0	1650.0	412.5
KNO_3	167.0	950.0	1900.0	475.0
$CaCl_2 \cdot 2H_2O$	84.7	220.0	110.0	
CaCl2	333.0			
$MgSO_4 \cdot 7H_2O$	447.3	1223.0	611.5	
$MgSO_4$	181.0			
Na_2SO_4			180.0	
KH_2PO_4	68.0	680.0	170.0	340.0
$NaH_2PO_4 \cdot H_2O$	14.9			
KCl	88.3			
Na_2 EDTA	3.7	37.3		18.7
$FeSO_4 \cdot 7H_2O$	2.8	27.8		13.9
$Fe_2(SO_4)_3$	2.3			
KI	0.76	0.83	0.83	0.42
H_3BO_3	2.0	6.2	6.2	3.1
$MnSO_4 \cdot H_2O$	3.1		16.9	
$MnCl_2 \cdot 4H_2O$	2.0	19.8		9.9
$ZnSO_4 \cdot 7H_2O$	2.3	9.2	8.6	4.6
$Na_2MoO_4 \cdot 2H_2O$	0.03	0.25	0.25	0.13
$CuSO_4 \cdot 5H_2O$	0.003	0.025	0.025	0.013
$CoSO_4 \cdot 7H_2O$	0.003	0.03		0.015
$CoCl_2 \cdot 6H_2O$			0.025	
FeNa EDTA			36.7	
Organics				
i-Inositol	39.7	100.0	100.0	50.0
Thiamine · HCl	0.19	1.0	0.4	0.5
Glycine	1.35		2.0	
Niacin	0.45			
Pyridoxine · HCl	0.1		0.5	
Nicotinic acid			0.5	
Casein Hydrolysate			1000.0	
MES	976.0	976.0		
Sucrose	95,840.0	79,419.0	863.25	
Kinetin			1.0	

(continued)

Table 2 *(continued)*

Constituent	Medium I, mg/L	Medium II, mg/L	Tobacco shoot medium, mg/L	Root medium, mg/L
Organics				
1-NAA	0.3	0.3		
6-BAP	0.075	0.075		
Agarose		5000.0		
Agar			8000.0	

Medium I, medium II, and root medium are modifications of those described by Shepard and Totten *(27)*. All salts and organics through casein hydrolysate are made as 1000X stocks, except $CaCl_2 \cdot 2H_2O$ (260X) and KCl (663X)·Na_2EDTA, $FeSO_4 \cdot 7H_2O$, and $Fe_2(SO_4)_3$ are mixed in a single stock solution in that order. Stocks are stored frozen at –20°C; any material precipitated by freezing is resolublized after thawing. An incomplete medium I containing all components through 2(*N*-Morpholino) ethanesulfonic acid [MES] is prepared in 1-L batches using stock solutions and solid MES. The incomplete medium is stored frozen in 100-mL aliquots at –20°C until the day GCP are to be isolated. Sucrose and hormones are added to complete wash and culture media on the day of protoplast isolation as described in **Subheading 3.3., step 27**. The final sucrose concentration in medium I is 0.28 *M*, in medium II, 0.23 *M*, and in shoot differentiation medium, 0.025 *M*. Concentration of MES in media I and II is 5 m*M*. Tobacco shoot medium is commercially available from Carolina Biological Supply, Gladstone, OR.

2.5. Primary Callus

1. Medium II in Petri dishes (1.5 cm deep × 10 cm diameter; ca. 20 mL/dish; **Table 2**).
2. Lighted incubator.

2.6. Secondary Callus

1. Tobacco shoot medium (**Table 2**).
2. Lighted incubator.

2.7. Plant Regeneration

1. Tobacco shoot medium (**Table 2**).
2. Root medium (**Table 2**) in Magenta vessels (Sigma Chemical Co., St. Louis, MO); 75 mL/vessel.
3. Lighted incubator.
4. 0.16-L small pots; potting soil.

3. Method

3.1. Plants

1. Germinate seeds of *N. glauca* at high density on the surface of moistened, autoclaved potting soil in small pots (0.16 L; *see* **Note 1**).

2. Germinate seeds and maintain seedlings on a 16-h light/8-h dark cycle in an environmental chamber. Mean (± SE) temperature during the light cycle is 28 ±2°C; mean temperature during the dark cycle is 21 ±2°C. Relative humidity in the chamber is 65–75%. Water seedlings with tap water daily. The photosynthetic photon flux density (PPFD) at seedling height is 50–70 μmol/m^2/s of photons of photosynthetically active radiation (PAR).
3. After 4–6 wk of growth, transfer plants to 2-L plastic pots containing an autoclaved mixture of 60% soil/40% sand (v:v).
4. After another 4–6 wk of growth, cull plants to 2 plants/pot and allow to grow to a height of 0.2 to 0.3 m.
5. Transfer plants to 10-L plastic pots containing the same soil/sand mix. Grow on a table under high-intensity fluorescent lights. The PPFD at the top of the canopy is 800–900 μmol/m^2/s of PAR.
6. Water plants three times daily with tap water at 6 h intervals for 4 min and every other day with Hoagland's nutrient solution (**Table 1**). Mean temperature is 27 ±2°C during the 16 h light cycle and 23 ±2°C during the 8 h dark cycle. Relative humidity in the room is 45–65%.

3.2. Day Before Culture

1. Autoclave at 121°C, 15 psi for 20 min; exhaust on dry cycle:

   Four, 2-L Erlenmeyer flasks each containing 1.4 L of deionized water. Stopper flasks with cotton wrapped in cheesecloth. Place a paper towel over each stopper and secure on each side with autoclave tape. Place a double layer of aluminum foil on top of each stopper, and secure to the flask with autoclave tape.

   One casserole dish, empty: Cover the dish with a single layer of foil and secure with autoclave tape. One casserole dish containing: two pairs of fine-point forceps wrapped individually in foil, one 6-in. cotton swab, one glass plate, one 15 × 60 mm glass Petri dish bottom, and one single-edge razor blade. Cover the dish with a single layer of foil, and secure with autoclave tape.

   Two, 1-L plastic Nalgene beakers, empty: Cover beakers with a double layer of foil and secure with autoclave tape.

   One, 2-L plastic beaker, empty; one, 2-L plastic beaker containing three 125-mL, screw-cap disposable plastic Erlenmeyer flasks, (caps secured loosely and covered with foil) and two plastic funnels, one lined with 220 × 220 μm mesh nylon netting and the other with 30 × 30 μm mesh nylon netting. Secure netting to rim of each funnel with autoclave tape; wrap funnels in foil. Cover beakers with a double layer of foil secured with autoclave tape.

3.3. Isolation of Guard Cell Protoplasts

1. Turn on shaking water bath, check water level, bring temperature to 28°C.
2. Thaw 100 mL of incomplete medium I (**Table 2**).
3. For each experiment, harvest one flat leaf from insertion level 4 or 5 from the top of the plant with a blade length of 0.11–0.2 m and with a relatively thick cuticle. Harvest 0.5 to 1.5 h prior to the onset of the light cycle (*see* **Note 5**). Store leaves

in moist paper towels in a plastic bag in darkness until detachment of epidermis is initiated.

4. Weigh, mix, and adjust pH of solutions A, B, C, and D; filter-sterilize.
5. Turn on laminar flow hood; purge for 10–30 min.
6. Cover hands with sterile, latex gloves (*see* **Note 3**).
7. Wipe the sides and bottom of the laminar flow cabinet with full-strength disinfectant (*see* **Note 4**).
8. Transfer solutions, glassware, and instruments to the laminar flow cabinet. Back row: solutions; middle row, L to R: 2-L beaker, 2-L beaker, 1-L beaker, 1-L beaker; front row L: casserole dish with implements; front row R: empty casserole dish. Remove foil and contents from beakers. Pour 2 L of sodium hypochlorite into 2 L beaker at left. Fill the second 2-L beaker with 2 L of sterile, deionized water; fill each of the two 1 L-beakers with 1 L of sterile, deionized water.
9. Remove foil from both casserole dishes.
10. Spray leaf with 95% ethanol, and then lightly buff with tissues to remove some of the wax.
11. Immerse right hand in sodium hypochlorite for 3–5 s. Using the right hand, immerse the leaf in the sodium hypochlorite solution for 3–5 s. Dip left hand by the same method. Immerse leaf in each beaker of water for 3–5 s, starting with the 2-L beaker. Lay the leaf in the empty casserole dish, top (adaxial) side up.
12. Transfer a few milliliters of solution A to the small Petri dish in the other casserole dish; pour the remainder over the leaf.
13. Keeping the leaf beneath the solution and starting at the base of the leaf, break (tear) the leaf near the midrib. Bend the torn leaf section toward you at angles of 120–160° to the leaf surface, and "peel" the adaxial epidermis away from the mesophyll. (*See* **Note 5**; it may be necessary to repeat this procedure three to five times to get most of the epidermis from one-half of the leaf.)
14. Transfer each sheet of epidermis, and any attached leaf material to the glass plate in the neighboring casserole dish. Spread the epidermis on the plate with the side that normally faces the mesophyll upward. Brush the epidermis gently with the cotton swab to remove adhering mesophyll. Using the single-edge razor blade, slice the epidermis horizontally into strips, and then vertically to small pieces (ca. 5 × 5 mm). Using fine forceps, transfer pieces to the small Petri dish.
15. Repeat **steps 13** and **14** until most of the epidermis has been peeled from the leaf.
16. Decant the solution in the Petri dish into the casserole dish. Gather epidermal fragments in the Petri dish into a ball with forceps, and drop them down the middle of the neck of a 125-mL Erlenmeyer flask.
17. Add solution B to the flask, rinsing any epidermis adhering to the sides down into the flask with the solution. Cap.
18. Incubate for 15 min at 28°C in a shaking water bath at 175 rpm. Depending on leaf age, digestion may take more or less time.
19. Remove used materials from the hood. Leave 2-L beaker of sodium hypochlorite and 2-L beaker of water.
20. After 10 min of incubation, sterilize gloved hands as above.

21. In hood, position funnel with 220 × 220 µm mesh nylon net over mouth of second 125-mL Erlenmeyer flask.
22. After 15 min, remove the incubating flask from the water bath. In the laminar flow cabinet, remove cap, and pour contents of flask over the 220 × 220 µm mesh nylon net to collect "cleaned" epidermis.
23. Rinse epidermis on net with 100 mL of solution C.
24. Open second set of sterile forceps; collect epidermis in a ball, and transfer to third 125-mL Erlenmeyer. Rinse sides of flask during addition of solution D. Cap.
25. Incubate for 3 h and 15 min at 28°C with shaking at 25 excursions/min.
26. Remove used material from laminar flow cabinet.
27. During incubation, prepare wash and culture media. Dissolve 9.584 g of sucrose in a total volume of 100 mL using incomplete medium I (**Table 2**). Divide 60 and 40 mL into separate beakers. Adjust pH of 60-mL aliquot to 6.1; adjust pH of 40-mL aliquot to pH 6.8. Dissolve 0.012 g of 1-napthalene acetic acid (NAA) and 0.003 g of 6-benzylaminopurine (BAP) in 10 mL of 95% ethanol to make a hormone stock of 1.2 mg/L NAA and 0.3 mg/L BAP. Add 10 µL of hormone stock to 10 mL of medium at pH 6.1. In laminar flow cabinet, using a 60-mL syringe, filter medium containing hormones through a disposable syringe filter into a 15-mL sterile, plastic conical centrifuge tube. Cap.
28. Filter remaining media at pH 6.1 and 6.8 through separate 0.45-µm disposable filter units; place in laminar flow cabinet along with two 15- and three 50-mL sterile plastic conical centrifuge tubes with caps. Uncap two 50-mL centrifuge tubes. Transfer pH 6.8 medium from filter unit to one 50-mL tube and cap. Transfer pH 6.1 medium to a second 50 mL tube and cap.
29. In hood, unwrap and position funnel with 30 × 30 µm mesh nylon net over mouth of an open 50-mL conical centrifuge tube.
30. After 3 h and 15 min, remove the incubating suspension of epidermis from the water bath and swirl. In the laminar flow cabinet, remove cap from incubation flask, and pour contents over the nylon net.
31. With a 10-mL sterile, disposable pipet, divide filtrate equally between two 15 mL sterile plastic conical centrifuge tubes; cap.
32. Collect GCP by centrifuging the filtrate at 40*g* for 6–7 min.
33. After centrifugation, remove supernatant with 10-mL pipette and discard in waste container in hood.
34. With a fresh, sterile 10-mL pipet, add 8 mL of medium at pH 6.8 to each tube. Resuspend GCP in wash medium by rolling centrifuge tubes gently between the palms of the hands. Cap tubes, and centrifuge at 40*g* for 6–7 min.
35. Repeat **steps 33** and **34** two more times, with medium of pH 6.1 without hormones. After the second centrifugation in this medium, aspirate the supernatant in each tube to 0.5 mL. Use a 1-mL sterile pipet to resuspend GCP in one of the tubes and transfer them into the other (combined total volume of 1 mL).
36. Using a hemocytometer, count GCP to estimate cell density and dilute accordingly with medium (pH 6.1) to give a final value of 1.25×10^5 cells/mL.

3.4. Primary Cultures and Colony Formation

1. Initiate liquid cultures by pipeting 0.3 mL of the cell suspension to wells of eight-well microchamber culture slides. Add to each chamber 0.1 mL of the medium containing NAA and BAP.
2. Incubate chamber slides in sterile plastic Petri dishes (2.5 cm deep × 15 cm diameter) containing moist paper towels; seal edges of dishes tightly with Parafilm®.
3. Incubate cultures at 25–32°C (*see* **Note 6**) in darkness or under red light (15–20 μmol/m²/s of photons of PAR) on a 12-h light/12-h dark cycle. For the latter experiments, red light can be provided by filtering the fluorescent light of the growth chamber described above through a red Plexiglas® filter.

3.5. Primary Callus

1. After 8–10 wk of culture in microchamber slides, transfer cultured cells to medium II.
2. Seal dishes with Parafilm®, and incubate at 25°C under continuous white fluorescent light (21–27 μmol/m²/s of photons of PAR).

3.6. Secondary Callus

1. After 8–10 wk, transfer green callus tissue (**Fig. 2A**) to a commercial *N. tabacum* shoot differentiation medium (**Fig. 2B**, **Table 2**).
2. Incubate at 25°C under continuous white fluorescent light (14–23 μmol/m²/s of photons of PAR).

3.7. Plant Regeneration

1. After 8–10 wk of growth on shoot medium, transfer callus to fresh shoot differentiation medium.
2. When shoots are 0.5–1 cm in height (**Fig. 2C**), transfer them to root medium (**Table 2**) in Magenta vessels, and incubate at 25°C under continuous white fluorescent light (30 μmol/m²/s of photons of PAR).
3. When roots are sufficiently developed (6–8 wk), transplant plants to small pots, and grow under the conditions described above for seedlings.

4. Notes

1. Seeds may be obtained from Lena Dugal, Exotic Seeds from around the World, 1814 N.E. Schuyler, Portland, OR 97212; URL: http://www.rdrop.com/users/idaho/tobacco.htm; phone and FAX: 503–282–1235. Seeds are sprinkled on the top of the soil and misted daily with an atomizer to prevent flushing seeds deep into soil. Once seedlings are 5–10 mm in height they are watered directly.
2. Treatment at low pH appears to precipitate some inhibitory, insoluble material that does not return to solution when the pH is raised to 5.5. Each new lot of enzyme should be tested, and the concentration adjusted for any differences in activity from the previous lot. Pectolyase activity is increased dramatically with an increase of only 1 or 2°C in temperature. The time required for the first enzyme digestion may vary depending on leaf age, with younger leaves requiring less

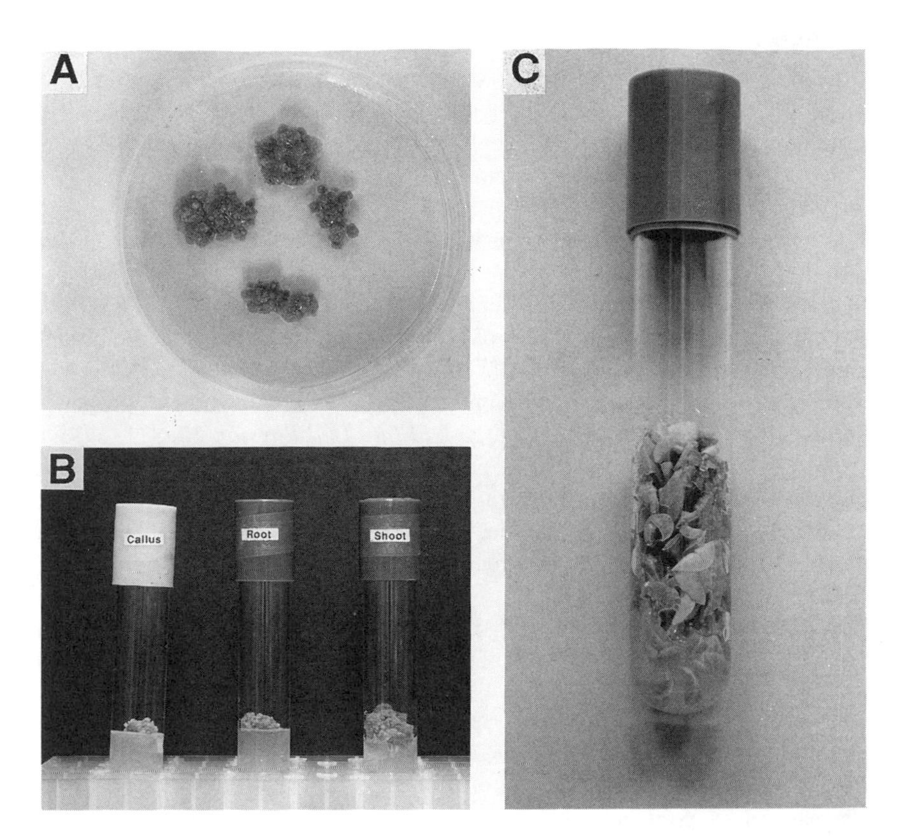

Fig. 2. Callus and shoots derived from cultured guard cell protoplasts of *N. glauca* (Graham), tree tobacco. **(A)** Primary callus, **(B)** secondary callus, and **(C)** shoots derived from secondary callus.

time. If leaves are too young, yields will be low, and peanut-shaped remnants of cells will appear in preparations.

3. Gloves are not powdered; they are air-tested for pinholes. Gloves are pulled over lab coat sleeves to cover any area of the arm or wrist that might be exposed.
4. We use Lysol Pine Action Cleaner, Household Products Division, Reckitt & Colman, Inc. (Montvale, NJ).
5. Plants are not used after they reached a height of 1 m. Leaf should be fully expanded with a well-developed waxy cuticle. Leaves are removed from darkness so that all stomata are closed. Osmotic potentials of guard cells of leaves with open stomata will not be uniform, and thus, solution D may not be hypertonic to all guard cells, reducing yields. Older leaves may give lower yields of GCP. Higher yields of epidermis from older (larger) leaves may saturate enzymes and reduce the number of GCP released over the 3-h, 15-min digestion period.
6. Temperature has a dramatic effect on survival; highest survival is at 32°C ***(20)***.

References

1. Cowan, I. R. (1982) Regulation of water use in relation to carbon gain in higher plants, in Encyclopedia of Plant Physiology, vol. 12B (Lange, O. L., Nobel, P. S., Osmond, C. B., and Ziegler, H., eds.), Springer, Heidelberg, pp. 589–613.
2. Zeiger, E. (1983) The biology of stomatal guard cells. *Annu. Rev. Plant Physiol.* **34,** 441–475.
3. Assmann, S. M. (1993) Signal transduction in guard cells. *Annu. Rev. Cell Biol.* **9,** 345–375.
4. Outlaw, W. H., Jr., Mayne, B. C., Zenger, V. E., and Manchester, J. (1981) Presence of both photosystems in guard cells of *Vicia faba* L.: implications for environmental signal processing. *Plant Physiol.* **67,** 12–16.
5. Shimazaki, K., Gotow, K., and Kondo, N. (1982) Photosynthetic properties of guard cell protoplasts from *Vicia faba* L. *Plant Cell Physiol.* **23,** 871–879.
6. Gotow, K., Kondo, N., and Syono, K. (1982) Effect of CO_2 on volume change of guard cell protoplasts from *Vicia faba* L. *Plant Cell Physiol.* **23,** 1063–1070.
7. Gotow, K., Shimazaki, K., Kondo, N., and Syono, K. (1984) Photosynthesis-dependent volume regulation in guard cell protoplasts from *Vicia faba* L. *Plant Cell Physiol.* **25,** 671–675.
8. Weyers, J. D. B., Fitzsimmons, P. J., Mansey, G. M., and Martin, E. S. (1983) Guard cell protoplasts—aspects of work with an important new research tool. *Physiol. Plant.* **58,** 331–339.
9. Gotow, K., Taylor, S., and Zeiger, E. (1988) Photosynthetic carbon fixation in guard cell protoplasts of *Vicia faba* L.—evidence from radiolabel experiments. *Plant Physiol.* **86,** 700–705.
10. Assmann, S. M., Simoncini, L., and Schroeder, J. I. (1985) Blue light activates electrogenic ion pumping in guard cell protoplasts of *Vicia faba. Nature* **318,** 285–287.
11. Shimazaki, K., Iino, M., and Zeiger, E. (1986) Blue light-dependent proton extrusion by guard-cell protoplasts of *Vicia faba. Nature* **319,** 324–326.
12. Shimazaki, K. and Kondo, N. (1987) Plasma membrane H^+-ATPase in guard-cell protoplasts from *Vicia faba* L. *Plant Cell Physiol.* **28,** 893–900.
13. Schroeder, J. I., Hedrich, R., and Fernandez, J. M. (1984) Potassium-selective single channels in guard cell protoplasts of *Vicia faba. Nature* **312,** 361–362.
14. Schroeder, J. I., Raschke, K., and Neher, E. (1987) Voltage dependence of K^+ channels in guard cell protoplasts. *Proc. Natl. Acad. Sci. USA* **84,** 4108–4112.
15. Ward, J. M., Pei, J-M., and Schroeder, J. I. (1995) Roles of ion channels in initiation of signal transduction in higher plants. *Plant Cell* **7,** 833–844.
16. Fairley-Grenot, K. and Assmann, S. M. (1991) Evidence for G-protein regulation of inward K^+ channel current in guard cells of fava bean. *Plant Cell* **3,** 1037–1044.
17. Hentzen, A. E., Smart, L. B., Wimmers, L. E., Fang, H. H., Schroeder, J. I., and Bennett, A. B. (1996) Two plasma membrane H^+-ATPase genes expressed in guard cells of *Vicia faba* are also expressed throughout the plant. *Plant Cell Physiol.* **37,** 650–659.
18. Cupples, W., Lee, J., and Tallman, G. (1991) Division of guard cell protoplasts of *Nicotiana glauca* (Graham) in liquid cultures. *Plant Cell Environ.* **14,** 691–697.

19. Herscovich, S., Tallman, G., and Zeiger, E. (1992) Long-term survival of Vicia guard cell protoplasts in cell culture. *Plant Sci.* **81,** 237–244.
20. Roberts, C., Sahgal, P., Merritt, F., Perlman, B., and Tallman, G. (1995) Temperature and abscisic acid can be used to regulate survival, growth, and differentiation of cultured guard cell protoplasts of tree tobacco. *Plant Physiol.* **109,** 1411–1420.
21. Hall, R. D., Riksen-Bruinsma, T., Weyens, G., Lefébvre, M., Dunwell, J. M., and van Tunen, A. (1997) Sugar beet guard cell protoplasts demonstrate a remarkable capacity for cell division enabling applications in stomatal physiology and molecular breeding. *J. Exp. Bot.* **48,** 255–263.
22. Sahgal, P., Martinez, G., Roberts, C., and Tallman, G. (1994) Regeneration of plants from cultured guard cell protoplasts of *Nicotiana glauca* (Graham). *Plant Sci.* **97,** 199–208.
23. Hall, R. D., Pedersen, C., and Krens, F. A. (1994) Regeneration of plants from protoplasts of sugarbeet (*Beta vulgaris* L.), in *Plant Protoplasts and Genetic Engineering,* vol. V (Bajaj, Y. P. S., ed.), Springer-Verlag, Berlin, pp. 16–37.
24. Hall, R. D., Verhoeven, H. A., and Krens, F. A. (1995) Computer-assisted identification of protoplasts responsible for rare division events reveal guard-cell totipotency. *Plant Physiol.* **107,** 1379–1386.
25. Hall, R. D., Riksen-Bruinsma, T., Weyens, G. J., Rosquin, I. J., Denys, P. N., and Evans, I. J. (1996) A high efficiency technique for the generation of transgenic sugar beets from stomatal guard cells. *Nature Biotechnol.* **14,** 1133–1138.
26. Hoagland. D. R. and Arnon, D. I. (1938) The water culture method for growing plants without soil. *Circular of the California Agricultural Experiment Station,* No. 347.
27. Shepard, J. F. and Totten, R. E. (1975) Isolation and regeneration of tobacco mesophyll cell protoplasts under low osmotic conditions. *Plant Physiol.* **55,** 689–694.

23

In Vitro Fertilization with Isolated Single Gametes

Erhard Kranz

1. Introduction

An experimental system has been established to isolate, handle, and fuse single gametes, which enables studies of gamete interaction, of gametic hybridization, and of events that occur immediately after gamete fusion. Also, starting with gamete fusion, the development of a single zygote into an embryo and finally into a plant can be followed in vitro. With experimental access to single gametes, to gamete fusion, and to events after fertilization, the consequences and significance of such early events on plant formation can be studied.

In contrast to animal and lower plant systems, which use naturally free-living gametes, in vitro fertilization with higher plants presupposes the isolation of gametes. Sperm cells have to be isolated from pollen grains or tubes, egg and central cells from an embryo sac, which also normally contains two synergids, and some antipodal cells. Moreover, the embryo sac is generally embedded in the nucellar tissue of the ovule. Double fertilization is the fusion of one sperm with the egg to create the embryo and the fusion of the other sperm with the central cell to form the endosperm *(1)*. In vivo this occurs deep within ovule tissues in the embryo sac and normally with the help of one of the two synergids. In vitro, these fusions are performed without any surrounding cells. Also, by use of a microculture system, zygote, embryo, and plant development take place in the absence of mother tissue, as is the case with endosperm formation. Therefore, in vitro fertilization with isolated, higher plant gametes is different from that in vivo.

In vitro fertilization includes the combination of three basic microtechniques:

1. The isolation and selection of male and female gametes.
2. The fusion of pairs of gametes.
3. Single cell culture.

From: *Methods in Molecular Biology, Vol. 111: Plant Cell Culture Protocols*
Edited by: R. D. Hall © Humana Press Inc., Totowa, NJ

Cells of the embryo sac are usually isolated using mixtures of cell-wall-degrading enzymes in combination with a manual isolation procedure (for review, *see* **2**). It is also possible to isolate female gametes of, for example barley, wheat, and rapeseed using only mechanical means ***(3–5)***. The same can be achieved in maize ***(6)***, but it is more effective to treat the maize ovule tissue with cell-wall-degrading enzymes for a short period prior to the manual isolation step in order to soften the nucellar tissue. Sperm cells are obtained from pollen grains (tricellular pollen) or tubes (bicellular pollen) mainly by osmotic bursting or by squashing or grinding of the material (for review, *see* **2**). Egg cells, central cells, and synergids can be isolated only in small numbers. However, the possibility of selection, individual transfer, and handling of single-gametes enables studies at the single cell level, for example, on cell physiology, adhesion, and putative recognition events with one egg and one sperm. Furthermore, with microtechniques originally developed for somatic protoplasts ***(7,8)***, defined gamete fusion is possible ***(9)***.

Isolated gametes, generative cells, nongametic cells of the embryo sac, and somatic cells have been fused electrically ***(6,10–12)*** and chemically, by calcium ***(13,14)*** and by polyethylene glycol ***(15)***. Fusiogenic media, for example, those including calcium, might be used to determine conditions, which promote in vivo membrane fusion. The fusion method using electrical pulses for cell fusion is described here in detail, because it is well established and an efficient part of the in vitro fertilization procedure. Early steps in zygote development can be analyzed in microdroplet culture under defined conditions without feeder cells ***(6,16,17)***. However, to achieve sustained growth of in vitro zygotes, cocultivation with feeder cells is necessary ***(3,10,12)***. Embryogenesis and plant regeneration from isolated male and female gametes fused in vitro have, so far, been achieved exclusively using nurse culture and electrofusion techniques ***(18)***. Moreover, because of the efficient production of in vitro zygotes by the latter method, the numbers of such cells allow molecular analyses to be performed at the single-cell level for gene expression studies of known genes involved in events during fertilization and zygote formation ***(19)***. Also by use of reverse transcriptase-polymerase chain reaction (RT-PCR) methods, cDNA-libraries have been generated from a small, but sufficient number of egg cells ***(20)*** and in vitro zygotes ***(21)*** to isolate egg-cell-specific and fertilization-induced genes. The potential of in vitro fertilization techniques for studies on fertilization mechanisms and development have been reviewed (for example, **22,23**). Skill and experience in the handling of individual cells are necessary. Unless otherwise mentioned, the procedures are described for maize *(Zea mays)*.

2. Materials

2.1. General Materials

1. Plant material grown in the greenhouse under standard conditions (*see* **Note 1**).
2. 3-cm plastic dishes.
3. Cover slips, siliconized at the edges with Repel-Silane (Merck, Darmstadt, Germany; Pharmacia Biotech, Uppsala, Sweden) and UV-sterilized.
4. Mineral oil, autoclaved (paraffin liquid for spectroscopy, Merck, Darmstadt, Germany).
5. Mannitol solution adjusted to 570 mosM/kg H_2O before autoclaving (concentration about 530 m*M*), after autoclaving 600 mosM/kg H_2O.
6. Mixture of cell-wall-degrading enzymes containing: 1.5% pectinase (Serva, Heidelberg), 0.5% pectolyase Y23 (Seishin, Tokyo), 1.0% cellulase Onozuka RS (Yakult Honsha, Tokyo), and 1.0% hemicellulase (Sigma) dissolved in bidistilled water, adjusted to 570 mosM/kg H_2O with mannitol at pH 5.0, and filter-sterilized before freezing. Store 10-mL aliquots at –20°C.
7. Glass needles with fine tips, preferably prepared using a microforge.
8. Capillaries, tip openings 100–300 µm (drawn by hand) and 20 µm (drawn by a puller).
9. Electrodes (platinum wire, diameter 50 µm) fixed to an electrode support, which is mounted under the condenser of the microscope.
10. Feeder: maize suspension cells (*see* **Note 2**).
11. Medium for in vitro zygotes and feeder cells (ZMS). MS medium ***(24)*** with the modifications NH_4NO_3 (165 mg/L) and organic constituents. These are: nicotinic acid (1.0 mg/L), thiamine-HCl (10.0 mg/L), pyridoxine-HCl (1.0 mg/L), L-glutamine (750 mg/L), proline (150 mg/L), asparagine (100 mg/L), myo-inositol (100 mg/L) ***(25)***, and 2,4-D (2.0 mg/L), adjusted to 600 mosM/kg H_2O with glucose, pH 5.5, and filter-sterilized ***(18)***.
12. "Millicell-CM" inserts (diameter 12 mm), Millipore, Bedford, MA.
13. Regeneration media (RMS): Filter-sterilized MS media ***(24)*** solidified with 4 g/L agarose (type I-A; Sigma) and the modifications: RMS1: medium without hormones and supplemented with 60 g/L sucrose; RMS2: same medium as RMS1, but containing 40 g/L sucrose; RMS3: medium without hormones and supplemented with 10 g/L sucrose, macro- and microsalts half-concentration ***(18)***.

2.2. Equipment

1. Laminar flow box.
2. Inverted microscope.
3. Sliding stage for the insertion of a cover slip and a plastic dish, self-made (*see* **Fig. 1**).
4. Micropump: computer-controlled dispenser/dilutor (Microlab-M, Hamilton, Darmstadt, Germany).
5. Electrofusion apparatus (CFA 500, Krüss, Hamburg, Germany).
6. Electrode support, self-made.
7. Positioning system (optional), especially useful for gently moving the electrodes along the *z*-axis by a step motor (type MCL, Lang, Huettenberg, Germany).

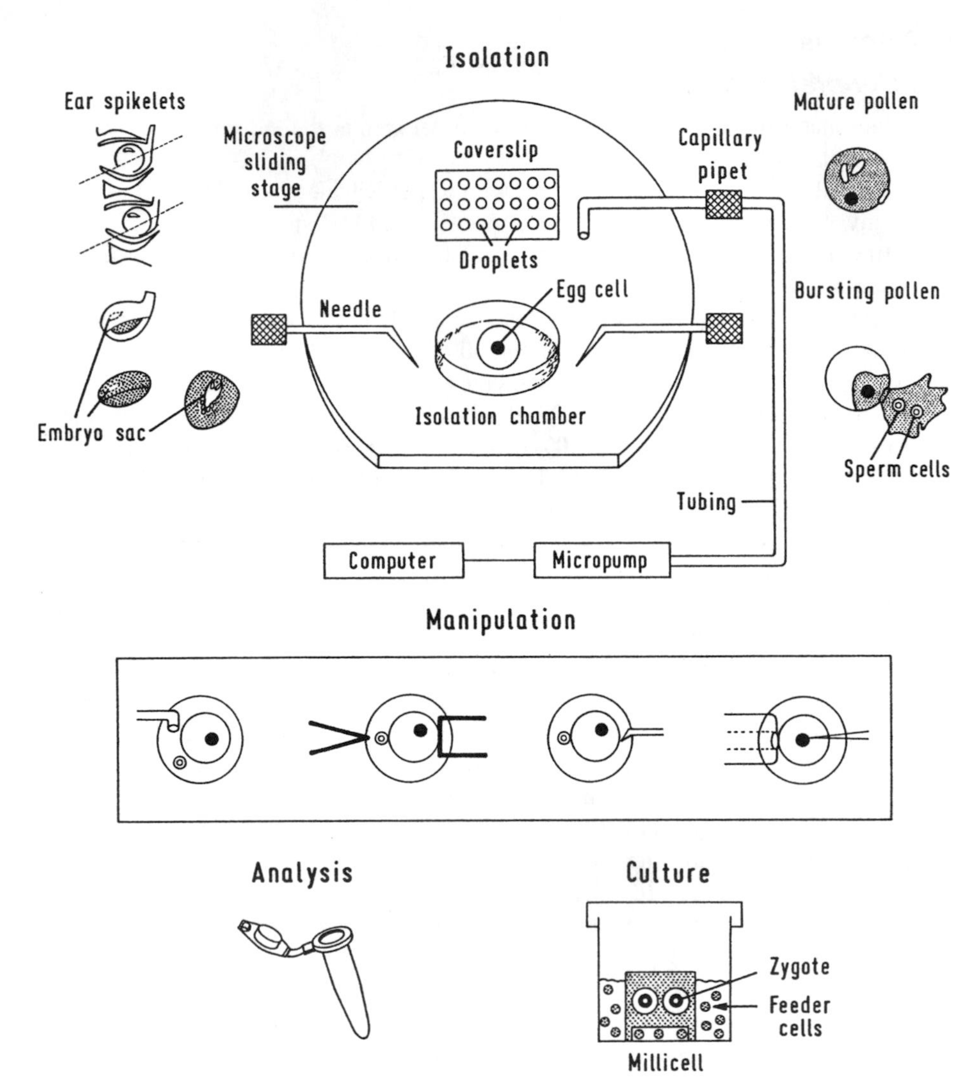

Fig. 1. Setup for isolation, individual selection, and transfer of maize gametes. Ear spikelets are cut as indicated (dotted lines). After removing integuments, cells of the embryo sac are isolated from nucellar tissue pieces in the isolation chamber (plastic dish) manually with needles following transfer with a capillary into microdroplets on a cover slip. Subsequently, sperm cells are selected in the isolation chamber after release from the pollen grains by osmotic shock and transferred into microdroplets. Manipulations are performed in these microdroplets under microscopic observation: physiological and biochemical studies, cell fusion, and microinjection for further analysis and culture.

3. Methods

3.1. Isolation of Gametes

1. Collect pollen from freshly dehisced anthers. Pollen can be stored for several hours at room temperature in plastic dishes containing a wet filter paper fixed to the lid to provide a moistened atmosphere.
2. Collect ears after silk emergence, and sterilize the outer leaves with ethanol (70%).
3. Dissect 20–30 nucellar tissue pieces from the ovules under a dissecting microscope (*see* **Fig. 1**). The embryo sac should be visible in the tissue pieces. Collect the tissue pieces in 1 mL mannitol solution (600 mosM/kg H_2O) in 3-cm plastic dishes, and add 0.5 mL enzyme mixture (*see* **Notes 3** and **7**). Incubate the mixture at room temperature for 30 min without shaking. After treatment, the dishes can be stored in the refrigerator at 6°C (*see* **Note 4**).
4. Overlay the siliconized and UV-sterilized cover slip with 300 µL autoclaved mineral oil, and inject 2 µL mannitol droplets (600 mosM/kg H_2O) in three rows, each with 10 droplets using a microcapillary and a micropump. Take care that the droplets spread over the glass surface.
5. Isolate egg cells and the other cells of the embryo sac directly in the incubation dish with glass needles under microscopic observation, and transfer the cells by a microcapillary (tip opening 100–200 µm for egg cells and 300 µm for central cells) into the microdroplets using the micropump (*see* **Notes 1** and **4**, and **Fig. 1**).
6. Overlay about 1000 pollen grains in a 3.5-cm diameter plastic dish with 1.5 mL mannitol solution (600 mosM/kg/H_2O). After grain bursting, transfer the sperm cells (tip opening of the capillary 20 µm) by use of the micropump into the microdroplets containing egg cells (*see* **Notes 5** and **6**, and **Fig. 1**).

3.2. Fusion of Gametes

1. Fix two electrodes (50-µm diameter platinum wire) to an electrode support. Before use, sterilize the ends of the electrodes in a weak flame. Mount the support under the condenser of the microscope. Adjust the electrodes to the crosshairs position and lower these onto the cover slip and into one droplet.
2. Align and fix the two gametes at one electrode. Prepare and adjust the electrodes carefully as demonstrated in **Fig. 1**. By moving the microscope stage, first move one egg cell toward the electrode. Finally, the egg cell is fixed to the electrode by dielectrophoresis (1 MHz, 70 V × cm^{-1}). Using the same procedure, the sperm cell is fixed to the egg cell. Now the final distance of the electrodes is adjusted to be approximately twice the sum of the diameters of the cells.
3. Egg–sperm fusion is induced by applying a single, or a maximum of three negative DC-pulses (50 µs; 0.9–1.0 kV × cm^{-1}) (*see* **Note 7**).
4. The fusion products are removed from the electrode by gently moving the sliding stage. Move the electrodes out of the droplet (*see* **Notes 8** and **9**).

3.3. Zygote and Embryo Culture, Plant Regeneration

1. In a "Millicell-CM" insert, place 100 µL ZMS medium, and insert it into a 3.5-cm plastic dish, containing 1.5 mL of a feeder suspension (*see* **Notes 2** and **9**). Transfer the fusion products into the insert using a microcapillary. During the next day, the dish can be placed on a rotary shaker (50–70 rpm). This may be advantageous when suspension feeder cells are used. Culture conditions are 26 ± 1.0°C, a light/dark cycle of 16/8 h, and a light intensity of about 50 µmol/m^2/s.
2. About 10–14 d after gamete fusion in liquid culture, transfer embryos (minimal size of 0.4-mm length) by use of a Pasteur pipet onto solid regeneration medium without hormones and supplemented with 60 g/L sucrose (1.5 mL RMS1 medium) in a 3.5-cm plastic dish for a first passage of 2 wk (*see* **Note 10**). When a coleoptile and roots are formed, transfer the structures for another 1–2 wk onto 1.5 mL of RMS2 medium containing 40 g/L sucrose. Transfer plantlets into a glass jar containing 50 mL of RMS3 medium (10 g/L sucrose, macro- and microsalts half-concentration, solidified with 4 g/L agarose). After about another 2 wk, transfer the maize plants (leaf lengths about 15–20 cm) to soil.

4. Notes

1. To obtain a high yield of egg cells, use only healthy and well-growing donor plants. Using maize, routinely 20–40 egg cells can be isolated/experimenter/d, and the same number of fusion products can be obtained. Under optimal conditions, up to 60 zygotes can be created by 1 person/d. The procedure is line-independent.
2. Feeder cells should grow actively. It can be useful to subculture suspension cells twice a week before use. Microspore cultures also can be used as a feeder system ***(3)***. Although the feeding effect can also be achieved with nurse cells of a different species than the zygote, make sure that culture conditions meet the requirements of both the zygotes and the feeder cells. For example, when suspension cells are used for feeding, these have to be adapted (usually several passages before use) to a higher osmolality than normally used for these cells to meet the requirement for egg protoplasts and zygotes.
3. Depending on the quality of the material, the use of mannitol solution (650 mosM/kg H_2O after autoclaving) in this step can be advantageous in improving cell fusion.
4. When ovule tissue is treated with cell-wall-degrading enzymes, the concentration of the enzymes should be kept low, and treatment should be short to avoid spontaneous fusion of the cells of the embryo sac.
5. Sperm cells are short-lived after isolation. Therefore, it is important to optimize conditions that guarantee the maintenance of good quality of these cells during a suitable time necessary to perform cell fusion. For example, whereas sperm cells of maize can be used for fusion within 30 min after isolation ***(10,11)***, sperm cells of wheat are useful for fusion for only few minutes after isolation ***(6,12)***. If possible, use oval- or spindle-shaped sperm cells for fusion, because these fuse more efficiently with egg cells than round ones (***6***; and *see* **Fig. 2**). Occasionally, the

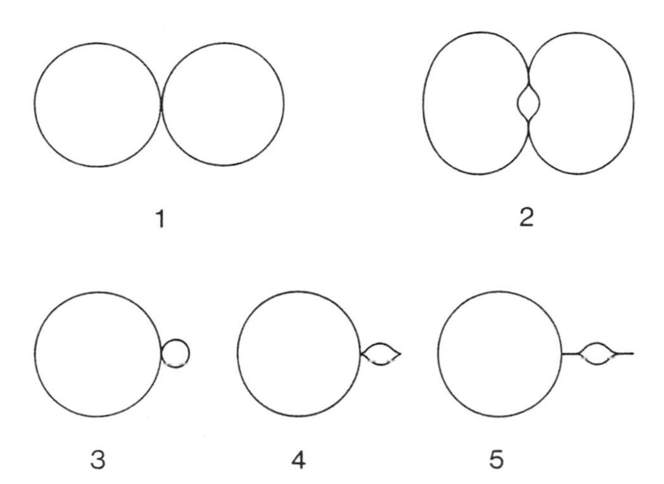

Fig. 2. Influence of turgor and shape of protoplasts on the effectiveness and speed of cell fusion. 1, Protoplasts (pairs of two somatic, two egg or one somatic and one egg protoplast) with high turgor are attached only locally at one point resulting in high frequency of fusion. 2, Same protoplast pairs, but with low turgor are attached at a broader area creating space between the protoplasts, which results in only a low frequency or no fusion. 3, Alignment of an egg and sperm protoplast each with a high turgor, results in high frequency of fusion. The speed and effectiveness of fusion is lower with low turgor. Speed of fusion and effectiveness is high using oval-shaped sperm cells ***(4)*** or after alignment of the pointed end of one of the tail-like extensions of the spindle-shaped sperm cell on the egg membrane ***(5)***.

two sperm cells spontaneously fuse after release from the pollen grain. To avoid the use of such a fusion product, select sperm cells only when two are visible in close proximity to each other and separated from others.

6. For efficient cell fusion, the turgor of the cells should be high enough (*see* **Fig. 2**). Avoid shifts from low to higher osmolalities between isolation and fusion media. Use the same medium preparation. Take care that the microdroplets always are covered by oil. Otherwise, the osmolality of the medium can dramatically increase with deleterious effects to the cells.
7. With well-prepared electrodes, the fusion frequency will be nearly 100%. When no cell fusion occurs, lower the distance of the two electrodes and pulse again. Sometimes some remnant cell-wall material is attached to the egg membrane, which prevents fusion. This can be overcome by rotation of the egg cell to establish a new fusion site. Another reason for nonfusion can be that the turgor of the protoplasts is too low (*see* **Fig. 2**). If this is the case, transfer the protoplasts into a fusion medium with a slightly lower osmolality. When a lot of protoplasts burst after the applied pulse, the osmolality of the fusion medium is too low or the electrodes are not well adjusted.

For the fusion, combination of cells of the same size, for example, egg + egg or egg + somatic protoplast, use dielectric alignment conditions of 1 MHz, 30 V/cm and DC pulses of 0.3 kV × cm^{-1}, 50 μs.

8. There are only two reports describing calcium-mediated egg–sperm fusion ***(13,14)***. It has to be seen, if this time-consuming method can be used efficiently for routine work. The alignment of the two gametes can be performed using a glass needle in a microdroplet. When adhesion has occurred, egg and sperm cells can be fused in mannitol solution (about 400–500 mosM/kg H_2O) containing 1–50 m*M* calcium ($CaCl_2$). As with electrical fusion, the turgor of the protoplasts is important. Thus, determine the optimum osmolality of the fusion medium.
9. Single egg cells and zygotes can be encapsuled in droplets of agarose-containing mannitol solution (ultralow gelling temperature agarose, type IX, Sigma, 1%). After cell transfer, the droplets are solidified for 10 min at 6°C. The droplets can be fixed on small plastic pieces, which can be useful, for example, for fixation, electron microscopical studies, and transportation of single cells to another laboratory for further analyses ***(16,17)***. To follow development at a fixed position, zygotes can be embedded in agarose-containing medium, previously filled into the "Millicell-CM" inserts (100 μL, ultralow gelling temperature agarose, type IX, Sigma, 1%).
10. To obtain secondary embryogenesis, transfer embryos from liquid culture medium (ZMS) onto solid RMS1 medium supplemented with 2,4-D (2 mg/L) for two to three passages.

References

1. Goldberg, R. B., de Paiva, G., and Yadegari, R. (1994) Plant embryogenesis: Zygote to seed. *Science* **266,** 605–614.
2. Theunis, C. H., Pierson, E. S., and Cresti, M. (1991) Isolation of male and female gametes in higher plants. *Sex. Plant Reprod.* **4,** 145–154.
3. Holm, P. B., Knudsen, S., Mouritzen, P., Negri, D., Olsen, F. L., and Roué, C. (1994) Regeneration of fertile barley plants from mechanically isolated protoplasts of the fertilized egg cell. *Plant Cell* **6,** 531–543.
4. Kovács, M., Barnabás, B., and Kranz, E. (1994) The isolation of viable egg cells of wheat (*Triticum aestivum* L.). *Sex. Plant Reprod.* **7,** 311–312.
5. Katoh, N., Lörz, H., and Kranz, E. Isolation of viable egg cells of rape (*Brassica napus* L.). *Zygote* **5,** 31–33.
6. Kranz, E., von Wiegen, P., and Lörz, H. (1995) Early cytological events after induction of cell division in egg cells and zygote development following in vitro fertilization with angiosperm gametes. *Plant J.* **8,** 9–23.
7. Koop, H.-U. and Schweiger, H.-G. (1985) Regeneration of plants after electrofusion of selected pairs of protoplasts. *Eur. J. Cell Biol.* **39,** 46–49.
8. Spangenberg, G. and Koop, H.-U. (1992) Low density cultures: microdroplets and single cell nurse cultures. *Plant Tissue Culture Manual* A10 (Lindsey, K., ed.), Kluwer Academic Publishers, Dordrecht, pp. 1–28.

9. Kranz, E. (1992) *In vitro* fertilization of maize mediated by electrofusion of single gametes. *Plant Tissue Culture Manual,* E1 (Lindsey, K., ed.), Kluwer Academic Publishers, Dordrecht, pp. 1–12.
10. Kranz, E., Bautor, J., and Lörz, H. (1991) *In vitro* fertilization of single, isolated gametes of maize mediated by electrofusion. *Sex. Plant Reprod.* **4,** 12–16.
11. Kranz, E., Bautor, J., and Lörz, H. (1991) Electrofusion-mediated transmission of cytoplasmic organelles through the *in vitro* fertilization process, fusion of sperm cells with synergids and central cells, and cell reconstitution in maize. *Sex. Plant Reprod.* **4,** 17–21.
12. Kovács, M., Barnabás, B., and Kranz, E. (1995) Electro-fused isolated wheat (*Triticum aestivum* L.) gametes develop into multicellular structures. *Plant Cell Rep.* **15,** 178–180.
13. Kranz, E. and Lörz, H. (1994) *In vitro* fertilisation of maize by single egg and sperm cell protoplast fusion mediated by high calcium and high pH. *Zygote* **2,** 125–128.
14. Faure, J.-E., Digonnet, C., and Dumas, C. (1994) An *in vitro* system for adhesion and fusion of maize gametes. *Science* **263,** 1598–1600.
15. Sun, M., Yang, H., Zhou, C., and Koop, H.-U. (1995) Single-pair fusion of various combinations between female gametoplasts and other protoplasts in *Nicotiana tabacum. Acta Botanica Sinica* **37,** 1–6.
16. Faure, J.-E., Mogensen, H. L., Dumas, C., Lörz, H., and Kranz, E. (1993) Karyogamy after electrofusion of single egg and sperm cell protoplasts from maize: Cytological evidence and time course. *Plant Cell* **5,** 747–755.
17. Tirlapur, U. K., Kranz, E., and Cresti, M. (1995) Characterization of isolated egg cells, *in vitro* fusion products and zygotes of *Zea mays* L. using the technique of image analysis and confocal laser scanning microscopy. *Zygote* **3,** 57–64.
18. Kranz, E. and Lörz, H. (1993) In vitro fertilization with isolated, single gametes results in zygotic embryogenesis and fertile maize plants. *Plant Cell* **5,** 739–746.
19. Richert, J., Kranz, E., Lörz, H., and Dresselhaus, T. (1996) A reverse transcriptase-polymerase chain reaction assay for gene expression studies at the single cell level. *Plant Sci.* **114,** 93–99.
20. Dresselhaus, T., Lörz, H., and Kranz, E. (1994) Representative cDNA libraries from few plant cells. *Plant J.* **5,** 605–610.
21. Dresselhaus, T., Hagel, C., Lörz, H., and Kranz, E. (1996) Isolation of a full-length cDNA encoding calreticulin from a PCR library of *in vitro* zygotes of maize. *Plant Mol. Biol.* **31,** 23–34.
22. Kranz, E. and Dresselhaus, T. (1996) *In vitro* fertilization with isolated higher plant gametes. *Trends Plant Sci.* **1,** 82–89.
23. Rougier, M., Antoine, A. F., Aldon, D., and Dumas, C. (1996) New lights in early steps of *in vitro* fertilization in plants. *Sex. Plant Reprod.* **9,** 324–329.
24. Murashige, T. and Skoog, F. (1962) A revised medium for rapid growth and bio-assays with tobacco tissue cultures. *Physiol. Plant.* **15,** 473–497.
25. Olsen, F. L. (1987). Induction of microspore embryogenesis in cultured anthers of Hordeum vulgare. The effects of ammonium nitrate, glutamine and asparagine as nitrogen sources. *Carlsberg Res. Commun.* **52,** 393–404.

V

Protocols for Genomic Manipulation

24

Protocols for Anther and Microspore Culture of Barley

Alwine Jähne-Gärtner and Horst Lörz

1. Introduction

The establishment of true breeding lines is a critical step for variety development in most crop plants. Traditionally, plant breeders have achieved homozygosity for all genes by using the time- and labor-intensive methods of self-fertilization or backcrossing. The phenomenon of parthenogenesis allows the use of androgenetic in vitro methods in order to shorten the time needed for the production of homozygous plants. It is possible to switch microspore development from the normal gametophytic pathway into the sporophytic pathway. The process of microspore embryogenesis makes it possible to regenerate homozygous plants originating from single cells. In barley, there is a high percentage—up to 90% dependent on the genotype—of dihaploid regenerants owing to a single autoendoreduplication of the genome during the first division of the microspore. Accordingly, there is no need for an application of agents, such as, e.g., colchicine, to induce chromosome doubling.

The most common androgenetic in vitro method is the culture of intact anthers, which is a relatively simple and fast method, and requires only minimal facilities. Anther culture is widely used for the production of homozygous plants. Research on barley anther culture was initiated by Clapham *(1)*, and since then, this technique has been optimized extensively. Many efforts have been made in order to develop efficient and reliable culture systems. Several important modifications have enabled plant breeders to use anther culture very successfully as a routine method in breeding programs for the production of pure lines. The development of efficient anther culture methods and their applications have been the subject of several reviews *(2–5)*.

From: *Methods in Molecular Biology, Vol. 111: Plant Cell Culture Protocols*
Edited by: R. D. Hall © Humana Press Inc., Totowa, NJ

Alternatively, microspores can be mechanically isolated and cultured independent of the anther. They represent a unique experimental material, since it is possible to isolate large numbers of single, almost synchronously developing cells, which will eventually undergo embryogenesis. Therefore, isolated microspores are considered as ideal target cells for in vitro selection and transformation. Progress in the development of barley microspore culture has been very rapid, especially in the last few years. There are now several methods for the isolation and culture of barley microspores that have been reviewed in detail ***(5,6)***. Meanwhile, fertile, transgenic barley plants have been produced by particle bombardment of isolated microspores ***(7)***.

There are a huge number of publications on the optimization of methods for barley anther and microspore culture. A review of the literature shows that it is obviously necessary to refine, in each laboratory, the experimental parameters dependent on the local conditions.

It can be frequently observed that the most decisive step in barley anther and microspore culture is the growth of the donor plants. Only microspores of superior quality are a suitable starting material for successful culture experiments. The number of microspores capable of division and regeneration can vary widely within one variety because of the environmental conditions in which the plants are grown ***(8)***. The vigor of donor plants is influenced by several parameters, the most important of which are: photoperiod, light intensity and spectrum, temperature, and nutrition ***(6)***. For barley, high light intensity (20,000 lx) and low temperatures are favorable for the support of androgenesis ***(9)***. Plant growth takes longer under low temperatures. Therefore, the development of microspores is more homogeneous and it is easier to harvest the spikes at the ideal stage. Furthermore, it has been shown that the culture response is optimal when spikes are harvested at the beginning of the flowering period and declines with increasing plant age ***(10,11)***. For establishment of a reliable culture system, the donor plants should be grown under controlled conditions, either in a greenhouse or in a phytotron. Optimal donor plants should be free of diseases and pests. However, pest control procedures should be avoided, since they reduce explant quality and subsequent culture response ***(12,13)***. Here the growth conditions are described that could be realized with our facilities in the institute.

For successful microspore culture experiments, it is important to determine a suitable developmental stage for the microspores to be used. In barley, most scientists use the mid- to late uninucleate stage (**Fig. 1**). The habit of the tiller can—to a certain extent—be correlated to the developmental stage of the microspores. Usually, spikes can be preselected on the basis of the interligule length between the flag leaf and the second leaf, and on the thickness of the tiller.

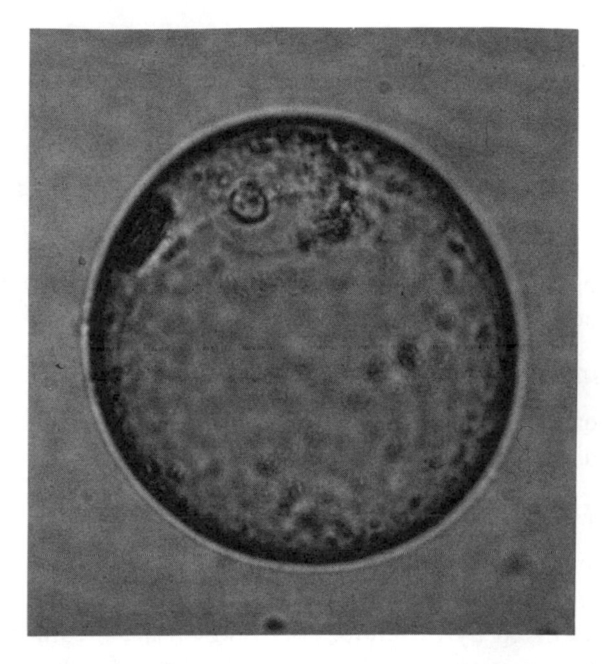

Fig. 1. Developmental stage of microspores, which is suitable for barley anther and microspore culture. Miduninucleate stage, the nucleus is close to the germ pore.

For switching the genetic program of barley pollen from the gametophytic to the sporophytic pathway, a signal is necessary. In barley, two procedures are being used routinely: cold pretreatment and starvation. The storage of spikes for 3–5 wk at 4°C, at relatively high humidity in the dark ***(14)*** is a common and simple method. Another, more labor-intensive procedure is the incubation of isolated anthers for 4 d on a medium containing 0.3 *M* mannitol instead of a metabolizable sugar at 25°C ***(15)***. Each of the respective methods has been reported to be more effective than the other ***(16–18)***, so that it can be recommended to compare both kinds of pretreatments. In our laboratory, the cold pretreatment has proven to be more successful and, furthermore, much easier to perform. It is therefore the method that is described in this protocol.

The optimization of culture media for barley anther and microspore culture has been the subject of many investigations ***(5)***. Major improvements have been achieved through modification of the nitrogen (N) supply ***(19)*** and the carbohydrate source ***(20)***. The medium that is presented here comprises all major improvements and has been found to be suitable for anther as well as microspore culture.

2. Materials

2.1. Donor Plants

1. The model genotype in barley anther and microspore culture is a two-rowed winter-type barley (*Hordeum vulgare* L. cv. Igri).
2. Controlled greenhouse or phytotron (14/12°C day/night, 16 h light, 10,000–16,000 lx, lamp type F96T12/CW/VHO 1500, 215 W, Philips, relative humidity 70–90%).
3. Vernalization chamber (2°C, 9 h light, 4000 lx, lamp type VIALOX NAV-T, 400 W, Osram, relative humidity 95%).

2.2. Pretreatment

For cold pretreatment, two-compartment Petri dishes (Greiner, Germany) are recommended.

2.3. Culture Media

The culture medium described in **Table 1** (*see* **Note 1**) is suitable for anther as well as for microspore culture. The pH of media is adjusted to 5.6–5.8, and solutions are sterilized by filtration. For solid culture medium, solutions are prepared double-concentrated and mixed with an equal volume of autoclaved 0.4% GelRite (Roth, Germany) in water. For anther culture, a final concentration of 20% Ficoll 400 (Sigma, Germany) can also be included to increase the buoyancy of the medium. Solid medium used for isolated microspore culture is identical to liquid medium, but the L-glutamine is omitted. For regeneration, the carbohydrate content of the medium is reduced to 30 g/L, and 6-BAP and L-glutamine are omitted.

Medium 1: solid medium (anther culture).
Medium 2: liquid medium supplemented with 20% Ficoll (anther culture).
Medium 3: liquid medium (microspore culture).
Medium 4: solid medium without L-glutamine (microspore culture).
Medium 5: solid medium with reduced maltose (30 g/L) and without 6-BAP and L-glutamine.

2.4. Anther and Microspore Isolation

The use of fine-tipped forceps, e.g., Aesculap BD321, is recommended for the removal of awns prior to cold pretreatment and for the preparation of intact anthers. For determination of the microspore stage and the examination of microspore cultures, a light microscope is needed.

For the isolation of microspores, the following materials are needed:

1. Forceps and scalpel.
2. Waring Micro Blendor (Eberbach Corporation, US) (Blender Cup autoclaved).

Table 1
Composition of the Medium Used for Barley Anther and Microspore Culture *(19,21)*

Macroelements (mg/L)	
NH_4Cl	76
KNO_3	1300
$CaCl_2 \cdot 2H_2O$	450
$MgSO_4 \cdot 7H_2O$	350
KH_2PO_4	200
Na_2EDTA	37
$FeSO_4 \cdot 7H_2O$	28
Microelements (mg/L)	
H_3BO_3	5.00
$MnSO_4 \cdot 4H_2O$	25.00
$ZnSO_4 \cdot H_2O$	7.50
KI	0.75
$Na_2MoO_4 \cdot 2H_2O$	0.25
$CuSO_4 \cdot 5H_2O$	0.025
$CoCl_2 \cdot 6H_2O$	0.025
Vitamins (mg/L)	
Myoinositol	100
Nicotinic acid	1
Pyridoxine HCl	1
Thiamine HCl	10
Amino acids (mg/L)	
L-Glutamine	420
Sugars (g/L)	
L-Maltose	50
Growth regulator (mg/L)	
6-BAP	1

3. Centrifuge (required speed: 714 rpm, e.g., Sigma 3K12).
4. 100-μm nylon sieve (autoclaved).
5. 10- and 50-mL sterile conical polystyrene tubes.
6. Hemocytometer.
7. 0.3 *M* mannitol (autoclaved).
8. 19% Maltose (autoclaved).

3. Methods

3.1. Growth of Donor Plants

1. Germinate seeds in a peat–soil mix in growth chamber.
2. After 2–3 wk, vernalize the Igri seedlings for a period of 8–10 wk.

3. Add fertilizer to the plants starting 8 wk after vernalization (0.7% Wuxal, Schering; N:P:K = 12:4:6) with each watering.
4. Two to 4 wk before the first spikes are harvested, transfer the plants to a controlled greenhouse (18/14°C day/night, 16 h light, at least 18,000–20,000 lx, lamp type SON-T Agro, Philips, relative humidity 70–90%).
5. After transfer, add one teaspoon of Nitrophoska blue (N:P:K:Mg = 12:12:17:2), and increase fertilizing to 1% Wuxal. The application of pesticides should be avoided as far as possible, and tillering plants should not be treated at all.
6. Select tillers on the basis of the thickness of the spike, the interligule length between the flag leaf, and the second leaf and emergence of the awns (*see* **Note 2**). In the case of Igri, the appropriate morphological stage for harvesting spikes occurs when the awns have emerged about 0.3–0.5 cm from the flag leaf. The distance between the penultimate leaf and the flag leaf is usually about 3–6 cm.

3.2. Pretreatment

1. Surface-sterilize freshly harvested spikes with an aerosol of 70% ethanol.
2. Remove the ensheathing leaves under sterile conditions.
3. Stage the microspores by removing an anther from the floret in the central part of the spike.
4. Squash the anther in a drop of water (*see* **Note 3**).
5. Examine the microspores microscopically. Only those spikes with microspores at the mid- to late uninucleate stage should be used for cold pretreatment.
6. Remove awns from the spikes under sterile conditions in a laminar flow bench. About five spikes can be transferred to one-half of a two-compartment Petri dish. A drop of water is placed in the other half in order to maintain a relatively high humidity.
7. Perform the cold pretreatment by wrapping the dishes in aluminum foil, and store them for 2–5 wk at 4–6°C.

3.3. Anther Culture

After cold pretreatment, anthers can be prepared from the spikes. During cold pretreatment, the anthers usually change their color from green to yellow. Anthers from the base and the top of the spike should not be used, since they are generally not at the appropriate stage.

1. Isolate anthers with fine-tipped forceps, and avoid damage to the anthers.
2. Place all anthers from one spike on solid medium (medium 1) or on Ficoll-containing medium (medium 2) (**Fig. 2A**).
3. Incubate the dishes in the dark for 4–5 wk at 26°C.
4. After 4 wk, assess the anther culture response by counting the number of responding anthers, i.e., the anthers that gave rise to embryogenic callus (**Fig. 2B**).
5. Transfer callus carefully to regeneration medium (medium 5), and incubate for about 3–4 wk in the light (16 h) at 24°C.

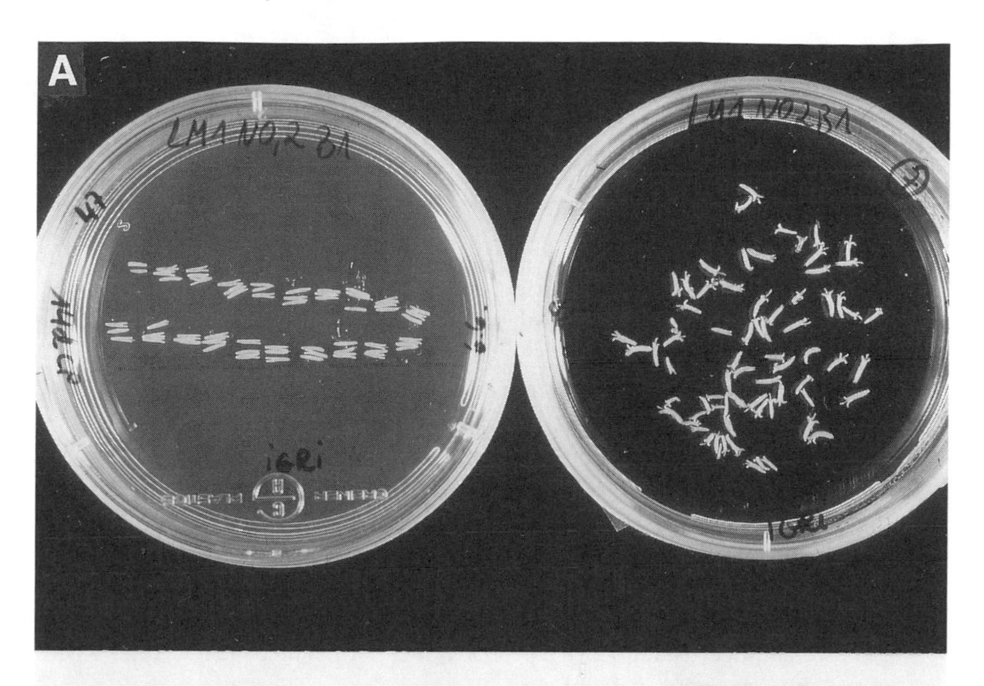

Fig. 2. Barley anther culture. **(A)** Freshly isolated anthers on solid (left dish) and on Ficoll medium (right dish). **(B)** Anther-derived callus after 3 wk on solid (left dish) and on Ficoll medium (right dish).

6. Transfer plantlets to soil, and determine the number of regenerants.
7. Grow regenerants initially in soil under glass or plastic boxes in order to acclimatize the plantlets.
8. Vernalize vigorous plants, and grow to maturity.

3.4. Microspore Culture

Up to 30 cold-pretreated spikes can be used for one isolation procedure *(19)*.

1. Cut the spikes with a scalpel into 1-cm segments, and discard the top and the bottom segment.
2. Transfer the segments with a forceps into a Waring Microblender, and add 20 mL of 0.3 *M* mannitol.
3. Isolate microspores by blending twice for 5 s at low speed.
4. Filter the crude microspore preparation through a 100-µm sieve, and retransfer the material remaining on the sieve back into the blender.
5. Repeat the whole procedure three times with 10 mL of 0.3 *M* mannitol.
6. Transfer the whole extract into a 50-mL tube, and centrifuge for 8 min at 85*g* (20°C).
7. Remove the supernatant with a pipet. Do not pour off, since the microspore pellet is soft and resuspends quickly.
8. Resuspend the pellet in 8 mL of 19% maltose, and transfer the suspension to a 10-mL tube.
9. Carefully place a 1-mL layer of 0.3 *M* mannitol on top.
10. Centrifuge for 10 min as described above. The fraction of viable microspores (**Fig. 3B**) is located in a band at the mannitol/maltose interphase (**Fig. 3A**).
11. Collect the band carefully with a 1-mL pipet, and transfer it to a 50-mL tube.
12. Add 20–30 mL of 0.3 *M* mannitol.
13. Determine the total number of microspores with a hemocytometer. Additionally, the viability of microspores can be determined with fluorescein diacetate *(22)*.
14. Centrifuge the microspore suspension as before, remove the supernatant carefully, and add liquid culture medium (medium 3) to final density of $2–5 \times 10^5$ microspores/mL.
15. Culture 1-mL aliquots in Petri dishes (3-cm Ø) or in 6-well plates.
16. Keep cultures in the dark at 26°C, and add after 1–2 wk, 1 mL of fresh liquid culture medium (medium 3).
17. Determine the percentage of proliferating microspores (**Fig. 3C**) after 2 wk.
18. After 3–4 wk, transfer the micospore-derived aggregates onto solid medium without L-glutamine (medium 4) and culture at 26°C in the dark.
19. Transfer cultures after 1 wk to the light for regeneration (16 h, 24°C).
20. Subculture at intervals of 2 wk, and transfer shoots to regeneration medium (medium 5).
21. Transfer regenerated plants to soil as described above.

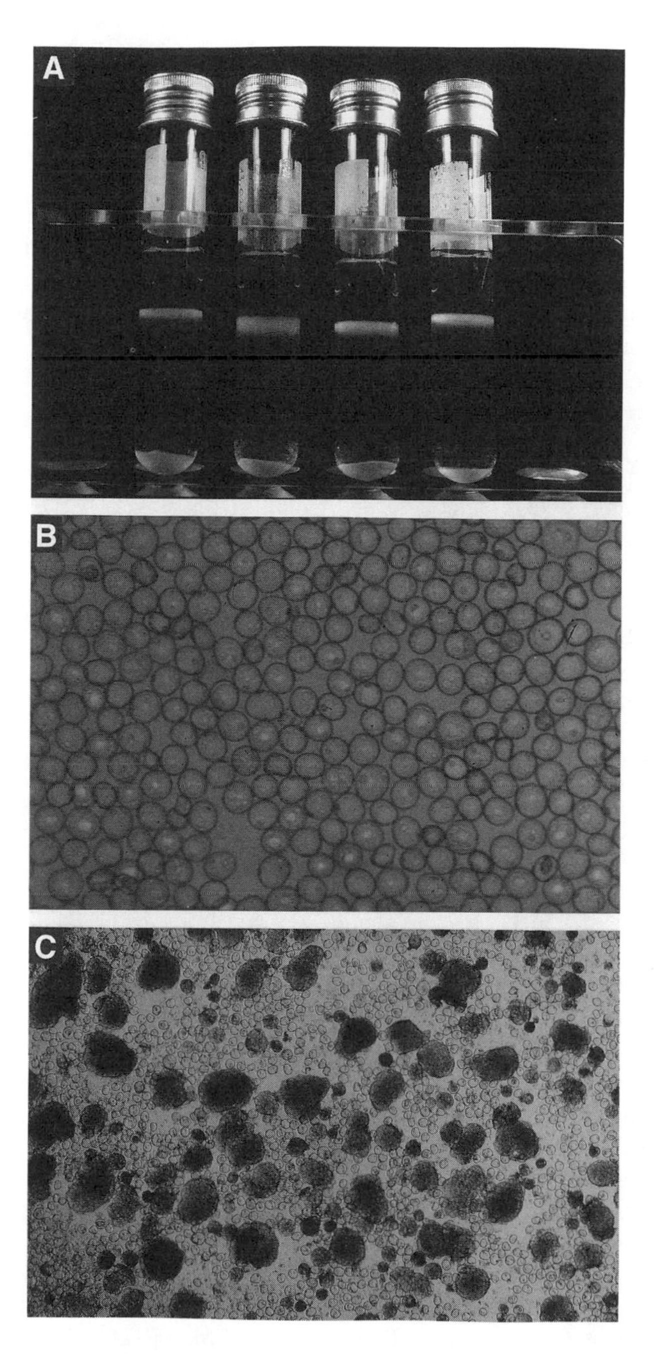

Fig. 3. Culture of isolated barley microspores. **(A)** Viable microspores collected in a band after centrifugation. **(B)** Freshly isolated microspores **(C)** Microspore-derived aggregates after 2 wk in culture.

4. Notes

1. Standard stock solutions can be used for media preparation. Stock solutions of vitamins should be kept at –20°C. For preparation of a 6-BAP stock solution (1 mg/mL), the hormone should first be dissolved in 1 mL 1 *M* KOH and then made up to volume with water.
2. These morphological parameters can give an indication on the microspore stage. However, these parameters depend on the actual growth conditions and the genotype. Therefore, microscopic examination of squashed anthers is strongly recommended in order to find a correlation between the morphological habit and the microspore stage.
3. In barley, staining with acetocarmine is usually not necessary; the nucleus can be recognized without staining.

References

1. Clapham, D. (1973) Haploid *Hordeum* plants from anthers *in vitro*. *Z. Pflanzenz.* **69**, 142–155.
2. Dunwell, J. M. (1985) Anther and ovary culture, in *Cereral Tissue and Cell Culture* (Junk, W., ed.), Martinus Nijhoff, Dordrecht, The Netherlands, pp. 1–44.
3. Luckett, D. J. and Davey, N. L. (1992) Utilisation of microspore culture in wheat and barley improvement. *Aust. J. Bot.* **40**, 807–828.
4. Pickering, R. A. and Devaux, P. (1992) Haploid production: approaches and use in plant breeding, in *Biotechnology in Agriculture No. 5 Barley: Genetics, Biochemistry, Molecular Biology and Biotechnology*(Shewry, P. R., ed.), C.A.B. International, Wallingford, pp. 519–547.
5. Jähne-Gärtner, A. and Lörz, H. Anther and microspore culture of barley. Submitted.
6. Jähne, A. and Lörz, H. (1995) Cereal microspore culture. Review article. *Plant Sci.* **109**, 1–12.
7. Jähne, A., Becker, D., Brettschneider, R., and Lörz, H. (1994) Regeneration of transgenic, microspore-derived, fertile barley. *Theor. Appl. Genet.* **89**, 525–533.
8. Heberle-Bors, E. (1989) Isolated pollen culture in tobacco: plant reproductive development in a nutshell. *Sex. Plant Reprod.* **2**, 1–10.
9. Foroughi-Wehr, B. and Mix, G. (1979) *In vitro* response of *Hordeum vulgare* L. anthers cultured from plants grown under different environments. *Env. Exp. Bot.* **19**, 303–309.
10. Gaul, H., Mix, B. Foroughi-Wehr, B., and Okamoto, M. (1976). Pollen grain development of *Hordeum vulgare. Z. Pflanzenz.* **76**, 77–80.
11. Wheatley, W. G., Marsolais, A. A., and Kasha, K. J. (1986) Microspore growth and anther staging in barley anther culture. *Plant Cell Rep.* **5**, 47–49.
12. Kasha K. J., Ziauddin, A., and Cho, U. H. (1990) Haploids in cereal improvement: anther and microspore culture, in *Gene Manipulation in Plant Improvement* II (Gustafson, J. P., ed.), Plenum, New York, pp. 213–235.
13. Jähne A., Lazzeri, P.A, Jäger-Gussen, M., and Lörz, H. (1991) Plant regeneration from embryogenic cell suspensions derived from anther cultures of barley (*Hordeum vulgare* L.). *Theor. Appl. Genet.* **82**, 74–80.

14. Huang, B. and Sunderland, N. (1982) Temperature-stress in barley anther culture. *Ann. Bot.* **49**, 77–88.
15. Wei, Z. M., Kyo, M., and Harada, H. (1986) Callus formation and plant regeneration through direct culture of isolated pollen of *Hordeum vulgare* cv. "Sabarlis." *Theor. Appl. Genet.* **72**, 252–255.
16. Olsen, F. L. (1991) Isolation and cultivation of embryogenic microspores from barley (*Hordeum vulgare* L.) *Hereditas* **115**, 255–266.
17. Roberts-Oehlschlager, S. L. and Dunwell, J. M. (1990) Barley anther culture: pretreatment on mannitol stimulates production of microspore-derived embryos. *Plant Cell. Tiss. Org. Cult.* **20**, 235–240.
18. Hoekstra, S., van Zijderveld, M. H., van Bergen, S., van der Mark, F., and Heidekamp, F. (1994) Genetic modification of barley for end use quality, in *Improvement of Cereal Quality by Genetic Engineering* (Henry, R. J. and Ronalds, J. A., eds.), Plenum, New York, pp. 139–144.
19. Mordhorst A. P. and Lörz, H. (1993) Embryogenesis and development of isolated barley (*Hordeum vulgare* L.) microspores are influenced by the amount and composition of nitrogen sources in culture media. *J. Plant Physiol.* **142**, 484–492.
20. Hunter, C. P. (1987) Plant generation method. European patent application No. 87200773.7.
21. Jähne, A., Lazzeri, P. A., Jäger-Gussen, M. and Lörz, H. (1991) Plant regeneration from embryogenic cell suspensions derived from anther cultures of barley (*Hordeum vulgare* L.). *Theor. Appl. Genet.* **82**, 74–80.
22. Widholm, J. M. (1972) The use of fluoresceine diacetate and phenosafranine for determining viability of cultured plant cells. *Stain Technol.* **47**, 189–194.

25

Microspore Embryogenesis and In Vitro Pollen Maturation in Tobacco

Alisher Touraev and Erwin Heberle-Bors

1. Introduction

Microspores have two developmental options when they are isolated from anthers and cultured in vitro. In a rich medium, they can develop into mature pollen grains that are fertile on pollination in vivo *(1,2)*. Their development closely resembles pollen formed in vivo, whereas the in vitro culture conditions simulate the changing environment a microspore/pollen grain experiences inside the anther. The microspore divides by an asymmetrical division, and the daughter cells differentiate into mature pollen grains, with their characteristic cell-cycle arrest (vegetative cell in G1, generative cells in G2). This simple pathway makes in vitro pollen maturation an excellent system to study developmental processes, such as cell fate determination and cellular differentiation *(3)*. In addition, in vitro pollen maturation has also been used for pollen selection *(4)* and plant transformation *(5)*. Another potential application is the rescue of sterile or self-incompatible pollen for "self"-pollinations.

When microspores or young pollen grains are stressed, e.g., by starvation or heat, gametophytic development stops, and a novel type of microspore/pollen grain is formed, which on transfer to nonstress conditions (rich medium without growth factors at ambient temperature) divides continuously and eventually develops into an embryo, i.e., a sporophyte *(6,7)*. Starvation induces microspore embryogenesis in tobacco *(6)*, wheat *(8)*, barley *(9)*, and rice *(10)*, whereas heat shock works in the *Brassicas* *(11)*, tobacco *(6)*, and wheat *(8)*.

On colchicine treatment of these haploid embryos, the resulting doubled haploids are completely homozygous, and a population of doubled haploids represents the genetic variability of male meiosis in a hybrid from which the microspores are taken. Such doubled-haploid "recombinant inbreds" are valu-

From: *Methods in Molecular Biology, Vol. 111: Plant Cell Culture Protocols*
Edited by: R. D. Hall © Humana Press Inc., Totowa, NJ

able tools for gene mapping, and they allow us to speed up breeding processes *(12,13)*. In molecular breeding, they help to transfer rapidly and fix transgenes from a easily transformable genotype into a cultivar, whereas the delivery of DNA into embryogenic microspores further advances genetic improvement of crop plants by producing homozygous transgenics in one step *(14,15)*.

At present, tobacco is the only system in which both the gametophytic and sporophytic pathway are well established. Both are highly efficient and can be easily reproduced in any lab.

2. Materials

2.1. Equipment

1. Laminar air flow.
2. Autoclave.
3. Two incubators or growth rooms (25 and 33°C).
4. Growth room, 16 h light, 25°C.
5. Microwave oven.
6. Clinical centrifuges and centrifuge tubes.
7. Growth chambers for donor plants (optional).
8. UV/light microscope.
9. Inverted microscope.
10. Domestic fridge and freezer.
11. Balance.
12. pH meter.
13. Millipore (or Corning) filtration units (0.2-μm membrane).
14. Magnetic stirrer.

2.2. Glassware, Culture Vessels, and Miscellaneous Items

1. Glass beakers (for 100–1000 mL), flasks (100–500 mL), funnels (10–1000 mL).
2. 6-, 12-, and 24- well plastic Petri dishes (tissue-culture tested from Falcon or Corning).
3. Glass vials with cap (17-mL, 26-mm diameter, Merck), and 12-mL conical sterile plastic centrifuge tubes (Kabe Labortechnik, Nürnberg, Germany).
4. Magnetic rod (18 mm in length), which moves freely on bottom of the glass vial.
5. Metal sieve (60-μm pore size, 30-mm diameter, Sigma) fitting to the top of a 100-mL Erlenmeyer flask with wide neck.
6. Parafilm (American National Can.).
7. Pasteur pipet (plastic, disposable from local producer).
8. Gilson pipetman (for 20, 200, and 1000 μL).
9. Plastic Petri dishes (tissue-culture-treated, Falcon or Corning) 100 × 15 mm, 100 × 20 mm, 60 × 15 mm, 35 × 10 mm.
10. Paper bags to prevent cross pollination.
11. Labels with thread, pen.

2.3. Chemicals

2.3.1. General

1. Agar (Sigma, US).
2. Activated charcoal (Sigma, US).
3. Ethanol ***(70%)***.
4. DAPI stain (Partec solution or Sigma [powder]).
5. Chemicals used in tissue culture media (Sigma, US, *see below*).

2.3.2. Media Compositions Used in This Chapter

1. B medium ***(16)***:

Chemicals	mg/L	Molarity
KCl	1490	20 m*M*
$CaCl_2 \cdot 2H_2O$	147	1 m*M*
$MgSO_4 \cdot 7H_2O$	250	1 m*M*
Mannitol	54,700	300 m*M*
Phosphate buffer	—	1 m*M*

Adjust pH to 7.0 by addition of KOH. Filter-sterilize.

2. AT3 medium (unpublished)

Chemicals	mg/L	Molarity
KNO_3	1950	19 m*M*
$(NH_4)_2SO_4$	277	2 m*M*
KH_2PO_4	400	2.9 m*M*
$CaCl_2 \cdot 2H_2O$	166	1.1 m*M*
$MgSO_4 \cdot 7H_2O$	185	0.7 m*M*
Fe-EDTA	10 mL from a 3.67 g/L stock solution	0.1 m*M*
B_5 vitamins	1 mL from 1000X stock solution (*see* Appendix)	
MS microsalts	1 mL from 1000X stock solution (*see* Appendix)	
MES	1950	10 m*M*
Glutamine	1256	8.5 m*M*
Maltose	90,000	250 m*M*

Adjust pH to 6.2 by addition of KOH. Filter-sterilize.

3. T1 medium ***(11)***:

Chemicals	mg/L	Molarity
KNO_3	1001	10 m*M*
$Ca(NO_3)_2 \cdot 4H_2O$	236	1 m*M*
$Mg(SO_4) \cdot 7H_2O$	247	1 m*M*
H_3BO_3	10	0.16 m*M*
Uridine	244	1 m*M*
Cytidine	127	0.5 m*M*
Glutamine	438	3 m*M*
Lactalbumine hydrolysate	10,000	—

Sucrose	171,150	500 m*M*
Phosphate buffer	—	1 m*M*

Adjust pH to 7.0 by addition of KOH. Filter-sterilize.

4. P medium ***(17)***:

Chemicals	mg/L	Molarity
Sucrose	400.5	1.17 *M*
Proline	11.5	100 m*M*

Adjust pH to 7.0 by addition of KOH. Filter-sterilize.

5. PEG8000 medium ***(18)***:

Chemicals	mg/L	Molarity
$CaCl_2$	111	1 m*M*
KCl	75	1 m*M*
$MgSO_4 \cdot 7H_2O$	200	0.8 m*M*
H_3BO_3	100	1.6 m*M*
$CuSO_4$	Trace amount	30 µ*M*
Casein hydrolysate	300	—
PEG 8000	125,000	—
Sucrose	50,000	146 m*M*
MES	2900	15 m*M*

Adjust pH to 5.9 by addition of KOH. Autoclave.

6. GK medium ***(19)***:

Chemicals	mg/L	Molarity
$MgSO_4 \cdot 7H_2O$	200	0.8 m*M*
H_3BO_3	200	3.2 m*M*
$CuSO_4$	Trace amounts	30 µ*M*
Sucrose	50,000	146 m*M*

Adjust pH to 5.8 by addition of KOH. Autoclave.

7. MS medium: Murashige and Skoog medium (***20*** and *see* Appendix) supplemented with 1% activated charcoal, 1% sucrose, and 0.8 % agar. The pH is adjusted to 5.8 before autoclaving.

3. Methods

3.1. Preparations

Glassware and distilled water are sterilized by autoclaving at 120°C, 15 psi for 20 min. The concentrated media stocks containing salts and organic supplements can be prepared in advance and stored in a freezer for several months. Media for washing and culture of microspores must be filter sterilized.

3.2. Growth of Donor Plants

Many tobacco genotypes and species, including *Nicotiana tabacum* L. and *Nicotiana rustica* L., have been shown to be good donor plants for microspore cultures. Emphasis should be on conditions for profuse flower formation and high male fertility (good-quality pollen). Here we give a protocol established in our lab for *N. tabacum* L. cv. Petit Havana SR1 ***(21)***. Donor plants are grown in a growth chamber (25°C, 16 h light) with regular supply of fertilizers and routine watering. Continuous flowering can be achieved by regular harvest of open flowers. Thus, one generation of tobacco plants can be used for 4–6 mo without significant decrease in microspore culture response.

3.3. Flower Bud Sterilization

Flower buds with an approximate length of 10–11 mm, representing unicellular microspores (which can be confirmed by DAPI staining; *see* **Note 1**, and **Fig. 1A,B**) are excised. The buds are sterilized by dipping them in 70% ethanol for a maximum of 5 min (*see* **Note 2**).

3.4. Microspore Isolation

1. Under sterile conditions, squeeze the anthers out of the flower buds into a glass vial (17 mL), containing approx 3 mL of medium B (***16***; *see* **Note 3**). Anthers from 20–30 buds can be squeezed/vial.
2. Remove the medium, and replace with 6 mL of fresh medium B.
3. Put a sterile magnetic bar into the vial, and stir for 2–3 min at maximum speed until the mixture of microspores/anther debris and medium becomes milky (*see* **Note 4**).
4. Collect the suspension of released microspores and anther debris with a Pasteur pipet, and filter the suspension through a 60-μM metal sieve (*see* **Note 5**).
5. Centrifuge the filtrate for 2–3 min at 1200 rpm (about 200*g*; *see* **Note 6**).
6. Remove the top green layer of the two-layered pellet, which contains anther wall debris by using a 200- or 1000-µL Gilson pipette tip (*see* **Note 7**).
7. Resuspend the whitish pellet composed of microspores in 2–10 mL of medium B and centrifuge again. Repeat this procedure at least two to three times until there is no greenish layer above the whitish pellet. The microspores can be used for embryogenesis (*see* **Subheading 3.5.**) or pollen maturation (*see* **Subheading 3.6.**).

3.5. Induction of Microspore Embryogenesis and Regeneration of Haploid Tobacco Plants

1. Resuspend the final whitish pellet in medium B (microspores from one flower bud in 2 mL of medium B give the optimal density for a good response), and pour into small Petri dishes (35 × 10 mm, Falcon or Corning) with 1.5 mL of suspension/dish (*see* **Note 8**).
2. Seal Petri dishes with Parafilm, and incubate at 33°C for 5–6 d in the dark (*see* **Note 9**, *see* **Fig. 1C**).
3. Collect the suspension from the Petri dishes in 12-mL test tubes, and pellet by centrifugation at 1000 rpm (175*g*) for 5 min (*see* **Note 10**).
4. Discard the supernatant, resuspend the pellet in AT3 medium, and pour back into the original Petri dish with 1.0 mL/Petri dish.

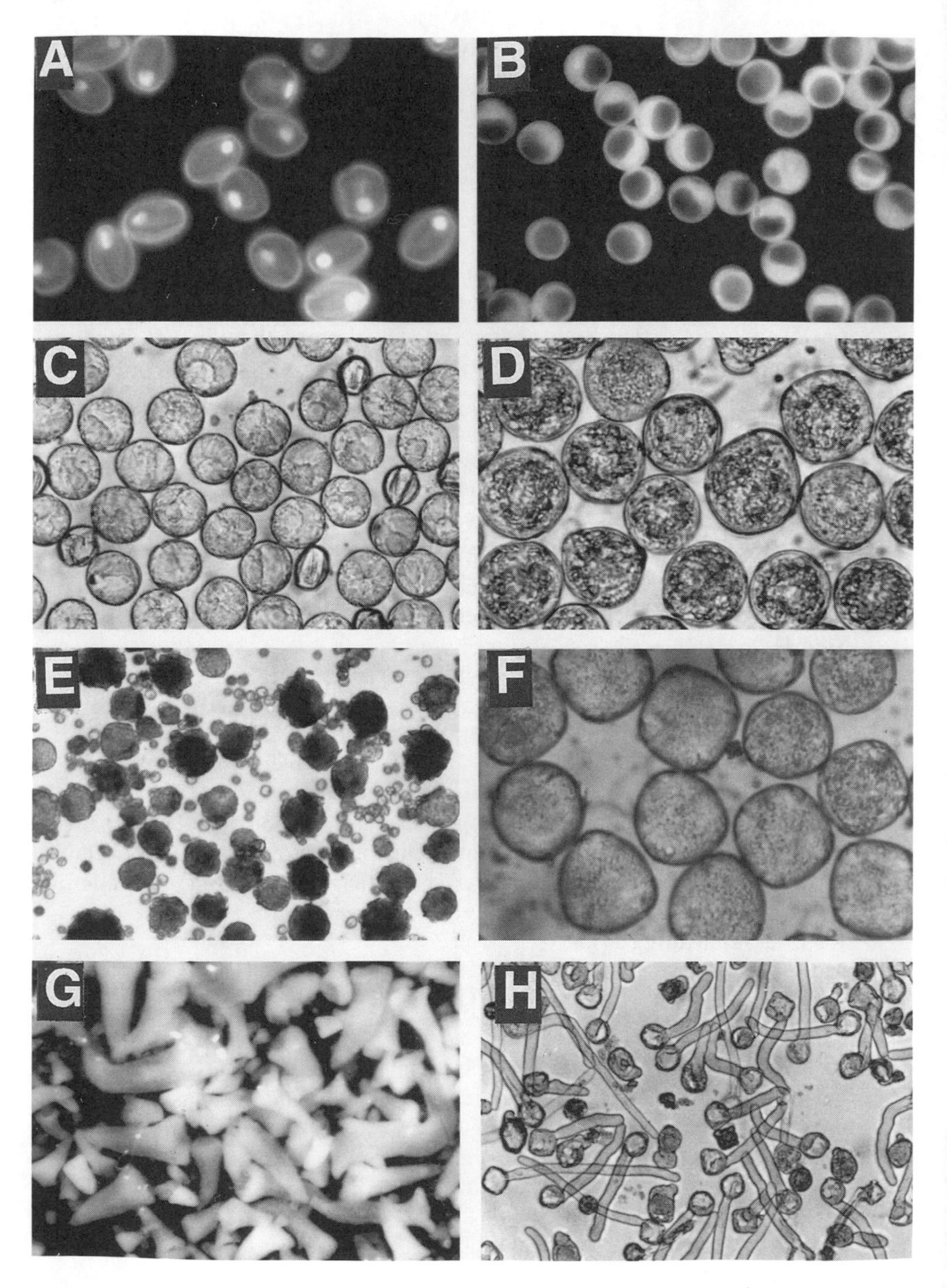

Fig. 1.

5. Seal the Petri dish with Parafilm, and incubate for 1.5–2 mo at 25°C in the dark (*see* **Fig. 1E,G**).
6. Transfer seedlings onto agar solidified MS medium (***20***; and *see* Appendix) with 1% activated charcoal and 1% sucrose, and move to light.

3.6. Maturation of Tobacco Microspores In Vitro

1. Resuspend the final whitish pellet in medium T1 (***17***; microspores from one bud in 1 mL of medium give optimal maturation) and pour into small Petri dishes (35 × 10 mm, Falcon or Corning) with 1 mL suspension/dish.
2. Seal Petri dishes with Parafilm, and incubate at 25°C for 4 d in the dark (*see* **Note 11** and **Fig. 1D**).
3. Dilute cultures with 1 mL of medium P, and incubate for 1 d (day 5; *see* **Note 12**).
4. Dilute cultures with another 1 mL of medium P, and incubate for one more day (day 6; *see* **Note 13**). Fully matured pollen grains (*see* **Fig. 1F**) can be used for in vitro germination (*see* **Subheading 3.7.** and **Fig. 1H**) or for in situ pollination (*see* **Subheading 3.8.**).

3.7. Germination of In Vitro Matured Tobacco Pollen

1. Collect the in vitro matured pollen grains in 12-cm test tubes.
2. Centrifuge the suspension at 1500 rpm (275*g*) for 1–2 min.
3. Discard the supernatant, and resuspend the pellet in a volume of PEG8000 medium to give a density of 10,000 pollen grains/mL (***18***; *see* **Note 14**).
4. Pour the suspension into 6- or 10-cm Petri dishes (cultures should be diluted by a factor of 2–3 in order to make observation of germination easier).
5. Seal Petri dishes with Parafilm, and incubate at 25°C in the dark.
6. Observe and evaluate germination frequency and tube length after 5–10 h or longer (*see* **Note 15**).

3.8. In Situ *Pollination with In Vitro Matured Tobacco Pollen*

1. Collect in vitro matured pollen grains in 12-cm test tubes.
2. Centrifuge suspension at 1500 rpm (275*g*) for 1–2 min.

Fig. 1. *(previous page)* Embryogenesis and maturation from isolated tobacco microspores in vitro. **(A)** and **(B)** freshly isolated uninucleate tobacco microspores stained with DAPI **(A)** and FDA **(B)**; **(C,E,G)** in vitro microspore embryogenesis: **(C)** microspores after 6 d starvation in sucrose- and nitrogen-free medium B; **(E)** globular embryos after 30 d of culture of embryogenic microspores in medium AT3; **(G)** torpedo and cotyledonary embryos after 1.5 mo culture of microspores in AT3. **(D,E,F)** in vitro microspore maturation : **(D)** starch formation in microspores after 2 d of incubation in maturation medium T1; **(E)** fully mature pollen after 6 d of culture of microspores in vitro ; **(H)** germination of in vitro matured pollen after 5 h incubation in medium PEG8000.

3. Discard the supernatant, and resuspend the pellet in 1–2 mL of medium GK (*see* **Note 16**).
4. Centrifuge the suspension at 1500 rpm (275*g*) for another 1–2 min.
5. Discard the supernatant, and resuspend the pellet in 0.5 mL of medium GK.
6. Transfer the suspension into small PCR Eppendorf tubes (e.g., 500-μL tubes).
7. Centrifuge the suspension at 1500 rpm (275*g*) for 1–2 min.
8. Discard the supernatant, and resuspend the pellet in a volume of medium GK to give a density of 1.250 pollen grains/μL (*see* **Note 17**).
9. Take 4 μL of pollen suspension using a 10- or 20-μL Gilson pipet, and apply as a drop carefully onto the stigma of a flower, which has been emasculated 1 d before (*see* **Note 18**).
10. Label the pollinated flowers, and bag them 5–6 h after pollination when the drop of pollen suspension on the stigma has dried.
11. Harvest seed pods, and count seeds after about 1 mo.

4. Notes

1. Diamidino-2-phenylindole (DAPI) is a specific fluorescent stain to visualize DNA. It can be purchased as a ready-to-use solution from Partec (Germany) or as a powder from Sigma (US). The powder is dissolved in buffer ***(22)*** or in 50% ethanol. The nuclei of tobacco microspores are stained well in both solutions, and can be observed with a range of UV filter sets under a fluorescent microscope. Alternatively, staging can be performed by acetocarmine or acetoorcein stain, but the sensitivity of these stains is less than that of the DAPI stain.
2. In general, surface sterilization of buds with 70% ethanol is sufficient to obtain sterile cultures, but sometimes one can also use sodium or calcium hypochlorite as sterilization agents after the ethanol treatment.
3. Medium B is simple, cheap, osmotically well balanced, and used in our lab for all tobacco microspore isolation procedures. Depending on skill and the isolation procedure used, the viability of isolated microspores should be 90–95% in this medium.
4. The duration and time of stirring depend on the vial used and on the type of stirrer, and can easily be optimized empirically. The magnetic bar should move inside the glass vial freely. In case of small-scale isolations (<5 buds), anther maceration with a glass rod can be used successfully ***(23)***, and in the case of large-scale isolations (more than 50 buds), a Waring blendor is the best choice ***(24)***.
5. Unicellular and early bicellular tobacco microspores have average diameters of 20–23 μ*M*, and the pore size of the sieve used for filtration may vary from 40 to 60 μm.
6. Microspores can resist centrifugation speeds even higher than 600*g*, but high speed may affect subsequent viability of the microspores.
7. The microspores pellet much faster than the anther debris, and therefore, one can see two clear pellet layers (one greenish, one whitish) after centrifugation. The greenish layer can be removed easily after some practice.

8. Microspore density is an important parameter for optimal induction of embryogenesis. The calculation of optimal density for tobacco microspores is given in refs. ***(17)*** and ***(25)***, and has been found to be: microspores from one bud to 2–4 mL of medium.
9. This is the key step to block the gametophytic pathway of development and divert microspores into the sporophytic pathway. Embryogenic microspores at the end of the starvation treatment should have increased in size and should show a "star"-like phenotype under the inverted microscope with the nucleus being in a more central position as compared to the starting stage and cytoplasmic strands radiating out from it through the vacuole (*see* **Fig. 1C**).
10. The embryogenic microspores formed are very sensitive to damage, and low speed must be used to pellet them gently.
11. Depending on the developmental stage and synchrony of the starting population (ratio of unicellular to early bicellular microspores), the incubation time in medium T1 can be shortened to 3 d ***(17)***. Starch accumulation is seen in immature pollen grains after 2–3 d of culture in medium T1, which is a good criteria for optimal and synchronous maturation of microspores (*see* **Fig. 1D**).
12. This step and the following step are simulating the dehydration of maturing pollen.
13. A mature pollen grain is 30–32 μm in size and has a slightly rectangular shape, whereas earlier stages are fully spherical. The exine should be fully intact. No protrusions from the germ pores should be seen. The number and size of starch grains are smaller when compared to a slightly earlier stage.
14. The germination medium based on the use of PEG8000 ***(12)*** gives a very high frequency of germination of in vitro matured tobacco pollen (*see* **Fig. 1H**) as compared to medium GK ***(19)*** or GV ***(17)***.
15. In vitro germination usually starts 4–5 h after incubation, but the best time to evaluate germination frequency is after 9–10 h.
16. Do not use the PEG8000 medium for *in situ* pollination experiments. The in vitro matured pollen when suspended in PEG8000 medium and loaded onto a stigma, germinate rapidly, but do not penetrate the style and form a white cover on the surface of the stigma.
17. It has been shown ***(5,19)*** that the minimum number of in vitro matured pollen grains applied onto one stigma should not be <1000 in order to obtain any seed set. A minimum of 5000 is needed to obtain optimum seed set.
18. The receptivity of the stigma of an emasculated flower bud is a major limiting factor in obtaining seed set. The wounding, which accompanies emasculation 1 d before *in situ* pollination, induces the formation of flavonols in the stigma and style, which are important growth factors for the germinating pollen grains ***(19,26)***. The stigmatic surface should be wet, and a drop of stigmatic diffusate should be visible at the moment of pollination. A drop of up to 6 μL can be applied onto a stigma, but usually 4 μL drops of pollen suspension are used ***(5)***.

References

1. Benito Moreno, R. M., Macke, F., Alwen, A., and Heberle-Bors, E. (1988) *In-situ* seed production after pollination with *in-vitro*-matured, isolated pollen. *Planta* **176,** 145–148.
2. Stauffer, C., Benito Moreno, R. M., and Heberle-Bors, E. (1991) In situ pollination with in vitro matured pollen of *Triticum aestivum. Theor. Appl. Genet.* **81,** 576–580.
3. McCormick, S. (1993) Male gametophyte development. *Plant Cell* **5,** 1265–1275.
4. Touraev, A., Fink, Ch., Stöger, E., and Heberle-Bors, E. (1995). Pollen selection: a transgenic reconstruction approach. *Proc. Natl. Acad. Sci. USA* **92,** 12,165–12,169.
5. Touraev, A., Stöger, E., Voronin, V., and Heberle-Bors, E. (1998) Plant male germ line transformation. *Plant J.* **12(4),** 949–956.
6. Touraev, A., Ilham, A., Vicente, O., and Heberle-Bors, E. (1996) Stress induced microspore embryogenesis from tobacco microspores: an optimized system for molecular studies. *Plant Cell Rep.* **15,** 561–565.
7. Touraev, A., Vicente, O., and Heberle-Bors, E. (1997) Initiation of microspore embryogenesis by stress. *Trends Plant Sci.* **2,** 297–302.
8. Touraev, A., Indrianto, A., Wratscko, I., Vicente, O., and Heberle-Bors, E. (1996) Efficient microspore embryogenesis in wheat (*Triticum aestivum* L.) induced by starvation at high temperatures. *Sex. Plant Reprod.* **9,** 209–215.
9. Hoekstra, S., van Zijderveld, M. H., Louwerse, J. D., Heidekamp, F., and van der Mark, F. (1992) Anther and microspore culture of *Hordeum vulgare* L. cv. Igri. *Plant Sci.* **86,** 89–96.
10. Ogawa T., Fukuoka, H., and Ohkawa, Y. (1994) Induction of cell division of isolated pollen grains by sugar starvation in rice. *Breeding Sci (Jpn.)* **44,** 75–77.
11. Pechan, P. M. and Keller, W. A. (1988) Identification of potentially embryogenic microspores in *Brassica napus. Physiol. Plant.* **74,** 377–384.
12. Morrison, R. A. and Evans, D. A. (1988) Haploid plants from tissue culture: new plant varieties in a shortened time frame. *Bio/Technology* **6,** 684–690.
13. Ferrie, A. M. R., Palmer, C. E., and Keller, W. A. (1995). Haploid embryogenesis, in *In Vitro Plant Embryogenesis* (Thorpe, T. A., ed.), Kluwer, Dordrecht, pp. 309–344.
14. Jähne, A., Becker, D., Brettschneider, R., and Lörz, A. (1994) Regeneration of transgenic, microspore-derived, fertile barley. *Theor. Appl. Genet.* **89,** 525–533.
15. Stöger, E., Fink, C., Pfosser, M., and Heberle-Bors, E. (1995) Plant transformation by particle bombardment of embryogenic pollen. *Plant Cell Rep.* **14,** 273–278.
16. Kyo, M. and Harada, H. (1986) Control of the developmental pathway of tobacco pollen *in vitro. Planta* **168,** 427–432.
17. Tupy, J., R'hová, L., and Zársky, V. (1991) Production of fertile tobacco pollen from microspores in suspension cultures and its storage for *in situ* pollination. *Sex. Plant Reprod.* **4,** 284–287.
18. Read, S. M., Clarke, A. E., and Bacic, A. (1993) Stimulation of growth of cultured *Nicotiana tabacum* W38 pollen tubes by poly (ethylene glycol) and $Cu_{(II)}$ salts. *Protoplasma* **177,** 1–14.

19. Ylstra, B., Touraev, A., Benito Moreno, R. M, Stöger, E., van Tunen, A. J., Vicente, O., Mol, J. N. M., et al. (1992). Flavonols stimulate development, germination, and tube growth of tobacco pollen. *Plant Physiol.* **100,** 902–907.
20. Murashige, T. and Skoog, F. (1962) A revised medium for rapid growth and bioassays with tobacco tissue cultures. *Physiol. Plant* **15,** 473–497.
21. Maliga, P., Breznovits, A., and Marton, L. (1973) Streptomycin-resistant plants from callus culture of haploid tobacco. *Nature* **244,** 29–30.
22. Vergne, P., Delvallée, I., and Dumas, C. (1987) Rapid assessment of microspore and pollen development stage in wheat and maize using DAPI and membrane permeabilization. *Stain Technol.* **62,** 299–304.
23. Benito Morcno, R. M., Mackc, F., Hauscr, M.-T., Alwen, A., and Heberle-Bors, E. (1988) Sporophytes and male gametophytes from *in vitro* cultured, immature tobacco pollen, in *Sexual Reproduction in Higher Plants* (Cresti, M., Gori, P. and Pacini, E., eds.), Springer, Heidelberg, NY, pp. 137–142.
24. Swanson, E. B., Coumans, M. P., Wu, S. C., Barsby, T. L., and Beversdorf, W. D. (1987) Efficient isolation of microspores and the production of microspore-derived embryos from *Brassica napus. Plant Cell* Rep. **6,** 94–97.
25. Garrido, D., Charvat, B., Benito Moreno, R. M., Alwen, A., Vicente, O., and Heberle- Bors, E. (1991) Pollen culture for haploid plant production in tobacco, in *A Laboratory Guide for Cellular and Molecular Plant Biology* (Negrutiu, I. and Gharti-Chhetri, G., eds.), Birkhäuser, Basel, pp. 59–69.
26. Vogt, T., Pollak, P., Tarlyn, N., and Taylor, L. P. (1994). Pollination- or wound-induced kaempferol accumulation in petunia stigmas enhances seed production. *Plant Cell* **6,** 11–23.

26

Embryo Rescue Following Wide Crosses

Hari C. Sharma

1. Introduction

Crosses between species of the same or different genera are called wide crosses (WCs). They are useful in plant improvement for gene transfer and haploid production, and in general biology, for genome mapping, study of chromosome behavior, and phylogenetic relationships ***(1–8)***. Broadening the genetic base of modern crops is necessary to prevent genetic vulnerability. If genes are not present in crop species, we resort to mutagenesis, transformation, protoplast fusion, and wide crosses. The first two technologies are not routine for a number of reasons ***(9,10)***. The utility of protoplast fusion may lie in transformation or production of hybrids, which are sexually impossible ***(11,12)***. Nevertheless, wide crossing will remain an important method of genome manipulation for introducing unique variation to crop species. In recent years, it has become imperative that genetic material from WCs be exploited in breeding ***(10,13)***.

In the manipulation, exploitation, and study of alien genomes through WCs, the first required step is to make a hybrid. The alien genome can then be studied and utilized to various degrees from the whole genome and single chromosomes to specific gene transfer by manipulating the hybrid genome through either natural recombination or specialized techniques depending on the genomic relationship of the two species ***(14)***. However, crossability barriers often make hybridization difficult. These barriers, caused by gene action, genome incompatibility, ploidy levels of species, and the environment, lead to no or poor seed set on crossing, retard or stop hybrid seed development, or lead to sterility of the hybrid plant ***(15,16)***. However, reproductive barriers are not always absolute ***(17)***. Molecular analysis shows that genomic changes in wide hybrids only partially coincide with those of speciation ***(18)***, and methods to

From: *Methods in Molecular Biology, Vol. 111: Plant Cell Culture Protocols*
Edited by: R. D. Hall © Humana Press Inc., Totowa, NJ

Fig. 1. 14-d-old normal seed of wheat (left) and *Agropyron junceum* (right), and aborting seed of wheat x *A. junceum* wide cross with only watery endosperm (middle).

overcome crossability barriers have been devised ***(7,19–21)***. In nature, hybridization is suppressed owing to less favorable conditions for fertilization, hybrid seed development, and survival of F_1 sterile plants, but under artificial conditions, genotypic variation for crossability, multiple pollinations, application of growth regulators, and in vitro techniques can be exploited to make the WC possible. In making the hybrid, seed abortion is the major barrier, largely caused by the failure of proper endosperm development and its inability to nourish the embryo (**Fig. 1**). The technique of rearing the embryo from an aborting seed into a plant in vitro is called embryo rescue (ER) or embryo culture (EC). Thus, in ER following WCs, synthetic medium supplies the life support for the hybrid embryo. ER following WCs is one of the most successful and beneficial plant cell/tissue-culture techniques ***(7,20,22,23)***. The technique has been used routinely, but procedures and modifications are rarely or incompletely described in publications. The few protocols that are available deal with EC in general, and not ER in WCs ***(22–25)***. Kaltsikes et al. ***(26)*** described a method for the production of wheat × rye hybrids by EC. Collins and Grasser ***(27)*** reviewed interspecific hybridization and EC in *Trifolium,* Mok et al. ***(28)*** in *Phaseolus,* and Williams ***(29)*** in forages.

There are different ways of ER following WCs: direct ER where the seed is dissected, the embryo excised and cultured to germinate into a plant ***(30)***; ovule culture, followed ***(31)*** or not followed ***(32)***, by EC; ovary culture followed by ovule culture or EC ***(22)***; segmenting the cultured embryo for further culturing

(33); inducing callus and plantlet regeneration from embryos ***(34)***; culturing embryos onto nurse endosperm on medium ***(35)***; and spike culture followed by EC ***(36)***. Compared to direct ER, other variations are more cumbersome and require transfers to different media, and are useful if direct ER is unsuccessful. In some cases where comparative studies have been made, direct ER was more successful than some of these variations ***(37,38)***. Ovule culture has worked largely in cotton ***(39)***, but is not useful when maternal tissue or endosperm is inhibitory to the embryo ***(40,41)***. It has utility where embryo abortion occurs early or when dissection is not possible. Postpollination application of growth regulators can prolong seed development on the mother plant. Nurse culture may be useful when no suitable medium is found. Direct ER following WC is a much simpler technique. A survey of literature from 1984 to 1996 by the author, based on Center for Agri. and Biosci. (CAB), has shown that direct ER has been used far more frequently than these variations.

However, detailed, stepwise procedures that can be followed and performed by a new researcher are lacking. This chapter provides an updated overview and complete step-by-step protocol of ER from seeds that result from WCs.

2. Materials

1. Seeds of parental species of wheat *(Triticum aestivum)* and wheatgrass *(Agropyron caninum)*, greenhouse, growth chamber or field facilities, pots, and soil to grow plants.
2. Pollinating bags, surgical scissors, forceps, tags, paper clips, pencil, and coin envelops to make WC.
3. General laboratory glassware, including flasks, beakers, graduated cylinders, pipets, pipet rubber bulbs, stir bars for making solutions, and media.
4. Refrigerator, laminar flow cabinet, distilled water unit, autoclave, analytical balance to weigh from 0.1 mg up to a few grams, pH meter, stirrer with hot plate, and aluminum foil.
5. Chemicals listed in **Table 1** plus inositol, sucrose, agar, and hormone(s) as needed for media.
6. Acid (0.5 *N* HCl) and alkali (0.5 *M* NaOH) solutions to adjust pH of the media.
7. Clorox bleach and cheesecloth to surface-sterilize WC seeds.
8. Gas flame or a burner, spray bottles with 95 and 70% ethanol, and Coplin jar containing 70% ethanol.
9. Dissecting microscope or stereomicroscope with a light source, glass Petri dishes with moistened filter papers, sharp surgical forceps, scalpel, narrow spatula, needle with handle for dissecting seeds, and isolating embryos.
10. Culture tubes for medium, and slanting test tube racks (150 × 25 mm glass tubes with slip-on caps [**Fig. 2**], available from Carolina Biological Supply Co. or Sigma Chemical Co.).
11. Incubator with temperature and light controls.

Table 1
Composition of Stock Solutions and Final Concentrations of Ingredients in Media

Component	Stock sr. no.	Stock vol., mL	Ingredient	Stock amount, g	Final conc., mg/L
Inorganic	1	500	$NH_4 NO_3$	82.5	1650
			KNO_3	95.0	1900
	2	500	$MnSO_4 \cdot 4H_2O$	1.115	22.3
			$ZnSO_4 \cdot 7H_2O$	0.43	8.6
			$CuSO_4 \cdot 5H_2O$	0.00125	0.025
			$MgSO_4 \cdot 7H_2O$	18.5	370
	3	500	$CaCl_2 \cdot 2H_2O$	22.0	440
			KI	0.0415	0.83
			$CoCl_2 \cdot 6H_2O$	0.00125	0.025
	4	500	KH_2PO_4	8.5	170
			H_3BO_3	0.31	6.2
			$Na_2MoO_4 \cdot 2H_2O$	0.0125	0.25
	5	500	$FeSO_4 \cdot 7H_2O$	1.39	27.8
			$Na_2 \cdot$ EDTA	1.865	37.3
Vitamins	—	500	Thiamine-HCl	0.02	0.4
Amino acids	a	250	Glycine	0.25	10
	b	250	L-Arginine-HCl	0.25	10
	c	250	L-Tyrosine	0.25	10

3. Methods

1. Germinate the seeds of the parental species (e.g., wheat and wheatgrass) to be crossed and grow plants to flowering. Stagger two to three plantings to increase the chances of simultaneous flowering of the two parental species and making the cross (*see* **Note 1**).
2. Emasculate the florets of the plants to be used as females when the anthers are yellowish green and have not dehisced. Pollinate using pollen of the other species when the stigma becomes feathery (receptive). Enclose the spikes in pollinating bags using paper clips (**Fig. 3**). Pollinate twice over 2 d to increase seed set. Make reciprocal crosses to enhance chances of success. Write information about the cross, including date of pollination, on a tag, and tie it to the tiller (*see* **Note 2**).
3. Score the number of florets pollinated and seed set after 4–5 d. Remove the pollinating bags, and cover the spikes bearing seed(s) with coin envelopes to block light and facilitate seed development.
4. At about day 10 after pollination, inspect the growing caryopses by opening the bracts with forceps, and tentatively decide when the embryos are to be excised for culture. Plan to prepare stock solutions and ER media accordingly (*see* **Note 3**).

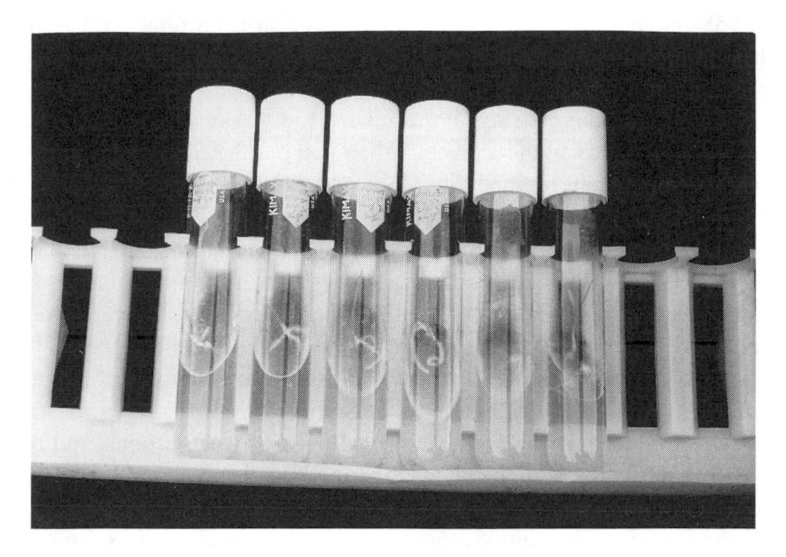

Fig. 2. Slip-on-type embryo culture tubes in a slanting test tube rack with germinating embryos on media demonstrating the embryo culture system.

Fig. 3. Wide crossing between wheat and *Agropyron intermedium.*

5. Turn on the laminar flow hood. Wipe it thoroughly with cloth well soaked in 95% ethanol (*see* **Note 4**).
6. Sterilize all glassware and water in the autoclave (121°C and 15 psi, 20 min). Sterilize the various dissecting tools listed in the **Subheading 2.** in the autoclave

or in 95% ethanol and keep in the sterile hood in Coplin jar with 70% ethanol with their points downward. Sterilize glass Petri dishes containing moistened filter papers. Wrap the glassware, tools, and Petri dishes in aluminum foil before putting in the autoclave. Bring all the autoclaved glassware and tools to the laminar flow hood. Spray with 95% ethanol as you put them in the hood.

7. Prepare stock solutions for modified MS media *(42)* following **Table 1** (*see* **Notes 5** and **6**). Prepare inorganic and vitamin stocks at strengths 50X and the amino acids stocks at 25X the final concentrations needed for the actual media. To prepare each inorganic stock, dissolve each salt in a minimum volume of distilled water in a beaker and then pool in the order shown to prevent precipitation. Make volume up to 500 mL by adding water. For Fe·EDTA complex, weigh 1.865 g of Na_2·EDTA, and put it into a beaker with 100 mL distilled water. Stir and apply gentle heat or give a longer time on a stirrer to dissolve. Dissolve 1.39 g of $FeSO_4 \cdot 7H_2O$ separately in the same way. Pool the two solutions, and bring volume to 500 mL with distilled water. Stir for 1 h to ensure complete mixing of the complex. Bring pH to 5.8. Store all stock solutions at 4°C (*see* **Note 7**).
8. From the above stock solutions, prepare the actual media with supplements. To prepare 1 L media, put 500 mL of distilled water in a 1 L beaker, and stir at moderate speed. Using separate pipets for each stock, add 10 mL each of inorganic stocks 1–5 (**Table 1**) in ascending order to avoid precipitation. Mix well before adding each solution. Add 10 mL of vitamin stock and then 10 mL each of the three amino acid stocks (a–c). Weigh and add 100 mg of inositol. Add 30–50 g of sucrose. Add 0.1–1.0 mg of kinetin (*see* **Note 8**). Make volume up to 800 mL. Adjust pH to 5.8 by adding dropwise, either 0.5 *N* HCl or 0.5 *M* NaOH. Make volume up to 1000 mL. Add 7 g of agar. Pour the media into a 1.5- or 2-L flask and close the flask mouth with aluminum foil. Autoclave for 3 min or boil on burner to melt agar (*see* **Note 9**). Distribute 20–25 mL/tube. Close the tubes with lids, and write name and date. Put the tubes in a metallic or any autoclavable container, and autoclave for 20 min to sterilize the media (*see* **Note 10**). After sterilization, close the lids well onto the culture tubes, since they may have lifted somewhat during autoclaving. Put the tubes on the slanting test tube rack to solidify. Wipe or spray with ethanol as you bring them into the laminar flow hood (*see* **Note 11**).
9. Harvest the seed by cutting the spike from the mother plant with a pair of scissors. Put the spikes in a beaker containing water and bring to the laboratory. Remove the seed with a pair of forceps and put in water to avoid drying out (*see* **Note 12**). Record the number of seeds harvested.
10. Prepare 15% (v/v) Clorox bleach (= 0.8% sodium hypochlorite = 0.37% chlorine) solution using distilled water, and fill a small sterile beaker. Put the seeds in the solution. If the beaker has been exposed outside the hood while pouring or preparing the solution, wipe the outer walls, and top of the beaker with ethanol after placing in the hood. Allow 20 min for surface-sterilization of the seeds in the solution. If the seeds float, keep dipping them under the liquid with forceps (*see* **Notes 13** and **14**). Clean the working area. Wipe hands with ethanol. Spray

and wipe the whole hood once again with ethanol during this time. After 20 min, decant the solution, blocking the seeds with forceps. Rinse three times with autoclaved water, shaking, and discarding the water each time. Finally, add some sterile water to keep seeds moist, and store, if necessary, at 5–8°C in darkness.

11. Organize the working area inside the laminar bench: Spray again the dissecting instruments (forceps, scalpel, needle and spatula) with ethanol and leave dipped in the Coplin jar containing 70% ethanol. Before every use, flame dry tools to burn off the alcohol and allow to cool. Wipe/spray the dissecting microscope thoroughly with ethanol paying particular attention to the stage and the surrounding area. Turn on the gas flame or burner. Place an autoclaved Petri dish containing a moistened filter paper on the dissecting microscope stage. Transfer a seed from the beaker onto the dish (*see* **Note 15**). Using the seed as the object, adjust the dissecting microscope at a magnification comfortable to your eyes (*see* **Note 16**). View the seed and locate the basal end containing the embryo (*see* **Note 17**). Hold the seed in the middle with a pair of flamed and cooled forceps. Make a shallow incision (slit) on the pericarp next to the holding point of the basal half of the seed with a scalpel. Using sharp forceps, open (peel) the pericarp directly above the embryo, exposing the embryo (**Fig. 4**). If during the dissecting operation, the filter paper in the Petri dish dries out, add a little water, to avoid dehydration of the embryo, by dipping the scalpel or forceps in the water beaker containing the sterile seeds, and then touching the filter paper. Working in a dry environment is otherwise difficult as the starch of the endosperm sticks to the forceps and hinders the operation (*see* **Note 18**).
12. Wipe the outside of a media tube with ethanol, especially its upper portion, close to the lid, and allow to evaporate. Open the media tube over the flame. Flame and cool the spatula, moisten it by dipping in water, and lift the embryo. Place the embryo onto the medium in the tube with flat side of scutellum in contact with the medium, i.e., embryonic axis upward (*see* **Note 17**). Close the tube with the lid after passing the inner side of the lid over the flame. Label the tube with information about the cross, embryo age, and date of culture. Put the tube back in the rack in the same position (*see* **Fig. 2**). Flame all the tools before starting on the next seed. Culture each embryo in a separate tube (*see* **Notes 19** and **20**).
13. Incubate the embryos at low temperature by placing the test tube rack(s) in the vernalization chamber or refrigerator (5–8°C) in the dark for 4–5 d and then transfer to the incubator at 24°C, 12 h photoperiod, and high humidity (*see* **Note 21**).
14. During germination within the first week, if any embryo becomes contaminated, resterilize for 2–5 min in disinfectant as for seeds.
15. After 2–3 wk, when the plantlet is well rooted, have two to three leaves, and its tip has reached the top of the culture tube, transplant it in a pot containing standard greenhouse soil with fertilizers and organic matter. Tease or invert and gently pull out the seedling from the medium, and wash carefully under running water to remove traces of agar (*see* **Note 22**).
16. Place the plants in a growth chamber or greenhouse and protect from direct sunlight. Cover the seedlings with a beaker or a clear plastic bag for a few days to

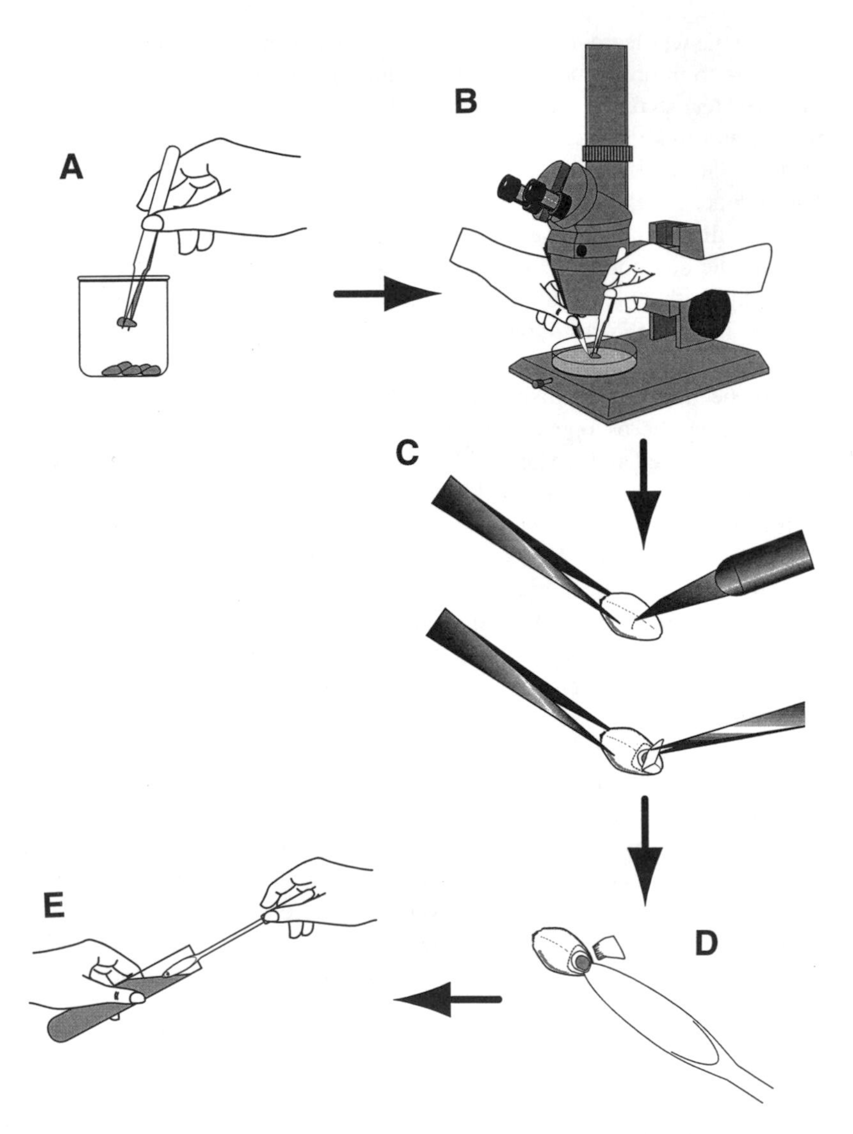

Fig. 4. Dissecting a wide cross-seed, and isolating and culturing the embryo on medium: Take the seed **(A)**, place in the dissecting dish below dissecting scope and hold with one hand using forceps, and operate with the other **(B)**. Incise with a scalpel and peel the pericarp with forceps **(C)**. Lift the naked embryo **(D)**, and place the embryo on the media in the tube **(E)**.

provide enough humidity. Uncover the plant after about a week when new leaves start appearing. When growth is resumed (**Fig. 5**), you have the hybrid(s) at hand (*see* **Notes 23** and **24**).

Fig. 5. Embryo rescued *A. caninum* x wheat F_1 hybrid seedling (middle) in soil after the plastic cover has been removed and good growth has started. On the sides are *A. caninum* (left) and wheat parent seedlings.

4. Notes

1. The example of ER following WC's in wheat *(T. aestivum)* is used for the protocol. Wheat is one of the crop species in which ER following WCs has proven most valuable. Seedlings of winter wheat and wheatgrasses (e.g., *Agropyron, Elymus* species) require vernalization (cold treatment, 5–8°C) for 8–10 wk to induce flowering. Grow the plants under optimum conditions in the greenhouse or growth chamber. Conditions in the field may be better, but unpredictable. A greenhouse environment with 21°C day, 18°C night temperatures and 10-h photoperiod for tillering, and 25°C day, 21°C night temperatures and 18-h photoperiod for flowering and crossing is good for wheat x wheatgrass crosses.
2. Since there is variation in the WC seed and embryo development in a spike, some or several of the seeds when dissected may not have embryos, not every embryo will be placed in the best appropriate position on the most appropriate media or grown under the most appropriate conditions, and some or several embryos may not grow. The researcher's first goal must therefore be to produce as many hybrid seeds as possible. Pollinating 1000 florets/cross is recommended. To facilitate fertilization and caryopsis growth, you may douse the floret after pollination with 50 ppm (50 mg/L) gibberellic acid (GA_3) solution. In several wheat x wheatgrass crosses, seed set is possible or higher when the alien species having the lower chromosome number is used as female.
3. Monitor the seed development very closely on a daily basis starting 1 wk after pollination. Delay the removal of seed from the mother plant as long as possible.

In wheat x wheatgrass crosses, usually the caryopses will begin to turn brown (drying) and are ready for ER, 12 d after pollination, and the embryo can be observed in the shriveling seed.

4. The hood should be turned on well before use to minimize chances of contamination. **Beware: 95% ethanol catches fire.** Use only 70% ethanol when the burner or the gas flame is on.
5. Buy plant cell culture or analytical-reagent-grade chemicals, and store in a dark and dry place following label instructions. Should you have different formulations of any salt, recalculate the amount accordingly. For example, in MS salts, one can use 370 mg/L $MgSO_4 \cdot 7H_2O$ or 181 mg/L $MgSO_4$ because their formula weights are 246.5 and 120.4, respectively.
6. Instead of preparing stocks, media can be obtained in prepared powder form, e.g., Sigma Chemical Co.'s product no. M5519, which is MS-based medium. We use this medium at 2.22 g/L with 50 g/L sucrose and 7 g/L agar, pH = 5.8 for ER following wheat x maize WCs (data unpublished).
7. At 4°C, the stock solutions should remain usable for a few months, but prepare fresh if microbial growth is seen. Unless you need a large quantity of media, prepare regularly small quantities of stocks.
8. The hormone(s) requirement will depend on the cross, growth, and age of hybrid seed, and if embryos do not germinate without hormonal stimulation. Media containing 1 mg kinetin/L for certain wheat x wheatgrass crosses facilitate embryo germination, but root primordia fail to elongate and lead to seedling starvation ***(43)***. Using a more labile form or a lower concentration of hormone(s) in such situations might solve the problem. Alternatively, transfer plantlets to hormone-free medium.
9. If agar is to be melted on a burner, keep shaking the flask constantly, or the agar will stick to the bottom of the container and will burn. Alternatively, autoclave the media in bulk in the flask and dispense into autoclaved tubes in the laminar hood after it has cooled to around 50°C.
10. All ingredients in the medium are heat-stable. If a heat-labile compound has to be added, then autoclave the rest of the medium in a 1.5- to 2-L flask, filter-sterilize the stock solution, and add to precooled (50°C) medium, and then distribute into sterile tubes. The stock solution concentration should be such that addition of no more than 5 mL will give the desired end concentration in the medium.
11. If the media is to be used within 1 wk, it is okay to keep it in the hood at room temperature. For a longer duration, store at 4°C to prevent desiccation.
12. There may be some accidental self-pollinated seed. Distinguish it by its normal and more advanced development, and discard. On the other hand, some of the shriveling hybrid seeds may be small, so open the bracts well to ensure that you do not overlook them.
13. Alternatively, wrap the seeds in cheesecloth, and immerse in the disinfectant with periodic agitation manually or by magnetic stirring to disperse air bubbles and facilitate distribution of the disinfectant.

Fig. 6. Drawing of a seed showing main structures encountered in ER.

Fig. 7. Drawing of an embryo showing its components.

14. If the seeds are from reciprocal matings or from more than one cross, then harvest, label, and surface-sterilize the seeds separately. You may surface-sterilize a few immature seeds from the parent plants for practice in excising embryos.
15. You can dissect the seed on the stereomicroscope stage itself, but this is not recommended. There will be dehydration of embryo without water and glare to eyes with water, and scratching of the stereomicroscope stage can result.
16. A lower magnification providing sharper and whole-seed view may be most desirable for manipulation.
17. To begin with, a new worker may want to practice with parental seeds. The cereal seed, such as wheat x wheatgrass hybrid seed, is composed of pericarp, endosperm, and embryo (**Fig. 6**). The pericarp is the seed-covering over the embryo and endosperm ***(44)***, and forms a hairy tip at the apex of the seed. The embryo is at the end opposite to the hairy tip. The scutellum forms the largest part of the embryo. It is attached to the embryonic axis (bulging side), and its lower side is flat (**Fig. 7**). With experience, just two pairs of forceps will do, and no scalpel is needed for incision; an opening can be made with the forceps used to peel off the pericarp. If the embryo, after one attempt, is uncovered only partly, a little more pericarp can be peeled away. It is a simple question of breaking the seed and getting the embryo out. Any variation in the operation is fine as long as the embryo is not injured. Sometimes, the naked embryo may not stay on the seed and may fall off.
18. It is useful to record embryo type (torpedo, globular, or heart shape), any characteristics of endosperm (starchy, watery, no endosperm), measure embryo size, or

take photograph(s). For measurements, the stereomicroscope should have an ocular micrometer, and for photography, it should be equipped with a camera. Also make a note if no embryo is found.

19. Lifting the embryo is easy if the spatula is moist. A somewhat dehydrated embryo will stick to the surface moisture of the spatula. Be careful not to injure the embryonic axis or push the embryo into the medium. If the embryo is too small to handle with the spatula, the scalpel or a dissecting needle with its tip moistened with culture medium or sterile water can be used to pick up the embryo. If proper placing of the embryo becomes a concern, then use media in Petri dishes rather than tubes. The media dish can be placed under the dissecting microscope while placing the embryo. Petri dishes can be sealed with parafilm. Petri dishes may be more economical, since they need less media and one can culture more than one embryo per dish. However, Petri dishes do not provide enough space for the seedling to grow, and the leaves may spread over the medium. Culturing more than one embryo per dish may be risky in that if one is contaminated, then it may contaminate the others. Furthermore, one seedling may grow too big, but cannot be transplanted because the others are still too young. There may also be some depletion of media components leading to weak seedlings.
20. After dissecting 5–10 seeds, replace the dissecting dish with a new one to avoid the possibility of contamination or losing the embryo in the torn filter paper.
21. Place a pan of water at the bottom of incubator if it does not have a humidity regulator.
22. To count chromosomes for an early confirmation of hybrid nature of the seedling, harvest 1–2 root tips, pretreat in 2% bromonaphthalene for 6-h overnight at 4°C, and fix in glacial acetic acid for 20 min at 4°C. Hydrolyze the root tips in 1 *N* HCl at 60°C for 10 min, stain in Feulgen stain for 10–15 min, and then squash in 1% acetocarmine ***(45)***.
23. If the parents were winter wheat and a wheatgrass, vernalization would be needed for flowering and backcrossing, or for studying inflorescence or pollen mother cells of the hybrids.
24. The number of embryos found and cultured, number of seedlings produced, and behavior of the remaining cultured embryos may be recorded. From the various data recorded, calculate % seed set, embryo recovery, and ER efficiency. This and other information will be useful for further actions.

Acknowledgments

Support of USAID (Grant No. HRN-5600-G-00–2032–00, PSTC, o/o Agri. and Food Security, Center for Economic Growth, Bureau for Global Programs, Field Support and Research), and Public Varieties of Indiana is gratefully acknowledged. This work is dedicated to my family, gurus, colleagues, and relatives who impacted my life. Purdue Univ. Agri. Res. Programs contribution no. 15408.

References

1. Chandler, J., Jan, C., and Beard, B. (1986) Chromosomal differentiation among the annual *Helianthus* species. *Syst. Bot.* **11,** 354–371.
2. Friebe, B., Mukai, Y., Dhaliwal, H., Martin, T., and Gill, B. (1991) Identification of alien chromatin in wheat germplasm by C-banding and *in situ* hybridization. *Theor. Appl. Genet.* **81,** 381–389.
3. Riera-Lizarazu, O., Rines, H., and Phillips, R. (1993) Retention of maize chromosomes in haploid oat plants from oat x maize crosses. *Agron. Abstract* **84,** 112.
4. Jiang, J., Friebe, B., and Gill, B. (1994) Recent advances in alien gene transfer in wheat. *Euphytica* **73,** 199–212.
5. Ozminkowski, R. and Jourdan, P. (1994) Comparing the resynthesis of *Brassica napus* by interspecific hybridization. *J. Am. Soc. Hort. Sci.* **119,** 808–815.
6. Bonhomme, A., Gale, M., Koebner, R., Nicolas, P., Jahier, J., and Bernard, M. (1995) RFLP analysis of an *Aegilops ventricosa* chromosome that carries a gene conferring resistance to leaf rust when transferred to hexaploid wheat. *Theor. Appl. Genet.* **90,** 1042–1048.
7. Sharma, H. (1995) How wide can a wide cross be? *Euphytica* **82,** 43–64.
8. Anderson, A., Crasta, O., Francki, M., Bucholtz, D., Sharma, H., and Ohm, H. (1997) Molecular and cytogenetic analysis of *Thinopyrum intermedium* translocations. *Plant Genome* **5,** 90.
9. Hayward, M., Bosemark, N., and Romagosa, I. (1993) *Plant Breeding; Principles and Prospects.* Chapman and Hill, London.
10. Jones, S., Murray, T., and Allen, R. (1995) Use of alien genes for disease resistance in wheat. *Annu. Rev. Phytopathol.* **33,** 429–443.
11. Evans, D., Flick, C., and Jensen, R. (1981) Disease resistance: incorporation into sexually incompatible somatic hybrids of the genus *Nicotiana. Science* **213,** 907–909.
12. Hansen, L. and Earle, E. (1995) Transfer of resistance to *Xanthomonas campestris* into *B. oleracea* by protoplast fusion. *Theor. Appl. Genet.* **91,** 1293–1300.
13. Sharma, H., Ohm, H., Goulart, L., Lister, R., Appels, R., and Benlhabib, O. (1995) Introgression and characterization of BYD resistance from *Thinopyrum intermedium* in wheat. *Genome* **38,** 406–413.
14. Clark, M. and Wall, W. (1996) *Chromosomes.* Alden, Oxford, UK. p. 345.
15. Stebbins, G. (1958) The inviability and sterility of interspecific hybrids. *Adv. Genet.* **16,** 637–655.
16. Linskens, H. (1972) The reaction of inhibition during incompatible pollination and its elimination. *Soviet Pl. Physiol.* **20,** 156–166.
17. Grant, V. (1981) *Plant Speciation.* Columbia University Press, p. 183.
18. Vershnin, A., Salina, E., and Svitashev, S. (1992) Is there a connection between genomic changes and wide hybridization? *Hereditas* **116,** 213–217.
19. Stalker, H. (1980) Utilization of wild species for crop improvement. *Adv. Agron.* **33,** 111–147.
20. Collins, G., Taylor, N., and DeVerna, J. (1984) *In vitro* approaches to interspecific hybridization and chromosome manipulation in crop plants. *Stadler Genet. Symp.* **16,** 323–383.

21. Baum, M., Laguda, E., and Appels, R. (1992) Wide crosses in cereals. *Annu. Rev. Pl. Physiol. Mol. Biol.* **43,** 117–143.
22. Bridgen, M. (1994) A review of plant embryo culture. *Hort. Sci.* **29,** 1243–1246.
23. Hu, C., Helena, M., and Zanettini, B. (1995) Embryo culture and embryo rescue for wide cross hybrids, in *Plant Cell, Tissue and Organ Culture: Fundamental Methods* (Gamborg, O. and Phillips, G. eds.), Springer-Verlag, Berlin, pp. 129–141.
24. Raghavan, V. (1980) Embryo culture. *Int. Rev. Cytol. Suppl.* **11B,** 209–240.
25. Monnier, M. (1990) Culture of zygotic embryos of higher plants, in *Methods in Molecular Biology, Plant Cell and Tissue Culture* (Pollard, J., and Walker, J., eds.), Humana Press, Clifton, NJ, pp. 129–139.
26. Kaltsikes, P. J., Gustafson, J., and Kaltsikes, P. I. (1986) Triticale: production through embryo culture, in *Biotechnology in Agriculture and Forestry* (Bajaj, Y. P. S., ed.), Springer-Verlag, Berlin, pp. 523–529.
27. Collins, G. and Grosser, J. (1984) Embryo culture, in *Cell Culture and Somatic Cell Genetics of Plants, Laboratory Techniques* (Vasil, I., ed.), Academic, New York, pp. 242–257.
28. Mok, D., Mok, M., Rabakoarihanta, A., and Shii, C. (1986) *Phaseolus:* wide hybridization through embryo culture, in *Biotech. in Agri. and Forestry* (Bajaj, Y., ed.) Springer-Verlag, Berlin, pp. 309–318.
29. Williams, E. (1987) Interspecific hybridization in pasture legumes. *Pl. Breed Rev.* **5,** 237–305.
30. Sharma, H. and Ohm, H. (1990) Crossability and embryo rescue enhancement in wide crosses between wheat and three *Agropyron* species. *Euphytica* **42,** 209–214.
31. McCoy, T. and Smith, L. (1986) Interspecific hybridization of perennial *Medicago* species using ovule-embryo culture. *Theor. Appl. Genet.* **71,** 772–783.
32. Shubhada, T., Paranjpe, S., Khuspe, S., and Mascarenhas, A. (1986) Hybridization of *Gossypium* species through *in ovulo* culture. *Pl. Cell, Tissue Organ Cult.* **6,** 209–219.
33. Guan, G., Li, P., and Wu, D. (1988) A study of backcrossing the interspecific F1 hybrid of *Lycopersicon esculentum* x *L. peruvianum* to *L. esculentum. Acta Hort. Sinica* **15,** 39–44.
34. Chen, Q., Jahier, J., and Cauderon, Y. (1992) Production of embryo callus-regenerated hybrids between *Triticum aestivum* x *Agropyron cristatum* with 1B chromosome. *Agronomie* **7,** 551–555.
35. Kruse, A. (1974) An *in vitro/in vivo* embryo culture technique. *Hereditas* **77,** 219–224.
36. Laurie, D. and Bennet, M. (1988) Cytological evidence of fertilization in hexaploid wheat x sorghum. *Plant. Breed.* **100,** 73–82.
37. Hossain, M., Inden, H., and Asahira, T. (1989) Interspecific hybrids between *Brassica campestris* and *B. oleracea* by embryo and ovary culture. *Memoirs College Agri. Kyoto Univ.* **135,** 21–30.
38. Sirkka, A. and Immonen, T. (1993) Comparison of callus culture at different times of embryo rescue. *Euphytica* **70,** 185–190.

39. Beasley, C., Ting, I., Linkins, A., Binbaum, E., and Delmer, D. (1974) Cotton ovule culture: progress and preview of potential, in *Tissue Culture and Plant Science* (Street, H., ed.), , Academic, New York, pp. 169–192.
40. Sharma, D., Chowdhury, J., Ahuja, U., and Dhankhar, B. (1980) Interspecific hybridization in the genus *Solanum. Z. Pflanzenzuchtng.* **85,** 248–253.
41. Hong, K., Om, Y., and Park, H. (1994) Interspecific hybridization between *Cucurbita pepo* and *C. moschata* through ovule culture. *J. Korean Soc. Hort. Sci.* **35,** 438–448.
42. Murashige, T. and Skoog, F. (1962) A revised medium for rapid growth and bioassays with tobacco tissue cultures. *Physiol. Plant.* **15,** 473–497.
43. Sharma, H. and Baenziger, P. (1986) Production, morphology and cytogenetic analysis of *Agropyron caninum (Elymus caninus)* x *Triticum aestivum* hybrids and backcross-1 derivatives. *Theor. Appl. Genet.* **71,** 750–756.
44. Briggle, L. (1967) Morphology, in *Wheat and Wheat Improvement* (Quisenberry, K., ed.), Am. Soc. Agronomy Pub., Madison, WI, pp. 92–115.
45. Sharma, H. (1982) A technique for somatic counts from root-tips of cereal seedlings raised by embryo culture. *Curr. Sci.* **51,** 143–144.

27

Mutagenesis and the Selection of Resistant Mutants

Philip J. Dix

1. Introduction

The prospect of utilizing plant tissue cultures as a means of generating and identifying novel genetic variants is one that has sparked the interest of researchers for many years. It has resulted in large numbers of research papers describing the selection and characterization of mutants, and several exhaustive reviews (e.g., ***1,2***), including a treatise giving detailed coverage of the whole field ***(3)***. By far the largest number of articles relate to the selection of resistant mutants, which are the most straightforward category to select in vitro, and which are the focus of the present chapter. One thing that is immediately apparent on superficially scanning the literature on these mutants is a high level of repetition. This is because the concept is simple: one challenges a large population of cells with a selective agent, and identifies "survivors" on the basis of their capacity to divide to form cell masses, or organized structures. Thus, the procedures used to select mutants resistant to a range of different antimetabolites have a certain uniformity. By the same token, selection for the same type of mutant (e.g., salt tolerant) in a range of species, involves a broadly similar strategy. On closer scrutiny of the same literature, however, it becomes apparent that far from there being a simple formula to obtain "results" for selection of every phenotype in every species, a great deal of refinement, sometimes major, sometimes subtle, has been necessary in each case. This is owing to the diversity of detail, both in the tissue-culture procedures available for each crop and in the cellular consequences of exposure to different antimetabolites. Awareness of this is important in appreciating the limitations of any set protocol for selection of mutants in vitro. In choosing species, culture systems, mutagenesis treatment, and selective agent for this chapter, the author is providing a specific workable system, which should lead to the selection of

From: *Methods in Molecular Biology, Vol. 111: Plant Cell Culture Protocols*
Edited by: R. D. Hall © Humana Press Inc., Totowa, NJ

mutants. Inevitably, however, variations in any of these will require refinements to the procedures. It is hoped that these protocols will provide a sound baseline on which those modifications can be based.

The procedures utilize *Nicotiana plumbaginifolia*, an amenable species with which the author has considerable experience, and two culture systems, cell suspension cultures and mesophyll protoplasts. They should be applicable without modification to related species, such as *Nicotiana tabacum*, tobacco, and be adaptable for use with other species with suitable cell suspension and/or protoplast culture systems. The author's laboratory has a special interest in the role of proline in stress tolerance, and therefore, in hydroxy-proline-resistant lines, which accumulate proline. It should be possible to substitute other antimetabolites subject to modifications to the procedures. The mutagenesis treatments described here relate to exposure to low concentrations of the two most commonly used nitroso-ureas, but short-term exposure to higher concentrations, and to ethyl methane sulfonate (EMS) is discussed in **Note 9**, whereas UV irradiation is covered in **Note 10**. For advice on other approaches and procedures for in vitro mutagenesis, the reader is referred to earlier articles ***(4,5)***.

2. Materials

1. MS plant salt mix (*see* Appendix).
2. Vitamins and plant growth hormones in aqueous stock solutions as follows:
   a. Thiamine-HCl, 1 mg/mL.
   b. Meso-inositol: no stock, used as powder.
   c. 2,4-Dichlorophenoxyacetic acid (2,4-D): 0.1 mg/mL.
   d. 6-Furfurylaminopurine (kinetin): 0.05 mg/mL.
   e. Benzylaminopurine (BA): 0.5 mg/mL.
   f. Naphthalene acetic acid (NAA): 1 mg/mL.
   g. Gibberellic acid (GA_3): 0.5 mg/mL.

   2,4-D, kinetin, NAA, and GA_3 will dissolve more readily if a small amount of 1 *M* NaOH is added dropwise to the solution until the hormone dissolves, whereas BA is most easily dissolved in 0.1 *M* HCl.
3. All components of K_3 medium and W5 salts (***6***; **Table 1**).
4. 4-Hydroxy-L-proline: 1 *M*.
5. *N*-Nitroso-*N*-ethyl urea (NEU) or *N*-nitroso-*N*-methyl urea (NMU): 10 m*M*.
6. 20% (v/v) Domestos, or other proprietary bleach, giving a final concentration of 1% sodium hypochlorite.
7. RM medium: 4.6 g MS plant salt mix, 20 g sucrose, and 6.5 g Difco Bacto-agar (or equivalent)/L. This and all other media are autoclaved for 20 min at 1.1 kg/cm^2.
8. RMP medium: as RM, but with 100 mg meso-inositol, 1 mg thiamine-HCl, 0.1 mg 2,4-D, and 0.1 mg kinetin/L.
9. RMOP medium: as RM, but with 100 mg meso-inositol, 1 mg thiamine-HCl, 1 mg BA, and 0.1 mg NAA/L.

Table 1
Components of Modified K_3 Medium Used for Isolation and Culture of *N. plumbaginifolia* Protoplasts[a]

Ingredient	mg/L	Ingredient	mg/L
$CaCl_2 \cdot 2H_2O$	900	KI	0.75
KNO_3	2400	$MnSO_4 \cdot 4H_2O$	6.7
$NaH_2PO_4 \cdot 2H_2O$	120	$Na_2MoO_4 \cdot 2H_2O$	0.24
NH_4NO_3	240	$ZnSO_4 \cdot 7H_2O$	2.3
$(NH_4)_2SO_4$	130	meso-inositiol	100
$CoCl_2 \cdot 6H_2O$	0.025	Nicotinic acid	1.0
$CuSO_4 \cdot 7H_2O$	0.025	Pyridoxine·HCl	1.0
$FeSO_4 \cdot 7H_2O$	27.85	Thiamine·HCl	10
Na_2EDTA	37.3	Xylose	250
H_3BO_3	3.0	pH	5.6

[a]Sugar is not included in the above table. Sucrose or glucose is added as described in the list under **Subheading 2.** It is convenient to make up 5 L of the medium double-strength, omitting only the sucrose or glucose, and storing at –20°C. It is generally recommended to filter-sterilize the medium, but for *N. plumbaginifolia* protoplasts, good results can be obtained with autoclaved medium.

10. *N. plumbaginifolia* seeds.
11. Culture room facilities: Preferred conditions: 25°C, 1500 lx, 16-h d. Cell suspension cultures are maintained on a rotary shaker at 100 rpm in the culture room.

3. Methods

These are dealt with under the appropriate subheadings below. They include the sterilization and germination of seeds and the initiation of shoot, callus, and cell suspension cultures; the determination of the appropriate selective level of hydroxyproline; mutagenesis and the selection of resistant colonies; and the regeneration of plants from them.

3.1. Initiation of Shoot, Callus, and Cell Suspension Cultures

1. Place a few (ca. 100–200) tobacco seeds in an Eppendorf tube.
2. Drop a few drops of 70% ethanol in the tube for a few seconds, and then add 20% Domestos (without removing the alcohol) for 10 min.
3. Spin for 30 s at top speed in a microfuge, pour off the Domestos, and replace with sterile distilled water.
4. Repeat **step 3**, providing a second wash in sterile distilled water.
5. Pour the seeds out on to three to four 9-cm Petri dishes, each containing 20 mL RM medium. Seal with Parafilm and incubate in the culture room until the seeds germinate (*see* **Note 1**).
6. Thin out seedlings at 0.5–1 cm in height, transferring to RM medium in sterile tubs or jars (e.g., Plantcon containers—Flow Laboratories). At a height of 4–5 cm,

transfer to individual containers. When doing so, remove the roots and the basal 2 mm of the stem with a single clean cut, and insert the cut surface into the fresh medium. This encourages vigorous growth of adventitious roots.

7. When the seedling approaches the top of the container, establish shoot cultures by dissecting out nodes (with 0.5- 1-cm internode remaining above and below) and transferring to RM medium in fresh containers. Axillary buds will grow, and the process can be repeated at 4- to 8-wk intervals.
8. To initiate callus cultures, remove leaves and cut into strips (3- to 4-mm wide) with sterile scalpel and forceps. Transfer to the surface of 20 mL RMP medium in 9-cm Petri dishes, 5 explants/dish, with the lower epidermis downward. Seal with Parafilm, and incubate at 25°C in culture room or an incubator (illumination not necessary).
9. When sufficient callus has developed from cut surfaces of the explants (4–6 wk), excise small pieces (0.3- to 0.5-cm diameter), transfer to fresh RMP medium, and culture as before. In this way, callus stocks can be maintained indefinitely by subculturing at monthly intervals.
10. To initiate cell suspension cultures, transfer about 1 g of callus to 50 mL of RMP medium (without agar) in a 250-mL Erlenmeyer flask. Development of finely dispersed cell suspensions may be accelerated by cutting the callus into pieces first. Incubate the cultures at 25°C on an rotary shaker at 100 rpm.
11. Subculture the newly formed suspensions after 3 wk by swirling each flask and pouring at 3–4x dilution into flasks of fresh RMP medium.
12. Subculture the cell suspension cultures in the same way but at about 5X dilution, at 10- to 14-d intervals. After several such culture passages, the suspensions should become more finely dispersed and rapidly dividing. At this stage, the culture passage can be shortened to 7 d to eliminate lag and stationary phases of the growth curve. (For further comments on the suspension culture procedure, *see* **Note 2**.)

3.2. Determination of Selective Level of the Antimetabolite (See Also Note 3)

1. Prepare RMP media containing the following concentrations of hydroxyproline (mol wt = 131): 0, 0.1, 0.3, 1, 3, 10, 30, and 100 m*M* (*see* **Note 4**). The media are supplemented with 0.65% Difco Bacto-agar and dispensed, 20 mL/dish, into 9-cm Petri dishes. Hydroxyproline, from freshly prepared filter-sterilized stock solutions (prepared in liquid RMP medium), is added to flasks of autoclaved medium held at about 45°C. Since addition of stock will in some cases result in substantial dilution of the agar in the medium, calculated adjustments in the initial volume of liquid medium in such flasks may be required. Two stock solutions are recommended: 300 m*M* (39.3 mg/mL) for 100- and 30-m*M* final concentrations, and 100 m*M* (13.1 mg/mL) for the remainder. The volumes of media and stock solution required to give final volumes of 250 mL (sufficient for 12 dishes) are summarized in **Table 2**.
2. Place small (20- to 30-mg) pieces of healthy callus, taken from culture about 3 wk after subculture, on the surface of the media described above. Five inocula should

Table 2
Preparation of Media Containing a Range of Hydroxyproline Concentrations

Final concentration, mM	Stock solution, mM	Volume autoclaved medium[a], mL	Volume autoclaved solution added, mL
100	300	167	83
30	300	225	25
10	100	225	25
3	100	242.5	7.5
1	100	250	2.5
0.3	100	250	0.75
0.1	100	250	0.25
0	—	250	0

[a]Each flask contains agar for 0.65% concentration in a final volume of 250 mL (i.e., 1.63 g).

be evenly spaced on each dish. Use 10 dishes/hydroxyproline concentration to give a total of 50 calli/treatment. Seal dishes with Parafilm, and incubate at 25°C.

3. After 4–6 wk, weigh the calli, determine the mean and standard deviation for each treatment, and after subtracting the initial inoculum weight, calculate the percentage of fresh weight increase compared to the control. Plot these values against hydroxyproline concentration (log scale).
4. Select the lowest concentration giving complete inhibition of growth (likely to be 10 or 30 mM) as the selective concentration for mutant isolation.

3.3. Mutagenesis and Selection of Resistant Cell Lines

The following protocol commences with cell suspension cultures. If protoplasts are preferred, *see* **Note 5**.

1. To 10 flasks of freshly subcultured cell suspension cultures, on a 7-d passage cycle, add filter-sterilized NEU or NMU to give a final concentration of 0.1 mM (**important:** *see* **Note 6**).
2. Culture the treated cell suspensions and also nonmutagenized controls in the usual way for 7 d.
3. For each suspension culture, prepare RMP medium: levels of all components and agar for 250 mL, and selective levels of hydroxyproline (again refer to **Table 1**). In this case, however, make each flask up to a final volume of 200 mL, not 250 mL (omit 50 mL distilled water when preparing the medium). Place autoclaved flasks of medium in a water bath at 45°C, to keep agar molten, for 30–60 min prior to plating the cells.
4. Filter the contents of each flask of cells through two layers of sterile muslin to remove large aggregates (this step may be omitted if the suspension is very fine), and pour the suspension into one of the flasks of molten medium. Mix by swirl-

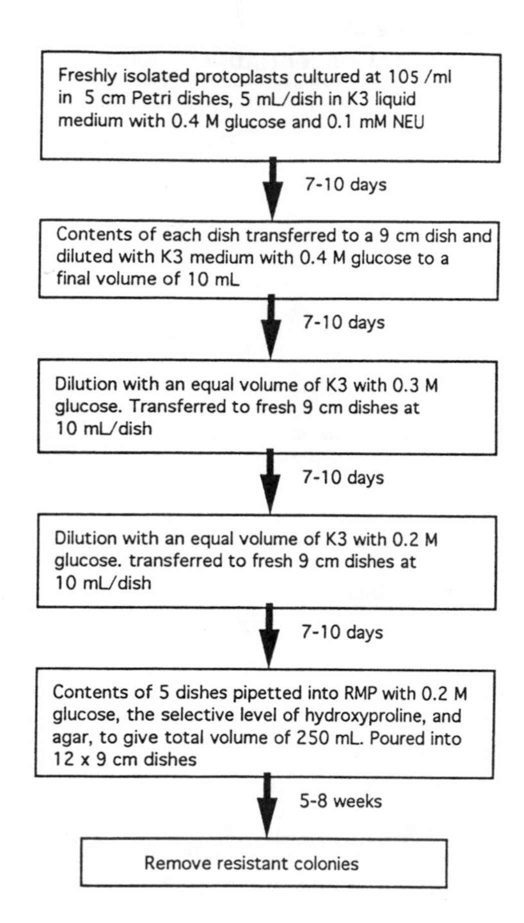

Fig. 1. A flow diagram for isolation of hydroxyproline-resistant colonies from mesophyll protoplasts of *N. plumbaginifolia*.

ing and pour into 12 sterile 9-cm Petri dishes. Over 100 dishes containing the selective level of hydroxyproline and 12 control dishes without hydroxyproline should be obtained.

Cell aggregates arising from mesophyll protoplasts (*see* **Note 5** and **Fig. 1**) are plated in the same way, except that filtration through muslin is not necessary, and the plating medium includes 0.2 *M* glucose or sucrose. In this case, pipet (using a sterile Pasteur pipet) the contents of five dishes of protoplast-derived colonies into each flask of molten medium, swirl, and pour as above. The number of dishes of protoplast-derived colonies and, therefore, the number of flasks of medium needed, will depend on the initial yield and viability of protoplasts, which is difficult to standardize.

5. Allow the agar to set, seal each dish with a strip of Parafilm, and incubate at 25°C in the dark.
6. Examine the dishes at weekly intervals. After 5–8 wk, remove any growing colonies from selective plates, and transfer individually to dishes of fresh RMP medium supplemented with the selective level of hydroxyproline.
7. After 4–7 wk, transfer small pieces (20–30 mg) of callus, from lines continuing to grow on medium containing hydroxyproline to RMOP medium in Petri dishes. Seal with parafilm, and incubate in the culture room at moderate (1000–2000 lx) illumination.
8. After 5–8 wk, excise any adventitious shoots with normal leaves, and transfer to RM medium. Incubate for a further 4–6 wk, until a vigorous root system is established. Transfer calluses that did not produce adventitious shoots on RMOP medium to fresh RMB medium. Shoots may be regenerated on RMB, or after several alternating culture passages on RMB and RMOP (*see also* **Note 7**).
9. Shoot cultures can either be maintained in vitro by transferring nodal cuttings to fresh medium as already described, or transferred to the greenhouse for growth to maturity and genetic analysis. In the latter case, gently remove agar from the roots, and plant individually in 12-cm pots 2/3 full of potting compost. To avoid desiccation, cover tightly with cling film for 1–2 wk, until the plantlets are established. When growing vigorously, transfer them to larger pots, where they should flower within 2–3 mo.
10. Check resistance of selected plants by initiating callus from shoot cultures and testing for growth on medium containing selective levels of hydroxyproline as already described (*see* **Note 8**).

4. Notes

1. In the event of poor germination, a gibberellic acid treatment can be included. Sterilized seeds are placed in an aqueous solution of GA_3 (0.5 mg/mL) for 1 h before transferring to Petri dishes.
2. The object of this protocol is to provide a straightforward set of instructions that will give a good chance of obtaining the desired mutants. For this reason, detailed monitoring of the growth of the suspension cultures has been excluded in favor of a simple empirical approach that has proven effective in our hands. If quantitative data on mutation frequency or direct comparisons between a number of experiments are required, it may be desirable to monitor the growth of the suspensions to ensure that comparable cultures are always used, and to perform cell counts prior to plating to ensure that a uniform cell density (about 10^4 cells/mL) is used in all experiments. Such measurements can pose their own problems in highly aggregated suspension cultures, but some useful guidelines and procedures can be found in ref. ***(7)*** and ***(8)***.
3. Callus is the most convenient material for producing a dosage–response curve, as described in the protocol. In the case of hydroxyproline, there is a good correlation between the sensitivity of callus and that of plated cells or protoplasts. For some chemicals, this correlation can be less precise, and it may be wise, if suffi-

cient material is available, to use as additional selective levels concentrations above and below that determined by the callus test. A more rigorous alternative is to measure dose–response by plating cells or protoplasts at the complete range of antimetabolite concentrations, using exactly the same procedure described in **Subheading 3.3.** In this case, instead of fresh weight, colony number after a suitable period (4–6 wk for plated cell suspensions, 6–8 wk for plated protoplasts) should be plotted against antimetabolite concentration.

4. Storage times of culture medium: All media described here, including those containing hydroxyproline, can be stored for 3–4 wk, preferably in the cold room without serious deterioration. Medium in Petri dishes, however, is subject to concentration by evaporation and should be used within 2 wk of preparation. Even for this period, it is important that they are sealed in bags (e.g., those in which the sterile dishes are obtained from suppliers) and kept in the cold room. NMU and NEU should be added to medium only immediately before use.
5. If mesophyll protoplasts are the desired starting material, the protocol for isolation, mutagenesis, and culture of the protoplasts is exactly as described for the isolation of chloroplast mutants in Chapter 28, this volume. When the colonies are at the stage where they are ready to be incorporated into solid medium, they can be handled in exactly the same way as mutagen-treated cell suspension cultures in the current protocol, except that 0.2 *M* glucose or sucrose should be included in the RMP medium used for plating. The steps in the isolation of proline resistant mutants from mesophyll protoplasts are outlined in the flow diagram (**Fig. 1**).
6. Both NMU and NEU, which are closely related, are dangerous carcinogens, and must be handled with extreme care. We recommend the use of a respirator, and protective gloves and apron during all manipulations involving mutagens. It is important to avoid skin contact. All working surfaces, balances, and the like, where spillage might have occurred, should be washed down immediately after use. Placing large sheets of absorbent paper backed with foil on the laminar flow work surface helps to contain any spillage. It is a good idea to exclude other workers from the work area while manipulations with mutagen are in progress. In the event of skin contact, and routinely after use, wash hands in soap and water gently, avoiding excessive rubbing of the skin. Do not use NMU in alkaline solutions, because it is very unstable. After filter-sterilizing mutagen solutions, do not remove the syringe from the millipore unit immediately, since the pressure that has built up in the syringe can result in the release of an aerosol of the mutagen. Mutagen solutions should be inactivated before disposal. Add an excess of 5% NaOH in the fume hood, and leave open overnight, before pouring down the sink and chasing with a large volume of tap water. Contaminated apparatus should also be treated with 5% NaOH, but in this case, after an overnight soaking, a second wash (>1 h) is recommended, followed by a thorough rinse in running tap water. Both mutagens are unstable in aqueous solution, with the advantage that the low concentrations used here do not need to be washed from the cells. The rate of breakdown is highly pH-dependent, and it is important that mutagen solu-

tions should be prepared using culture medium at its normal pH (5.6). Further information about these mutagens can be found in ref. ***(9)***. It is also possible to select spontaneous mutants without mutagenic treatment, but a larger number of cultures may be required.

7. Shoot regeneration from the resistant cell lines is a critical phase of the procedure. The efficiency with which regeneration can be induced by manipulating the hormonal composition of the medium (i.e., using RMOP or RMB) is variable, and in the case of cell suspension cultures, may be quite low. Regeneration can often be stimulated by the inclusion in the medium of silver ions, in the form of silver nitrate at 10 or 50 mg/L ***(10)***.
8. Resistance to the antimetabolite can also be checked in seedling progeny and callus derived from them.
9. Mutagens can also be supplied at higher concentrations for short time periods to freshly isolated protoplasts or cell suspension cultures. The method is fairly general for a range of mutagens, including NEU, NMU, and EMS. In all cases, preliminary experiments are needed to establish a suitable concentration and duration of treatment, one that gives a reduction to 10–50% colony formation in plated protoplasts, or cell suspension cultures compared to untreated controls. Recommended concentration ranges to investigate are 0.3–10 m*M* for NEU or NMU, or 0.1–3% v/v EMS. The duration of treatment should be about 60 min, after which the cells should be washed twice with fresh culture medium, before continuing to culture and plating in the usual way. EMS is a volatile liquid that needs to be dispensed into the liquid medium in a fume hood. Prior to disposal, solutions containing EMS should be inactivated by gradual addition to a large excess of 3 *M* KOH in 95% ethanol, heated under reflux, repeatedly stirred for 2 h before disposal down the drain, chased by a large volume of tap water.
10. UV mutagenesis is best applied to protoplasts 24 h after isolation. Dishes of protoplasts are placed under the UV source and the lids removed before it is turned on. A dose giving 10–50% subsequent colony formation compared to nonmutagenized controls should be used (typically in the range 200–2000 erg/mm^2).

References

1. Dix, P. J. (1986) Cell line selection, in *Plant Cell Culture Technology* (Yeoman, M. M., ed.), Blackwell Scientific, Oxford, pp. 143–201.
2. Dix, P. J. (1994) Isolation and characterisation of mutant cell lines, in *Plant Tissue Culture—Theory and Applications* (Vasil, I. K. and Thorpe, T. A., eds.), Kluwer Academic, Dordrecht, pp. 119–138.
3. Dix, P. J. (ed.) (1990) *Plant Cell Line Selection.* VCH, Weinheim.
4. Negrutiu I. (1990) *In vitro* mutagenesis, in *Plant Cell Line Selection* (Dix, P. J., ed), VCH, Weinheim, pp. 19–38.
5. Dix, P. J. (1993) Use of chemical and physical mutagens *in vitro*, in *Plant Tissue Culture Manual: Fundamentals and Applications* (Lindsey, K., ed.), Kluwer Academic, Dordrecht, F1, pp. 1–17.

6. Maliga, P. (1984) Cell culture procedures for mutant selection and characterisation in *Nicotiana plumbaginifolia*, in *Cell Culture and Somatic Cell Genetics of Plants,* vol. 1, *Laboratory Procedures and Their Applications* (Vasil, I. K., ed.), Academic, New York, pp. 552–5562.
7. King, P. J. (1984) Induction and maintenance of cell suspension cultures, in *Cell Culture and Somatic Cell Genetics of Plants,* vol. 1, *Laboratory Procedures and Their applications* (Vasil, I. K., ed.), Academic, New York, pp. 130–138.
8. Gilissen, L. W. J., Hänisch ten Cate, C. H., and Keen, B. (1983) A rapid method of determining growth characteristics of plant cell populations in batch suspension culture. *Plant Cell Rep.* **2,** 232–235.
9. Hagemann, R. (1982) Induction of plastome mutations by nitrosourea compounds, in *Methods in Chloroplast Molecular Biology* (Edelman, M., Hallick, R. B., and Chua, N. H., eds.), Elsevier/North Holland, Biomedical, Amsterdam, pp. 119–127.
10. Purnhauser, L., Medgyesy, P., Czakó, M., Dix, P. J., and Márton, L. (1987) Stimulation of shoot regeneration in *Triticum aestivum* and *Nicotiana plumbaginifolia* Viv. tissue cultures using the ethylene inhibitor $AgNO_3$. *Plant Cell Rep.* **6,** 1–4.

28

The Generation of Plastid Mutants In Vitro

Philip J. Dix

1. Introduction

The plastid genome (or "plastome") encodes a number of proteins associated with the structure and function of chloroplasts, as well as tRNAs and rRNAs associated with the plastid translational machinery *(1,2)*. Although there have been numerous studies on the genetics of algal chloroplasts, similar studies with higher plants have been hampered by the uniparental (maternal) pattern of transmission of chloroplasts observed in most species and also the shortage of suitable genetic markers *(3)*.

Two developments have added impetus to studies on higher plant plastid genetics. First, procedures have been developed *(4,5)* for the more efficient generation of plastome mutations. Second, breakthroughs in the genetic transformation of plastids *(6,7)* have led to an appreciation of the great biotechnological potential of expressing foreign genes in the plastids *(8)*. DNA delivery into plastids has been achieved through biolistics or PEG-mediated uptake into protoplasts, and selection of transformants is based on one of two alternate strategies *(9)*. Of these, the more attractive is the use of plastid mutations conferring insensitivity to antibiotics *(10)*. As well as rapidly generating homoplasmic transplastomic lines, these avoid the controversial use of bacterial antibiotic resistance genes *(9)* and indeed can restrict the foreign DNA introduced to solely the gene of interest.

Thus, plastid mutations are now of interest both for the fundamental information they can provide on plastid function and as a means for selecting genetically transformed plastids. The protocols given here describe two reliable procedures to select for mutations in plastid ribosomal RNA or protein genes conferring insensitivity to several antibiotics (streptomycin, spectinomycin, lincomycin), in Solanaceous species. They use the mutagen nitroso-methylurea

From: *Methods in Molecular Biology, Vol. 111: Plant Cell Culture Protocols*
Edited by: R. D. Hall © Humana Press Inc., Totowa, NJ

Table 1
Components of Modified K_3 Medium Used for Isolation and Culture of N. plumbaginifolia Protoplasts[a]

Ingredient	mg/L	Ingredient	mg/L
$CaCl_2 \cdot 2H_2O$	900	KI	0.75
KNO_3	2400	$MnSO_4 \cdot 4H_2O$	6.7
$NaH_2PO_4 \cdot 2H_2O$	120	$Na_2MoO_4 \cdot 2H_2O$	0.24
NH_4NO_3	240	$ZnSO_4 \cdot 7H_2O$	2.3
$(NH_4)_2SO_4$	130	myo-inositiol	100
$CoCl_2 \cdot 6H_2O$	0.025	Nicotinic acid	1.0
$CuSO_4 \cdot 7H_2O$	0.025	Pyridoxine · HCl	1.0
$FeSO_4 \cdot 7H_2O$	27.85	Thiamine · HCl	10
Na_2EDTA	37.3	Xylose	250
H_3BO_3	3.0	pH	5.6

[a]Sugar is not included in the above table. Sucrose or glucose is added as described in the list under **Subheading 2.** It is convenient to make up 5 L of the medium double-strength, omitting only the sucrose or glucose, and storing at –20°C. It is generally recommended to filter-sterilize the medium, but for *N. plumbaginifolia* protoplasts, good results can be obtained with autoclaved medium.

(NMU), which has been shown to be efficient in targeting the plastome ***(11)***. The procedure using protoplasts is effective for *Nicotiana plumbaginifolia* ***(4)*** and with modifications (*see* **Note 4**) can also be used to obtain herbicide resistant mutants. The leaf strip procedure ***(5)*** has been used successfully for several species (*Nicotiana tabacum, N. plumbaginifolia, Nicotiana sylvestris, Solanum nigrum,* and *Lycopersicon peruvianum*) and can also generate chlorophyll-deficient mutants on nonselective plates.

2. Materials

1. Shoot cultures of the species to be used (*see above*): The procedures for obtaining and maintaining these cultures, starting from seed, are as described in Chapter 27, this volume.
2. Protoplast enzyme solution: Modified K3 medium (***12***; **Table 1**) containing 0.4 *M* sucrose and 0.5% Driselase (Kyowa Hakko Kogyo Co., Tokyo, Japan) (*see* **Note 1**). K3 can be filter-sterilized or autoclaved in a pressure cooker, and stored for up to 4 wk, preferably in the cold room. Driselase must be added immediately before use, and the resulting enzyme solution filter-sterilized.
3. Protoplast wash solution, W5 ***(12)***: 150 m*M* NaCl, 125 m*M* $CaCl_2$, 5 m*M* KCl, 5 m*M* glucose, pH 5.6. W5 may be autoclaved and stored in the cold room for up to 2 mo.
4. Protoplast culture medium: Modified K3 supplemented with 0.4 *M* glucose, 0.1 mg/L 2,4-dichlorophenoxyacetic acid (2,4-D, from a 0.1 mg/mL stock pre-

pared with the dropwise addition of 0.1 *M* NaOH to dissolve the hormone), 0.2 mg/L 6-benzylaminopurine (BA, from a 0.5 mg/mL stock, prepared in 0.1 *M* HCl), and 1 mg/L naphthaleneacetic acid (NAA, from a 1 mg/mL stock prepared as for 2,4-D, above). Also, the same medium with glucose levels reduced to 0.3 and 0.2 *M*.

5. Regeneration medium (RMOP): Contains (per liter): 4.6 g MS salts (*see* Appendix), 20 g sucrose, 100 mg mesoinositol, 1 mg thiamine-HCl, 1 mg BA, 0.1 mg NAA (*see* **item 4** *above* for stock solutions of hormones), pH 5.6, solidified with 0.65% Difco Bacto-agar.
6. Leaf strip medium for *L. peruvianum*: Contains (per litre): 4.6 g MS salts, 20 g sucrose, 100 mg mesoinositol, 1 mg thiamine-HCl, 0.5 mg nicotinic acid, 0.5 mg pyridoxine-HCl, 0.2 mg indoleacetic acid (IAA), and 2 mg zeatin (Sigma). Stock solutions (1 mg/mL) of these hormones should be freshly prepared in 0.1 *M* NaOH (IAA) or HCl (zeatin), filter-sterilized, and added to molten, autoclaved medium (pH 5.6, 0.65% Difco Bacto-agar) before pouring.
7. Leaf strip medium for *S. nigrum*: As for *L. peruvianum,* except zeatin is replaced by 1 mg/L BA.
8. RM medium: Contains (per liter): 4.6 mg MS salts and 20 g sucrose, pH 5.6, solidified with 0.65% Difco Bacto-agar.
9. RM solution: as for RM medium, but without the agar.
10. 60-μm mesh nylon bolting cloth.
11. Mutagen: *N*-nitroso-*N*-methylurea (NMU) (*see* **Note 2**).
12. Protective clothing: Plastic or rubber apron and disposable gloves, industrial organic vapor cartridge respirator with cartridges, and prefilters (obtainable from "Sa-fir," East Hoathly, East Sussex, England).
13. Large sheets of absorbent paper backed with foil.
14. 5 *M* NaOH.
15. 5% (w/v) Streptomycin sulfate in H_2O, filter-sterilized.
16. 5% (w/v) Lincomycin hydrochloride in H_2O, filter-sterilized.
17. Culture room facilities: Preferred conditions: 25° C, 1500 lx, 16-h d.
18. 0.4 *M* glucose.

3. Methods

Mutant isolation is described below, using two different starting materials, leaf strips and mesophyll protoplasts. The use of leaf strips is technically more straightforward and is described first. Healthy leaves from shoot cultures of any of the species mentioned in **Subheading 1.** can be used in exactly the same way, the only differences being the culture media required. In our experience, *S. nigrum* gives substantially higher yields of streptomycin-, spectinomycin-, or lincomycin-insensitive mutants than the other species. Albino mutants can readily be obtained in the same experiments, as described in **Note 3**.

The rest of this section deals with the isolation of streptomycin- and lincomycin-insensitive mutants of *N. plumbaginifolia* from cultures initiated from mesophyll protoplasts. Modifications of this procedure, necessary to obtain mutants resistant to terbutryn and other photosynthetic herbicides, are described in **Note 4**.

3.1. Isolation of Antibiotic-Insensitive Mutants from Leaf Strips

1. Prepare mutagen solutions after carefully reading **Note 2**. Prepare a stock solution containing 80 mg NMU in 20 mL of RM solution, and filter-sterilize. For a 1-m*M* solution, add 2.6 mL of stock to 97.4 mL of autoclaved RM solution in a wide-necked 250-mL Erlenmeyer flask. For a 5-m*M* solution, add 12.9 mL–87.1 mL of RM solution. It is important that the RM solutions used have pH values of 5.5–6.0 to enhance the stability of the mutagen.
2. Remove leaves from shoot cultures and cut into small strips (2- to 3-mm wide by 5- to 15-mm long, depending on leaf size). Add 250 strips to 100 mL of each (1 and 5 m*M*) NMU solution and to 100 mL of RM solution (nonmutagenized control).
3. Incubate the flasks on a rotary shaker (about 50 rpm) at 25°C for 90 min.
4. Decant mutagen solution, and wash the leaf strips four to five times in a large excess of RM solution or sterile distilled water (pH adjusted to 5.5–6.0).
5. For each treatment, place 200 of the strips on the surface of selective medium in 40 plastic Petri dishes (5 strips/dish). The selective media are RMOP for *Nicotiana* species, and the media described in **Subheading 2.** for *S. nigrum* and *L. peruvianum,* in each case supplemented with 500 mg/L streptomycin sulfate (1000 mg/L can also be used for *Nicotiana* spp. and *S. nigrum*), 100 mg/L spectinomycin, or lincomycin hydrochloride. The medium is prepared by addition of a small aliquot from the concentrated, filter-sterilized stock solution of antibiotic to the autoclaved culture medium, cooled to about 50°C prior to pouring into dishes. Place the remaining 50 strips on the same medium without streptomycin. Seal all dishes with parafilm, and incubate in the culture room.
6. After about 40 d, the leaf strips on control dishes will show prolific shoot regeneration, whereas those on antibiotic-containing medium are bleached and show little morphogenesis. Green nodules will appear at the edges of some of the bleached leaf strips (from one or both mutagen treatments), and most of these will develop into shoots. When shoots have at least two leaves that are normal, not "vitreous" (glassy and translucent) in appearance, remove cleanly with a scalpel, and transfer to RM medium (embedding cut stem in the medium) for rooting. Continue to incubate in the culture room.
7. After 4–8 wk, plantlets with vigorous roots are obtained. Remove a single leaf, cut into strips, and test for insensitivity to the antibiotic on the same medium as that initially used for mutant selection. Typically, insensitivity is expressed in the mutants by the retention of chlorophyll and the appearance of numerous green adventitious shoots.
8. Rooted plants can either be propagated in vitro by nodal cuttings, or transferred to soil for growth to maturity and genetic analysis. Both these procedures are described in Chapter 28, this volume.

3.2. Isolation of Streptomycin or Lincomycin-Insensitive Mutants from Protoplast Cultures of N. plumbaginifolia

1. Remove healthy, fully expanded leaves from shoot cultures, slice finely using a sterile razor blade (in a holder) or scalpel and forceps, and transfer to sterile protoplast enzyme solution in 9-cm Petri dishes. Typically, 10 mL of solution in a dish should be sufficient for three to four leaves, and three such dishes should provide enough protoplasts for one experiment. Protoplast yields are variable, however, so it may be necessary to start with more material.
2. Incubate overnight (12–16 h) at 25°C in the dark.
3. Swirl the dishes several times to liberate protoplasts from leaf debris, remove the solution with a Pasteur pipet, and filter through 60-μm nylon mesh into a 100-mL Erlenmeyer flask.
4. Transfer the protoplast preparation to 10-mL capped glass centrifuge tubes (sterile), and spin at about 300*g* for 3 min.
5. Intact protoplasts float and form a tight green band at the surface of the medium. Remove this carefully with a Pasteur pipet, and transfer to a clean tube (not more than 1 mL/tube). Fill the tubes with W5 solution, cap, invert to ensure thorough mixing, and spin at about 50*g* for 2 min.
6. Pipet off the supernatant, and resuspend the protoplast pellet in a small volume of protoplast culture medium (K3) supplemented with 0.4 *M* glucose. Mix the contents of the tube, count the intact (spherical, with an uninterrupted plasma membrane) protoplasts, and dilute to 10^5/mL with the culture medium.
7. Transfer to 5 cm Petri dishes, 5 mL/dish, using either previously calibrated Pasteur pipets or automatic pipets with wide-bore tips.
8. To individual dishes, add (to a final concentration of 0.1 or 0.3 m*M*) NMU from a concentrated stock prepared in the culture medium and filter-sterilized. (**Important:** read **Note 2** carefully before using the mutagen.) Wrap all dishes with parafilm and incubate in the culture room at low light intensity (ca. 100 lx).
9. After 7–10 d, providing divisions can be observed in the protoplasts, dilute the protoplasts 2x with fresh K3 medium with 0.4 *M* glucose. To do this, pipet the contents of each dish into a 9-cm dish, and add 5 mL of medium. Seal and incubate as before. There is no need to wash out the mutagen, since it is unstable and breaks down within 48 h at the pH used.
10. Monitor the development of the protoplast-derived cell aggregates, and make dilutions at suitable intervals (*see* **Note 5**). In a good preparation, these intervals should be of 7–10 d. Each dilution should be 2x, and lead to a doubling of the number of dishes containing 10 mL culture. For the first dilution, use K3 medium with the glucose reduced to 0.3 *M*, and for the second, glucose reduced to 0.2 *M*. Within 10–14 d of the latter dilution, numerous microcolonies (about 1-mm diameter) should be visible, and the cultures are then ready for plating into solid medium.
11. For every four (more if there is a low colony density) dishes, prepare 500 mL of RMOP medium with 0.2 *M* glucose (instead of 2% sucrose), 0.65% agar, and 1000 mg/L streptomycin sulfate or lincomycin hydrochloride. The antibiotics are

added from the concentrated filter-sterilized stock solution to autoclaved medium, held at 45°C in a water bath.

12. Using a fine-tipped Pasteur pipet, remove the excess K3 medium from the cultures to be plated. This is best done by tilting the dishes slightly and allowing the microcolonies to settle, so that the medium can be removed from above them. Then, using a broad-tipped pipet, wash the contents of four dishes into the 500-mL molten medium by repeatedly transferring small amounts of the medium to the dishes and sucking up again, together with the colonies.
13. Mix well, and pour into 9 cm Petri dishes (about 20 mL/dish). Allow the agar to set, wrap the dishes with parafilm, and incubate in the culture room at 1000- to 1500-lx illumination.
14. Numerous white colonies should appear after 1–2 mo. Streptomycin- or lincomycin-insensitive colonies are green and are easily selected visually. Pick them off when large enough and transfer to the same medium, but with sucrose reduced to 0.1 *M*.
15. Resistant colonies will continue to grow and remain green. Subculture them onto RMOP medium, without the antibiotic, for shoot regeneration. Regenerated shoots are handled as described for mutants isolated from leaf strips.

4. Notes

1. Driselase powder is a crude preparation containing much insoluble material that can quickly block the millipore filter. This can be prevented by either spinning for a few minutes in a bench centrifuge, or filtering, to obtain a clean solution prior to filter sterilization.
2. NMU is a dangerous carcinogen and must be used with great care. We recommend the use of a respirator, and protective gloves and apron during all manipulations involving the mutagen. It is important to avoid skin contact. All working surfaces, balances, and the like, where spillage might have occurred, should be washed down immediately after use. Placing large sheets of absorbent paper backed with foil on the laminar flow work surface helps to contain any spillage. It is a good idea to exclude other workers from the work area while manipulations with mutagen are in progress. In the event of skin contact, and routinely after use, wash hands in soap and water gently, avoiding excessive rubbing of the skin. Do not use NMU in alkaline solutions because it is very unstable. After filter-sterilizing mutagen solutions, do not remove the syringe from the millipore unit immediately, since the pressure that has built up in the syringe can result in the release of an aerosol of the mutagen.

   Mutagen solutions should be inactivated before disposal. Add an excess of 5% NaOH in the fume hood, and leave open overnight, before pouring down the sink and chasing with a large volume of tap water. Contaminated apparatus should also be treated with 5% NaOH, but in this case, after an overnight soaking, a second wash (>1 h) is recommended, followed by a thorough rinse in running tap water. Additional advice on the use of these mutagens is given in ref. ***(11)***.
3. For all the species mentioned, albino mutants have also been obtained from leaf strips. These arise in response to the mutagenesis treatment on the dishes from

which the selective agent (antibiotic or herbicide) has been excluded. Among the normal green shoots differentiating from the leaf strips, some albino or variegated shoots are frequently obtained. Albino shoots can be rooted and maintained on RM medium in the same way as other shoot cultures. Albino shoots can be obtained from variegated ones by dissecting out white sectors and culturing on the appropriate regeneration medium.

4. The procedure for isolating antibiotic-resistant mutants from protoplast cultures of *N. plumbaginifolia* can be applied to the selection of mutants resistant to herbicides that inhibit photosynthesis, providing a selective medium permitting photomixotrophic growth is used. This is achieved by lowering to 0.3% the glucose level in the RMOP medium in which the microcolonies are plated. In order to reduce the osmotic stress resulting from plating in such a low-sugar medium, an additional dilution step (with K3 medium plus 0.1 *M* glucose) is introduced into the protoplast culture procedure. For triazine herbicides (e.g., terbutryn, atrazine, simazine), a suitable selective level is 10^{-4} *M*, and selection is based on the greening of colonies, exactly as in the case of antibiotic-resistant mutants. After retesting for resistance on selective medium, shoots are regenerated by transfer of small callus pieces to RMOP medium without the herbicide. Mutants resistant to metobromuron and bromoxynil have also been obtained in this way.
5. The instructions for the gradual dilution of protoplast cultures with fresh medium of decreasing glucose concentration are given as accurately as possible. The 7- to 10-d interval should work, but careful monitoring of the cultures is desirable. If the growth rate of the colonies seems to be slow, a longer interval must be used. On the other hand, rapid growth rates, especially if accompanied by browning or the appearance of dead cells, indicates a requirement for more rapid dilution.

References

1. Dyer, T. A. (1985) The chloroplast genome and its products, in *Oxford Surveys of Plant Molecular and Cell Biology*, vol 2 (Miflin, B. J., ed.), Oxford University Press, New York, pp. 147–177.
2. Borner, T. and Sears, B. B. (1986) Plastome mutants. *Plant Mol. Biol.* **4,** 69–92.
3. Medgyesy, P. (1990) Selection and analysis of cytoplasmic hybrids, in *Plant Cell Line Selection* (Dix, P. J., ed.), VCH, Weinheim, pp. 287–316.
4. Cséplö, A. and Maliga, P. (1984) Large scale isolation of maternally inherited lincomycin resistance mutations in diploid *Nicotiana plumbaginifolia* protoplast cultures. *Mol. Gen. Genet.* **196,** 407–412.
5. McCabe, P. F., Timmons, A. M., and Dix, P. J. (1989) A simple procedure for the isolation of streptomycin resistant plants in *Solanaceae. Mol. Gen. Genet.* **216,** 132–137.
6. Svab, Z., Hajdukiewitz, P., and Maliga, P. (1990) Stable transformation of plastids in higher plants. *Proc. Natl. Acad. Sci. USA* **87,** 8526–8530.
7. O'Neill, C. M., Horváth, G. V., Horváth, E., Dix, P. J., and Medgyesy, P. (1993) Chloroplast transformation in plants: polyethylene glycol (PEG) treatment of protoplasts is an alternative to biolistic delivery systems. *Plant J.* **3,** 729–738.

8. McBride, K. E., Svab, Z., Schaaf, D. J., Hogan, P. S. Stalker, D. M., and Maliga, P. (1995) Amplification of a chimeric *Bacillus* gene in chloroplasts leads to an extraordinary level of an insecticidal protein in tobacco. *Bio/Technology* **13,** 362–365.
9. Dix, P. J. and Kavanagh, T. A. (1995) Transforming the plastome: genetic markers and DNA delivery systems. *Euphytica* **85,** 29–34.
10. Kavanagh, T. A., O'Driscoll, K. M., McCabe, P. F., and Dix, P. J. (1994) Mutations conferring lincomycin, spectinomycin, and streptomycin resistance in *Solanum nigrum* are located in three different chloroplast genes. *Mol. Gen. Genet.* **242,** 675–680.
11. Hagemann, R. (1982) Induction of plastome mutations by nitroso-urea-compounds, in *Methods in Chloroplast Molecular Biology* (Edelman, M., Hallick, R. B., and Chua, N. H., eds.), Elsevier Biomedical, Amsterdam, pp. 119–127.
12. Maliga, P. (1984) Cell culture procedures for mutant selection and characterization in *Nicotiana plumbaginifolia,* in *Cell Culture and Somatic Cell Genetics of Plants,* vol 1 (Vasil, I. K., ed.), Academic, New York, pp. 552–562.

VI

Protocols for the Introduction of Specific Genes

29

Agrobacterium-Mediated Transformation of *Petunia* Leaf Disks

Ingrid M. van der Meer

1. Introduction

Agrobacterium-mediated transformation of plants is now applicable to many dicotyledonous and also several monocotyledonous plant species. It can be used to transform many different species based on various factors: the broad host range of *Agrobacterium* *(1)*, the regeneration responsiveness of many different explant tissues *(2)*, and the utility of a wide range of selectable marker genes *(3)*. In addition to tobacco, one of the first species that was routinely transformed using this method was *Petunia hybrida*.

P. hybrida is a very good model plant for the analysis of gene function and promoter activity. It is readily transformed, the culture conditions are easy fulfilled, generation time is 3–4 mo and one can grow up to 100 plants/m^2. Furthermore, its genetic map is well developed, and it contains active transposable elements *(4)*.

The protocol presented here is a simplified version of that of Horsch et al. *(5)*. The basic protocol involves the inoculation of surface-sterilized leaf disks with the appropriate disarmed strain of *Agrobacterium tumefaciens* carrying the vector of choice, which in this protocol confers kanamycin resistance. The plant tissue and *Agrobacterium* are then cocultivated on regeneration medium for a period of 2 or 3 d. During this time, the virulence genes in the bacteria are induced, the bacteria bind to the plant cells around the wounded edge of the explant, and the gene-transfer process occurs *(6)*. Using a nurse culture of tobacco or *Petunia* cells during the coculture period may increase the transformation frequency. This is probably owing to a more efficient induction of the virulence genes. After the cocultivation period, the growth of the bacterial population is inhibited by bacteriostatic antibiotics (cefotaxim or carbenicil-

From: *Methods in Molecular Biology, Vol. 111: Plant Cell Culture Protocols*
Edited by: R. D. Hall © Humana Press Inc., Totowa, NJ

lin), and the leaf tissue is induced to regenerate. The induction and development of shoots on leaf explants occur in the presence of a selective agent against untransformed plant cells, usually kanamycin. During the next 2–3 wk, the transformed cells grow into callus or differentiate into shoots via organogenesis. After 4–6 wk, the shoots have developed enough to be removed from the explant and induce rooting in preparation for transfer to soil. To speed up the rooting period, the shoots can be rooted without selection on kanamycin. In total, it takes about 2 mo, after inoculation of the leaf disks with *Agrobacterium* to obtain rooted plantlets that can be transferred to soil.

2. Materials

2.1. Bacteria Media

1. For the growth of *Agrobacterium,* use Luria broth (LB) medium:
   1% Bacto-peptone (Difco, Detroit, MI).
   0.5% Bacto-yeast extract (Difco).
   1% NaCl.
   Autoclave, and cool medium to at least 60°C. Add appropriate antibiotics to select for plasmids (50 mg/L kanamycin for pBin19 *[7]*).
2. LB agar: LB medium with 15 g/L agar (Difco). Autoclave, and cool medium to at least 60°C. Add appropriate antibiotics to select for plasmids (50 mg/L kanamycin for pBin19). Pour into sterile 20-mm Petri dishes.
3. *Agrobacterium* inoculation dilution medium: Murashige and Skoog (***8***, and *see* Appendix) salts and vitamins (4.4 g/L) (Sigma, Amsterdam). Autoclave.

2.2. Stock Solutions

For convenience, most stock solutions are prepared at 1000 times the concentration needed for the final media. The antibiotics are added to the media after autoclaving when the temperature of the media has cooled to 60°C.

1. Cefotaxime: 250 mg/mL (Duchefa, Haarlem, The Netherlands, or Sigma), filter-sterilize, keep at –20°C.
2. Kanamycin: 250 mg/mL (Duchefa or Sigma), filter-sterilize, keep at –20°C.
3. 6-Benzylaminopurine (6-BAP), (Sigma): 2 mg/mL. Dissolve 200 mg BAP in 4 mL 0.5 *N* HCl. Add, while stirring, drop by drop H_2O at 80–90°C and make up to 100 mL. Filter-sterilize.
4. 1-Naphtalene acetic acid (NAA) (Sigma): 1 mg/mL, dissolved in DMSO. No need to sterilize. Keep at –20°C. DMSO should be handled under a fume hood.

2.3. Plant Culture Media

1. Cocultivation medium: Murashige and Skoog (***8***, and *see* Appendix) salts and vitamins (4.4 g/L) (Sigma), 30 g/L sucrose, 2 mg/L 6-BAP, 0.01 mg/L NAA, adjust pH to 5.8 with 1 *M* KOH, add 8 g/L agar (Bacto Difco) and autoclave. Pour into sterile plastic dishes that are 20 mm high (Greiner, Kremsmunster, Austria).

2. Regeneration and selection medium: Murashige and Skoog salts and vitamins (4.4 g/L) (Sigma), 30 g/L sucrose, 2 mg/L 6-BAP, 0.01 mg/L NAA, adjust pH to 5.8 with 1 *M* KOH, add 8 g/L agar (Bacto Difco). Autoclave and cool media to 60°C, add 250 mg/L cefotaxime to kill off *Agrobacterium,* and the appropriate selective agent to select for transformed cells depending on the vector used (100 mg/L kanamycin for pBin19). Pour 25 mL into each sterile 20-mm high Petri dish (Greiner).
3. Rooting medium: Murashige and Skoog salts and vitamins (4.4 g/L) (Sigma), 30 g/L sucrose, adjust pH to 5.8 with 1 *M* KOH, and add 7 g/L agar (Bacto Difco). Autoclave and cool media to 60°C, add 250 mg/L cefotaxime to kill off *Agrobacterium,* and add the appropriate selective agent to select for transformed shoots depending on the vector used (100 mg/L kanamycin for pBin19). To speed up the rooting process, kanamycin may be omitted from the rooting medium. Pour in Magenta GA7 boxes (Sigma, 80 mL per box).

2.4. Plant Material, Sterilization, and Transformation

1. *P. hybrida* c.v. W115 (Mitchell), grown under standard greenhouse conditions.
2. *A. tumefaciens* strain LBA 4404 and *A. tumefaciens* LBA 4404 containing pBin19 (in which the gene of interest is inserted).
3. 10% Solution of household bleach containing 0.1% Tween or other surfactant.
4. Sterile H_2O.
5. Sterile filter paper (Whatman) and sterile round filters (Whatman, diameter 90 mm).

2.5. General Equipment

1. Sterile transfer facilities.
2. Rotary shaker at 28°C.
3. Cork borer (or a paper punch).
4. Magenta GA7 boxes (Sigma) and 20-mm high Petri dishes (Greiner).

3. Methods

3.1. Plant Material

Young leaves are used as the explant source for transformation. These explants can be obtained from aseptically germinated seedlings or micropropagated shoots, but in this protocol, they are obtained from greenhouse-grown material. The genotype of the source material is important in order to obtain high transformation rates. *P. hybrida* cv.W115 gives the best results and is most often used (*see* **Notes 1** and **2**). To obtain plant material suitable for transformation, seedlings should be germinated 4–6 wk prior to transformation.

Sow *P. hybrida* (W115) seeds in soil 4–6 wk prior to transformation, and grow under standard greenhouse conditions in 10 × 10 × 10 cm plastic pots. Use commercially available nutritive solution for house plants.

3.2. Leaf Disk Inoculation

1. Grow *A. tumefaciens* culture overnight in LB at 28°C on a rotary shaker (130 rpm) with appropriate antibiotics to select for the vector (50 mg/L of kanamycin for pBin19) (*see* **Note 3**). The *Agrobacterium* liquid culture should be started by inoculating 2 mL of liquid LB with several bacterial colonies obtained from an *Agrobacterium* streak culture grown on an LB agar plate at 28°C for 2–3 d. The streaked plate itself can be inoculated from the original –80°C frozen stock of the *Agrobacterium* strain (*see* **Note 4**). This stock is composed of a bacterial solution made from a 1:1 mixture of sterile glycerol (99%) and an overnight LB culture of the *Agrobacterium.*
2. Prepare the culture for inoculation of explants by taking 0.5 mL of the overnight culture and diluting 1 to 200 with Murashige and Skoog salts and vitamins medium (4.4 g/L) (Sigma) to a final volume of 100 mL. The *A. tumefaciens* inoculum should be vortexed well prior to use. Pour the *Agrobacterium* inoculum into four Petri dishes.
3. When the plants are 10–15 cm high, harvest the top leaves to provide explants (leaves 3–8 from the top). Lower leaves and leaves from flowering plants should not be used, since they usually have a lower transformation and regeneration response (*see* **Note 1**).
4. Prepare harvested leaves for inoculation by surface sterilization for 15 min in 10% solution of household bleach containing 0.1% Tween or other surfactant. Wash the leaves thoroughly three times with sterile H_2O. Keep them in sterile H_2O until needed (*see* **Note 5**). All procedures following the bleach treatment are conducted in a sterile transfer hood to maintain tissue sterility.
5. Punch out leaf disks with a sterile (1-cm diameter) cork borer (or cut into small squares to produce a wounded edge) in one of the Petri dishes containing the *Agrobacterium* inoculum (20–25 disks/Petri dish) (*see* **Note 6**). Avoid the midrib of the leaf or any necrotic areas. Cut 80–100 disks/construct (*see* **Note 7**).
6. Leave the disks in the inoculum for 20 min. After inoculation, the explants are gently sandwiched between two layers of sterile filter paper (Whatman) to remove excess inoculum.
7. (Optional) Prepare nurse culture plates by adding 3 mL of cell suspension culture (e.g., *P. hybrida* cv Coomanche or *Nicotiana tabacum* cv SR1) to Petri dishes containing cocultivation medium. Swirl the suspension to spread the cells over the surface of the medium and cover with a sterile Whatman filter paper (diameter 90 mm) (*see* **Note 8**).
8. Place 20 explants with the adaxial surface downward on each plate with cocultivation medium (either with or without nurse cells), and incubate for 2–3 d (*see* **Note 9**). Seal the plates with Nescofilm. The culture conditions for the leaf disks are as follows: a temperature of 25°C and a photoperiod of 14 h light (light intensity: 25–40 μmol/m^2/s)/10 h dark.

   The controls to check the transformation protocol are:

   Leaf disks inoculated with "empty" *Agrobacterium* (without pBin19 vector) on regeneration medium without selective agent (to check the regeneration).

Leaf disks inoculated with "empty" *Agrobacterium* (without pBin19 vector) on regeneration medium with selective agent (to check the efficiency of antibiotic selection).

9. After cocultivation, transfer the disks to regeneration and selection medium (seal the plates with Nescofilm), and continue incubation until shoots regenerate. Transfer the explants every 2–3 wk to fresh regeneration and selection medium (*see* **Note 10**).

3.3. Recovery of Transformed Shoots

1. After 2–3 wk, the first shoots will develop. Cut off the shoots cleanly from the explant/ callus when they are 1–1.5 cm long, and place them upright in rooting medium in Magenta boxes. The shoots should be excised at the base without taking any callus tissue (*see* **Note 11**). Take only one shoot from each callus on the explant to ensure no siblings are propagated representing the same transformation event (*see* **Note 12**). Shoots from distinctly different calli on the same explant are, however, likely to be derived from different transformation events and should be transferred separately. Give each shoot a code that allows it to be traced back to specific explants.

   Cefotaxime is kept in the medium to avoid *Agrobacterium* regrowth. The antibiotic used to select for the transgenic shoots can be added to the medium to select against escapes, although the rooting process is sped up when it is omitted.
2. (Optional) Before removing rooted shoots from sterile culture, transfer a leaf to selection medium to test for resistance to kanamycin. If the leaf is obtained from a tranformed plantlet it should stay green and form callus on selection medium. If it originates from an untransformed plantlet, it should become brown/white and die within a few weeks.
3. After 3–4 wk the shoots will have formed roots. Remove plantlets, wash agar from the base under a running tap, plant the transformants in soil, and transfer them to the greenhouse (*see* **Note 13**). To retain humidity, cover the pots with Magenta boxes or place them in a plastic propagation dome. The plants should then be allowed to come to ambient humidity slowly by gradually opening the dome or Magenta box over a period of 7 d (*see* **Note 14**).
4. Fertilize and grow under standard plant growth conditions.

3.4. Analysis of Transformants

Tissue culture can be used to confirm that the putative transgenic shoots produced are expressing the selectable marker gene (*see* **Subheading 3.3., step 2**). However, DNA analysis using Southern blotting or PCR will confirm whether regenerants have integrated the antibiotic resistance gene (and also a gene of interest if this was linked to it within the T-DNA). For DNA analysis using Southern blotting ***(9)***, leaf DNA can be isolated according to the protocol described by Dellaporte et al. ***(10)***. A more rapid method can be used to isolate genomic DNA as described by Wang et al. ***(11)*** when PCR is going to be used to analyze the presence of foreign DNA in the transformed plants (*see* **Note 15**)

Usually, one to five copies of the foreign DNA is integrated in the plant genome using the *Agrobacterium*-mediated transformation method. However, position effects may silence the expression of the introduced gene (*see* **Note 7**).

4. Notes

1. This transformation protocol works very well for the often-used *P. hybrida* varieties W115 (Mitchell) and V26. However, some *Petunia* lines show poor regeneration from leaf disk explants, and consequently, few or no transformants can be obtained from these plants. Uniform, clean, and young plants will perform best. It is important not to take leaf material from old, flowering plants.
2. Instead of greenhouse material, aseptically germinated seedlings or micropropagated shoots could also be used as explant source. Then, of course, there is no need to surface-sterilize the leaves.
3. The *Agrobacterium* strain that is most often used is LBA4404 (Clontech *[12]*). Also the more virulent strains C58 or AGL0 *(13)* can be used, but these can be more difficult to eliminate after cocultivation. During growth, the *Agrobacterium* culture will aggregate.
4. The streaked plate can be reused for approx 3 wk if kept at 4°C after growth.
5. Be very gentle with the plant material during sterilization, since the bleach will easily damage the leaves, especially wounded or weak, etiolated leaves. Damaged tissue should not be used.
6. Disks provide a very uniform explant and are conveniently generated with a cork borer or paper punch. However, square explants or strips can also be used. Avoid excessive wounding during the process. The cork borer should be allowed to cool before use after flaming.
7. This transformation protocol will yield approx 20 transgenic plants from 100 initial explants. The expression level of the construct of interest can be greatly influenced by a position effect owing to its site of integration into the host plant genome. This silencing of expression owing to position effects can occur in 20–40% of the transgenic plants, especially if weak promoters are used. Therefore, at least 20 transformants should be generated/construct.
8. The nurse culture is not essential for transformation, but can facilitate the process by increasing frequency and reducing damage to the explant by the bacterium. Any healthy suspension of tobacco or *Petunia* should work. The suspension cultures can be maintained by weekly transfer of 10 mL into 50 mL of fresh suspension culture medium.
9. The cocultivation time may have to be optimized for different *Agrobacterium* strains carrying different vectors.
10. If *Agrobacterium* continues to grow on the regeneration and selection medium (forming slimy gray-white bacterial colonies), 150 mg/L of vancomycin can also be added to the medium already containing cefotaxime.

11. Care should be taken that only the stem and none of the associated callus is moved to the rooting medium. Otherwise, no roots will develop.
12. It is common for multiple shoots to arise from a single transformed cell. Therefore, it is important to separate independent transformation events carefully so that sibling shoots are not excised.
13. It is important to wash away all of the agar medium from the roots and to transplant before the roots become too long. When there is still agar left, it could enhance fungal growth.
14. Gradual reduction in the humidity is necessary to harden off the plantlets in soil. The roots must grow into the soil, and the leaves must develop a protective wax cuticle. If the plantlets start to turn yellow and die from fungal contamination, the lid should be opened faster. If the plantlets begin to wilt, the lid should be opened slower.
15. Confirmation of transformation by PCR may not always be reliable, owing to possible carryover of the *Agrobacterium* into the whole plant. DNA analysis using Southern blot hybridization is a better way to confirm whether regenerants are true transformants. Furthermore, the number of inserts can be determined at the same time using this method.

References

1. Richie, S. W. and Hodeges, T. K. (1993) Cell culture and regeneration of transgenic plants, in *Transgenic Plants,* vol. 1 (Kung, S. and Wu, R., eds.), Academic, San Diego, pp. 147–178.
2. Jenes, B., Morre, H., Cao, J., Zhang, W., and Wu, R. (1993). Techniques for gene transfer, in *Transgenic Plants,* vol. 1 (Kung, S. and Wu, R., eds)., Academic, San Diego, pp. 125–146.
3. Bowen, B. A. (1993) Markers for gene transfer, in *Transgenic Plants,* vol. 1 (Kung, S. and Wu, R., eds.), Academic, San Diego, pp. 89–124.
4. Gerats, A. G. M., Huits, H., Vrijlandt, E., Maraña, C., Souer, E., and Beld, M. (1990) Molecular characterization of a nonautonomous transposable element (dTph1) of petunia. *Plant Cell* **2,** 1121–1128.
5. Horsch, R. B., Fry, J. E., Hoffman, N. L., Eichholtz, D., Rogers, S. C., and Fraley, R. T. (1985) A simple and general method for transferring genes into plants. *Science* **227,** 1229–1231.
6. Hooykaas, P. J. J. (1989) Transformation of plant cells via *Agrobacterium. Plant Mol. Biol.* **13,** 327–336.
7. Bevan, M. (1984) Binary *Agrobacterium* vectors for plant transformation. *Nucleic Acids Res.* **12,** 8711–8721.
8. Murashige, T. and Skoog, F. (1962) A revised medium for rapid growth and bioassays with tobacco tissue cultures. *Plant Physiol.* **15,** 473–497.
9. Maniatis, T., Fritsch, E. F., and Sambrook, J. (1982) *Molecular cloning: A Laboratory Manual.* Cold Spring Harbor Laboratory Press, Cold Spring Harbor, NY.
10. Dellaporte, S. L., Wood, J., and Hicks, J. B. (1983) A plant DNA minipreparation: version II. *Plant Mol. Biol. Rep.* **1,** 19–21.

11. Wang, H., Qi, M., and Cutler, A. J. (1993) A simple method of preparing plant samples for PCR. *Nucleic Acids Res.* **21,** 4153–4154.
12. Hoekema, A., Hirsch, P. R., Hooykaas, P. J. J., and Schilperoort, R. A. (1983) A binary plant vector strategy based on separation of *vir* and T region of the *Agrobacterium tumefaciens* Ti-plasmid. *Nature* **303,** 179–180.
13. Lazo, G. R., Stein, P. A., and Ludwig, R. A. (1991) A DNA transformation-competent Arabidopsis genomic library in *Agrobacterium. Bio/Technology* **9,** 963–967.

30

Transformation of Rice via PEG-Mediated DNA Uptake into Protoplasts

Karabi Datta and Swapan K. Datta

1. Introduction

For stable transformation of cereals through PEG-mediated DNA uptake into protoplasts, the two most critical requisites are the ability to isolate and culture protoplasts in large numbers, and the development of an efficient and reliable system for routine plant regeneration from protoplasts *(1–3)*. Based on early success with mesophyll protoplasts of some dicotyledonous species, extensive efforts were made to induce sustained division in protoplasts isolated from leaves or young shoots of different cereal plants. However, there is still no convincing evidence of sustained divisions in protoplasts isolated from leaves or shoots of any cereal. In contrast, protoplasts isolated from embryogenic suspension cultures could be induced to divide in culture *(4)*. Obtaining a fast-growing and highly embryogenic suspension culture is the most important factor for cereal plant regeneration from protoplasts *(4–8)*. Microspore cultures, and immature or mature embryos may be used to obtain embryogenic calli which can eventually be used to establish embryogenic cell suspensions ECS *(2, 7)*.

PEG-mediated gene transfer to rice protoplasts appears to be the most efficient, reliable, inexpensive, and simplest method, when it works *(2,8,9)*. In this system, from a suspension culture, a large population of protoplasts can be readily obtained for transformation enabling many chances of obtaining independent transformation events. Regeneration of transgenic plants is possible under suitable in vitro conditions through selection at an early stage of development. However, the tissue-culture response may vary depending on the plant genotype, handling, and the condition of the suspension cells. We have successfully used the procedure described below for gene transfer to rice for at least 10 Indica and several japonica cultivars.

From: *Methods in Molecular Biology, Vol. 111: Plant Cell Culture Protocols*
Edited by: R. D. Hall © Humana Press Inc., Totowa, NJ

2. Materials

2.1. Suspension Culture Establishment

2.1.1. Plant Material

Plant material is *Oryza sativa*—Indica-type rice cultivar, Chinsurah Boro II, IR72, Japonica type rice cultivar—Yamabiko: Immature inflorescence (panicles collected prior to the emergence from the flag leaf sheath), immature embryos (panicles collected 7–10 d after anthesis) and healthy mature seeds from the mentioned cultivars.

2.1.2. General Equipment

1. Laminar flow cabinet, inverted microscope, light microscope, incubator, gyrating shaker, autoclave, and distilled water plant.
2. Nescofilm, flasks, beaker, measuring cylinder, media bottles, forceps, scalpel, scissors, spirit lamp, and pipet bulb.
3. Plastic Petri dishes (50 mm), 24-well COSTA plates, disposable pipets, 2- and 10-mL capacity.
4. Greenhouse facilities.

2.1.3. Sterilizing Solutions

1. 70% Ethanol.
2. Sodium hypochlorite solution (e.g., 100 mL 1.8% sodium hypochlorite with two drops Tween 20).
3. Sterile distilled water.

2.1.4. Media

1. Microspore culture medium (R1)

Component	Quantity (mg/L)
NH_4NO_3	1650
KNO_3	1900
$CaCl_2 \cdot 2H_2O$	440
$MgSO_4 \cdot 7H_2O$	370
KH_2PO_4	170
KI	0.83
H_3BO_3	6.3
$MnSO_4 \cdot 4H_2O$	22.3
$ZnSO_4 \cdot 7H_2O$	8.6
$Na_2MoO_4 \cdot 2H_2O$	0.25
$CuSO_4 \cdot 5H_2O$	0.025
$CoCl_2 \cdot 6H_2O$	0.025
Na_2EDTA	37.3
$FeSO_4 \cdot 7H_2O$	27.8

Thiamine HCl	10
Glutamine	500
Casein hydrolysate	300
Myo-inositol	100
NAA	2
2,4-D	1
Sucrose	6% w/v
Ficoll 400	10% w/v

pH should be 5.6. Based on MS medium ***(10)***. Sterilize by filtration.

2. AA medium ***(11)***

Component	Quantity (mg/L)
$CaCl_2 \cdot 2H_2O$	150
$MgSO_4 \cdot 7H_2O$	250
NaH_2PO_4 H_2O	150
KCl	2950
KI	0.75
H_3BO_3	3.0
$MnSO_4 \cdot H_2O$	10.0
$ZnSO_4 \cdot 7H_2O$	2.0
$Na_2MoO_4 \cdot 2H_2O$	0.25
$CuSO_4 \cdot 5H_2O$	0.025
$CoCl_2 \cdot 6H_2O$	0.025
Nicotine acid	1.0
Pyridoxine-HCl	1.0
Thiamine-HCl	10.0
Inositol	100
Na_2EDTA	37.3
$FeSO_4 \cdot 7H_2O$	27.8
L-Glutamine	876
Aspartic acid	266
Arginine	174
Glycine	7.5
2,4-D	1.0
Kinetin	0.2
GA_3	0.1
Sucrose/maltose	20 g/L

pH should be 5.6. Sterilize by filteration.

3. Modified MS medium

Component	Quantity (mg/L)
NH_4NO_3	1650
KNO_3	1900
$CaCl_2 \cdot 2H_2O$	440
$MgSO_4 \cdot 7H_2O$	370

KH_2PO_4	170
KI	0.83
H_3BO_3	6.3
$MnSO_4 \cdot 4H_2O$	22.3
$ZnSO_4 \cdot 7H_2O$	8.6
$Na_2MoO4, 2H_2O$	0.25
$CuSO_4 \cdot 5H_2O$	0.025
$CoCl_2 \cdot 6H_2O$	0.025
$FeSO_4 \cdot 7H_2O$	27.8
Na_2 EDTA	37.3
Nicotinic acid	0.5
Pyridoxine-HCl	0.5
Thiamine-HCl	1.0
Glycine	2.0
Casein hydrolysate	300
Myo-inositol	100
2,4-D	1.5
Kinetin	0.5
Sucrose/maltose	30 g/L
Agar	8 g/L

pH should be 5.8. This is based on MS medium ***(10)***. Sterilize by autoclaving.

4. R2 medium ***(12)***

Component	Quantity (mg/L)
$NaH_2PO_4 \cdot 2H_2O$	240
KNO_3	4040
$(NH_4)_2SO_4$	330
$MgSO_4 \cdot 7H_2O$	247
$CaCl_2 \cdot 2H_2O$	147
$MnSO_4 \cdot H_2O$	0.50
$ZnSO_4 \cdot 7H_2O$	0.50
H_3BO_3	0.50
$CuSO_4 \cdot 5H_2O$	0.05
Na_2 MoO_4	0.05
Na_2 EDTA	37.3
$FeSO_4 \cdot 7H_2O$	27.8
Nicotinic acid	0.5
Pyridoxine-HCl	0.5
Thiamine-HCl	1
Glycine	2
Inositol	100
2,4-D	1
Sucrose/maltose	20 g

pH should be 5.8. Sterilize by autoclaving.

2.2. Protoplast Isolation, Transformation, and Regeneration

2.2.1. General Equipment

1. Centrifuge.
2. Temperature-controlled shaker.
3. 12-mL round-bottom screw-cap centrifuge tubes.
4. Petri dishes, 10 (deep model) and 3.5 cm (Falcon).
5. Nylon sieves—50 and 25 μm (Saulas, F93100 Montreuil, France)
6. Counting chamber (hemocytometer).

2.2.2. Solutions and Medium

1. Enzyme solution: 4% w/v cellulase onozuka RS, 1% w/v macerozyme R10 (both Yakult Honsha Co., Japan), 0.02% pectolyase-Y23 (Seishin Pharmaceutical Co., Japan), 0.4 *M* mannitol, 6.8 m*M* $CaCl_2$, pH 5.6; filter-sterilize, and store at –20°C.
2. Wash solution: 0.4 *M* mannitol, 0.16 *M* $CaCl_2$; autoclave.
3. MaMg solution (transformation buffer): 0.4 *M* mannitol, 15 m*M* $MgCl_2$, 1% (w/v) MES, pH 5.8; autoclave.
4. PEG solution (40% w/v): Dissolve 80 g PEG 6000 (Merck, Art 12033, 1000) in 100 mL distilled water containing 4.72 g Ca $(NO_3)_2 \cdot 2H_2O$ and 14.57 g mannitol. The PEG is dissolved by heating carefully in a microwave. Make the total volume 200 mL with distilled water. Divide the solution into two parts. In one part, the pH is adjusted to 10.0 using 1 *M* KOH and left overnight to stabilize. Then, adjust the final pH to 8.2 using the second part of the solution. Filter-sterilize the solution and store as 5-mL aliquots at –20°C (*see* **Note 1**).
5. P1 medium: R2 medium with 2 mg/L 2,4-D and 0.4 *M* maltose, pH 5.6; filter-sterilize (**Table 1**).
6. Agarose—protoplast medium: 600 mg (dry autoclaved) Sea plaque agarose melted in 30 mL P1 medium.
7. P2 medium: Soft agarose N_6 medium ***(13)*** with 6% maltose or sucrose, 2 mg/L 2,4-D, and 0.3 g Sea plaque agarose; autoclave (**Table 1**).
8. DNA for transformation - Plasmid DNA is used; 10 μg/sample of protoplast suspension (1.5×10^6 protoplasts) is used for transformation. Dissolve 10 mg calf thymus DNA (used as carrier DNA) in 5 mL distilled water. Shear by passage through an 18-gauge needle to give an average fragment size of 5–10 kb (check by running gel). Carrier and plasmid DNA is sterilized by precipitation and washing in 96% ethanol (also possible with 70% ethanol), and dried in a laminar flow hood. DNA is dissolved in sterile double-distilled water (2 μg/μL) and stored at 4°C (or at –20°C for longer period).
9. Nurse cells—OC cell line derived from seedlings of *O. sativa* L. C5924, provided by K. Syono of University of Tokyo (*see* **Note 2**).

Table 1
Composition of the Media Used

Component	P1 mg/L	P2 mg/L	P3 mg/L
$(NH_4)_2SO_4$	330.0	463.0	
KH_2PO_4	—	400.0	170.0
KNO_3	4044.0	2830.0	1900.0
NH_4NO_3	—	—	1650.0
$CaCl_2 \cdot 2H_2O$	147.0	166.0	440.0
$MgSO_4 \cdot 7H_2O$	247.0	185.0	370.0
Na_2EDTA	37.3	37.3	37.3
$FeSO \cdot 7H_2O$	27.8	27.8	27.8
$NaH_2PO_4 \cdot 2H_2O$	240	—	—
$2H_2O$	240	—	—
$MnSO_4 \cdot 4H_2O$	0.5	4.4	22.3
H_3BO_3	0.5	1.6	6.3
$ZnSO_4 \cdot H_2O$	0.5	1.5	8.6
KI	—	0.8	0.83
$CoCl_2 \cdot 6H_2O$	—	—	0.025
$CuSO_4 \cdot 5H_2O$	0.05	—	0.025
$Na_2MoO_4 \cdot 2H_2O$	0.05	—	0.25
Thiamine-HCl	1.0	1.0	1.0
Nicotinic acid	0.5	0.5	0.5
Pyridoxine-HCl	0.5	0.5	0.5
Glycine	2.0	2.0	2.0
Myo-inositol	100.0	100.0	100.0
Kinetin	—	—	2.0
NAA	—	—	1.0
2,4-D	2.0	2.0	—
Sucrose g/L[a]	136.92	60.0	30.0
Maltose g/L[a]	144.12	60.0	—
Agar g/L	—	—	8.0
Agarose g/L	—	3 or 6	—
	pH 5.6	pH 5.8	pH 5.8
	Filter-sterilized	Autoclaved	Autoclaved

[a]Either sucrose or maltose to be used.

3. Method

3.1. Growth of Donor Plants

1. Break the dormancy of clean and pure seeds of derived lines in the oven at 50°C for 3–5 d.
2. Sow the seeds in seed boxes and water sufficiently to wet the soil.

3. At 21 d after sowing, transplant the seedlings to 6-in. diameter pots (one seedling per pot) containing soil and fertilizer (2.5 g $[NH_4]_2$ SO_4, 1.25 g P_2O_5 and 0.75 g K_2O/pot).
4. Grow the plants in the glass house, and keep them well watered. The glass house should have full sunshine and sufficient ventilation to maintain daytime temperatures of 27–29°C with a 90% humidity level.

3.2. Establishment of Embryogenic Suspension Culture

3.2.1. From Microspores

1. Collect the tillers from the donor plants when most of the microspores are at the miduninucleate stage. In all cereals investigated, the early or miduninucleate stage of microspore development was found to give the best results ***(14)***.
2. Sterilize the selected spikes with 70% ethanol for 30 s, and then treat with sodium hypochlorite solution for 7 min followed by washing three times with sterile deionized water ***(15)***.
3. Float 30 anthers, each containing microspores at the miduninucleate stage on the surface of 10 mL of liquid microspore culture media (R1) in 50-mm sterile Petri dishes.
4. Culture in the dark at 25°C. These floating anthers shed microspores into the medium within 3–7 d.
5. After 4–6 wk, transfer the developing embryogenic calli (about 0.5 g) to 7 mL of AA medium ***(11)*** to establish the embryogenic cell suspension (**Fig. 1A**). At this stage, subculturing is necessary at least twice a week until a fine cell suspension is obtained. This cell suspension can then be subcultured every 7 d for long-time maintenance (at least 1 yr).

3.2.2. From Immature or Mature Embryos

1. Sterilize dehulled immature (12 d after anthesis) or mature seeds in 70% ethanol for 1 min, and then sodium hypochlorite (1.8%) solution for 30 min followed by washing three times with sterile deionized water.
2. Isolate the embryos from the sterile seeds, and place individually in 24-well COSTA plates with each well containing 1.5 mL of modified MS medium. The scutellar tissue should be upward, since callus growth occurs only with this orientation.
3. Incubate the embryos in the dark at 25°C. After 3–4 d, cut off the emerging shoots and roots, and subculture the remaining tissues onto fresh medium. At least two more subcultures are necessary at 2-wk intervals until yellowish-white soft embryogenic calli develop on the surfaces of the scutella.
4. Transfer this callus to 6 mL of liquid AA medium ***(11)*** in 50-mm Petri dishes, and incubate on a gyrating shaker at low speed (80 rpm) in diffuse light (3 μmol/m^2/s) at 25°C.
5. Subculture the callus every 7 d with continued manual selection of small and densely cytoplasmic cells, which are transferred to 20 mL of R2 medium in 100-mL Erlenmeyer flasks (*see* **Note 3**). These suspensions can be maintained for a long

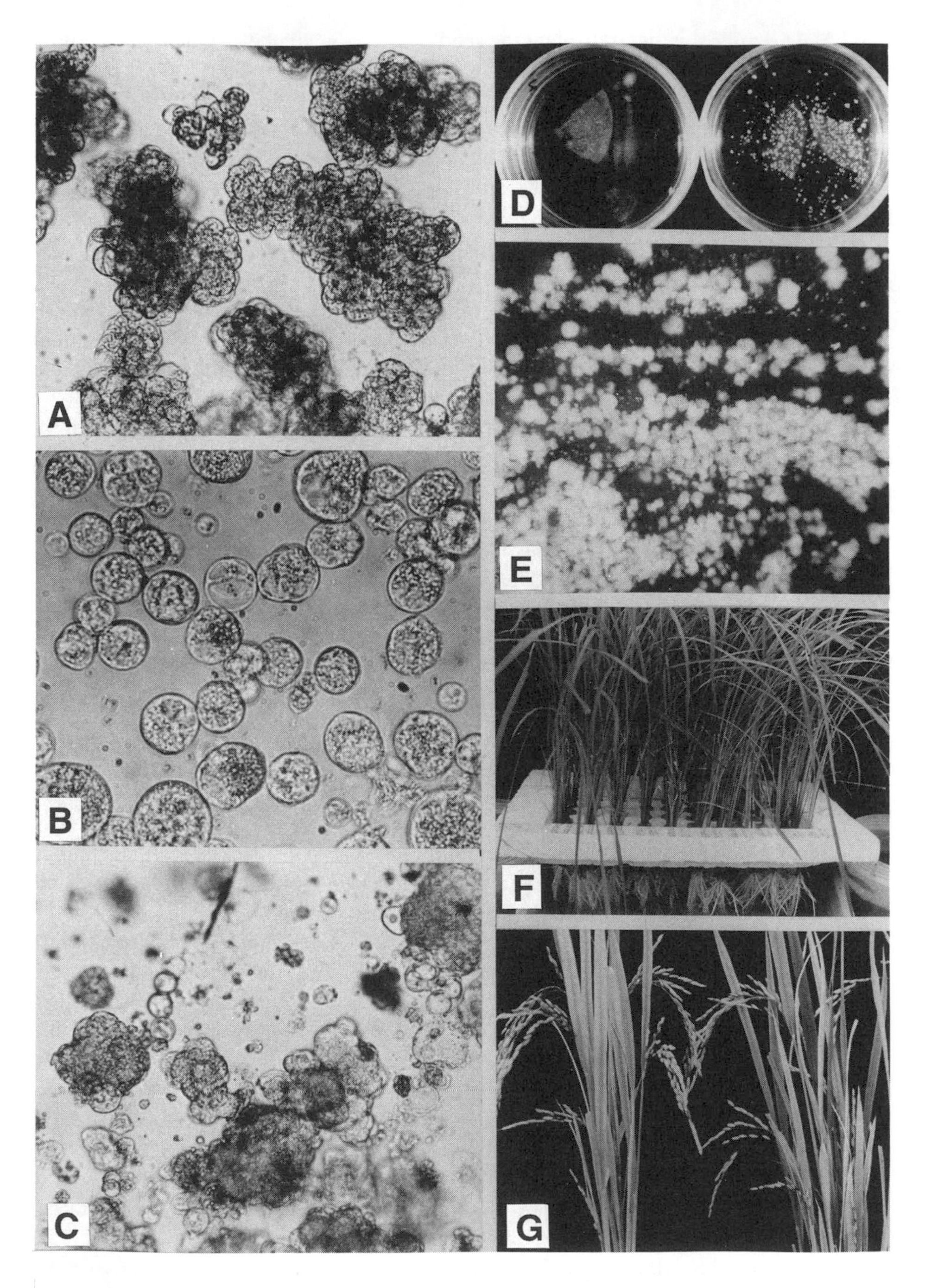

Fig. 1.

time by subculturing every 7 d, and incubating with gyratory shaking (80 rpm) at 25°C in dark or diffuse light.

3.3. Protoplast Isolation

1. Transfer (approx) 3–5 g of suspension cells 3–4 d after subculture into a deep 10-cm Falcon Petri dish. Allow the suspension cells to settle and remove the medium by pipeting off (*see* **Note 4**).
2. Add 20 mL of enzyme solution, seal with Nescofilm, and incubate without shaking at 30°C in the dark for 3–4 h (depending on protoplast release) (*see* **Note 5**).
3. Add an equal volume of wash solution to each Petri dish containing the protoplast enzyme mixture.
4. Remove the protoplast suspension with a 10-mL pipet, and pour through 50- and 25-μm sieves into a sterile glass beaker (100 mL).
5. Transfer the filtrate to 10-mL round-bottom screw-cap centrifuge tubes, and centrifuge for 10 min at 70*g* to separate off the enzyme solution.
6. Discard the supernatant, and resuspend the protoplast pellet in washing solution. Centrifuge and repeat once. Resuspend the pellet in 10 mL of washing solution. Count the density of protoplasts/mL using a hemocytometer. Freshly isolated protoplasts from ECS should appear densely cytoplasmic (**Fig. 1B**; *see* **Note 6**).

3.4. Direct Gene Transfer to Protoplasts Using PEG

1. Centrifuge the protoplasts in wash solution at 70*g* for 10 min (i.e., third washing). Remove the supernatant, and resuspend the protoplast pellet in transformation buffer (i.e., MaMg solution). Adjust the protoplast density to $1.5–2.0 \times 10^6$/0.4 mL.
2. Distribute the protoplast suspension (0.4 mL containing 1.5×10^6 protoplasts) into different centrifuge tubes using a 2 mL-pipet.
3. Add 6–10 μg of sterile plasmid DNA (preferably linearized) and 20 μg of calf thymus carrier DNA (transformation can be performed without calf thymus) to each 0.4-mL aliquot of the protoplast suspension (*see* **Note 7**).
4. Add 0.5 mL of the PEG solution dropwise, and mix gently. Incubate at room temperature (or preferably at 20°C) for 10 min.
5. Add slowly 10 mL of wash solution, mix, and centrifuge at 70*g* for 10 min to remove the PEG. Discard the supernatant.
6. Resuspend the protoplast pellet in 0.4 mL P1 medium.

Fig. 1. *(previous page)* **(A)** Fertile transgenic rice plants from transformed protoplasts obtained using the PEG method. ECS developed from microspore culture. **(B)** Freshly isolated, densely cytoplasmic, protoplasts from ECS. **(C)** Early division in protoplast culture. **(D)** Protoplasts in selection medium (left: Petri dish nontransformed cell, right: Petri dish with transformed calli developing from transformed protoplasts). **(E)** Protoplast-derived calli in bead-type culture without selection. **(F)** Putative transgenic rice plants in a plastic tray containing Yoshida solution after transfer to the transgenic greenhouse. **(G)** Transgenic fertile rice plants in the greenhouse.

3.5. Culture of Protoplasts

1. Melt agarose-protoplast medium in a microwave oven, and cool to 40°C.
2. Mix 0.6 mL of the agarose-protoplast medium with 0.4 mL of P1 medium containing protoplasts, and transfer into a 3.5-cm Petri dish. Incubate at 20°C for 1 h. The final density of the protoplasts should be around 1.5×10^6/mL.
3. Cut the solidified agarose gel into four segments, and transfer each to a 5-cm Petri dish containing 5 mL of P1 medium (bead-type culture) ***(16)*** with or without adding nurse culture ***(6,17)***. Incubate cultures in the dark at 28.5°C with slow shaking (40 rpm).
4. After 7 d, replace 2 mL medium with fresh P1 medium.
5. At the 10th d after protoplast isolation, remove (all) the nurse cells by transferring the segments to a fresh Petri dish containing 5 mL of P1 medium. Keep in dark at 28.5°C with slow shaking (30 rpm).

3.6. Selection

1. First selection: 14 d after transformation, add the selective agent to the medium (e.g., hygromycin B 25 mg/L, or kanamycin 50 mg/mL, G-418 25 mg/L, or phosphinotricin 20 mg/L, depending on the selectable marker gene used). Incubate as above, but without shaking. Early divisions of protoplasts are already present at this stage (**Fig. 1C**).
2. Second selection: after 2 wk, the selection pressure should be maintained in the same way by replacing 4 mL of medium with fresh medium supplemented with the same concentration of selective agent. Culture for another 2 wk. Putatively transformed, protoplast-derived colonies are visible at this stage (**Fig. 1D**). Nontransformed colonies without selection are shown in **Fig. 1E**.
3. Replace 4 mL of P1 medium with 2 mL of fresh P1 medium and 2 mL of suspension culture medium (R2 medium) without selective agent. Culture for 2 wk.
4. Third selection: Transfer the visible calli (putative transformed colonies) onto soft agarose medium (P2 medium) containing 0.3% agarose and the same concentration of selective agent. Incubate in the dark at 25°C for 2 wk (*see* **Note 8**).

3.7. Regeneration

1. Transfer the visible colonies to P2 medium containing 0.6% Sea plaque agarose without the selective agent (**Table 1**). Culture for another 2 wk under the same conditions.
2. Transfer the selected embryogenic colonies with a size of ca. 1.5–3.0 mm diameter with developing somatic embryos to P3 medium (**Table 1**). Incubate in darkness at 25°C until the development of embryos or embryogenic calli is observed (*see* **Note 9**).
3. Transfer the developing embryos onto the same medium, and culture in the light (24 µmol/m^2/s), with a 16-h photoperiod at 25°C to obtain plantlets.
4. Transfer the plantlets to MS medium without hormones until a well-developed root system is obtained. These plants can be transferred to the transgenic greenhouse if maintained under high humidity for the first 2 wk (*see* **Note 10**). Alternatively, plants can be transferred to a culture solution ***(18)*** in a plastic trays,

which may be placed in a phytotron or greenhouse (with 29°C light period, 21°C dark period), 14 h/day, light 160 μmol/m^2/s, daylight supplemented with Philips HPL N400W mercury lamps if necessary and 70–95% relative humidity (**Fig. 1F**; **Note 11**). The plants when transferred to soil (**Fig. 1G**) set seed after 3–4 mo.

4. Notes

1. For each transformation experiment (10 samples), use a new tube (5 mL) of PEG from the –20°C freezer to avoid possible toxic effects and changes of pH.
2. For nurse cells: any actively dividing rice cell suspension, even the same suspension culture used for protoplast isolation can be used. In our case, we use an OC cell line that is not regenerable, but still actively divides.
3. Some suspension lines (IR72) grow better in N_6LP medium, i.e., N_6/P1 basal medium ***(13)*** supplemented with 1 g/L proline, 860 mg/L glutamine, 1.5 mg/L 2,4-D, and 3% w/v sucrose or maltose ***(1,5)***.
4. Regenerable embryogenic cell suspension cultures (ECS) should be used for protoplast transformation. ECS can be obtained for all genotypes of japonica, indica, and IRRI-New plant-type rice. Usually, a 3- to 6-mo old suspension culture shows a better efficiency for protoplast isolation, transformation, and subsequent regeneration ***(19)***.
5. Enzymatic digestion of suspension cells should not exceed 5 h; if sufficient protoplasts are not released, the experiment for that day should be terminated.
6. The handling of protoplasts should be very gentle at all stages of the protocol.
7. For the selection, we use the selectable marker genes, such as the hygromycin resistance gene, kanamycin resistance gene, or a herbicide resistance gene. Sometimes a gene of interest may be linked to the selectable marker gene. For cotransformation, we add 10 μg of plasmid DNA containing the gene of interest, 10 μg of selectable marker gene, and 20 μg of calf thymus DNA. Supercoiled plasmid DNA (50 μg/mL/1.0 × 10^6 protoplasts even without calf thymus DNA) is suitable for transient gene expression, as examined 24 h after transformation.
8. A total period of selection pressure of <2 wk may produce escapes. Selection pressure for longer than 8 wk may affect the further growth and fertility of the plants, and may produce abnormal plants. Keep the selection pressure to the minimum time required to obtain the transformants. It is always better to remove the culture from the selection pressure to avoid later abnormality and sterility of the plants.
9. Take care of those embryogenic calli developing faster and separate them from the rest. Transfer them earlier for regeneration to plants. Faster-developing calli produce healthy (normal) plants. It is recommended to place a few (four to six) embryogenic calli together in a Petri dish/flask for regeneration.
10. Regenerated transgenic plants should be transferred to soil within 6 wk to minimize the tissue-culture effect.
11. In some countries, e.g., in Philippines, rice plants grow very well in the glass house/greenhouse conditions throughout the year without any artificial light.
12. Molecular analysis (Southern and Western analysis) should confirm the integration and expression of the transgene in the rice genome.

References

1. Datta, K., Potrykus, I., and Datta, S. (1992) Efficient fertile plant regeneration from protoplasts of the Indica rice breeding line IR72 (*Oryza sativa* L.) *Plant Cell Rep.* **11,** 229–233.
2. Datta, S. K., Peterhans, A., Datta, K., and Potrykus, I. (1990) Genetically engineered fertile Indica-rice plants recovered from protoplasts. *Bio/Technology* **8,** 736–740.
3. Peterhans, A., Datta, S. K., Datta, K., Goodall, G. J., Potrykus, I., and Paszkowski, J. (1990) Recognition efficiency of *Dicotyledoneae*-specific promoter and RNA processing signals in rice. *Mol. Gen. Genet.* **222,** 361–368.
4. Vasil, I. K. (1987) Developing cell and tissue culture systems for the improvement of cereal and grass crops. *J. Plant Physiol.* **128,** 193–218.
5. Datta, S. K., Datta, K., Soltanifar, N., Donn, G., and Potrykus, I. (1992) Herbicide-resistant Indica rice plants from IRRI breeding line IR72 after PEG-mediated transformation of protoplasts. *Plant Mol. Biol.* **20,** 619–629.
6. Kyozuka, J., Hayashi, Y., and Shimamoto, K. (1987) High frequency plant regeneration form rice protoplasts by novel nurse culture methods. *Mol. Gen. Genet.* **206,** 408–413.
7. Datta, S. K., Datta, K., and Potrykus, I. (1990) Fertile Indica rice plants regenerated from protoplasts isolated from microspore derived cell suspensions. *Plant Cell Rep.* **9,** 253–256.
8. Datta, S. K. (1995) Polyethylene-glycol-mediated direct gene transfer to indica rice protoplasts and regeneration of transgenic plants, in *Gene Transfer to Plants* (Potrykus, I. and Spangenberg, G., eds.), Springer-Verlag, New York, pp. 66–74.
9. Datta, S. K. (1996) Genetic transformation of rice from protoplasts of haploid origin, in In Vitro *Haploid Production in Higher Plants,* 2 (Jain, S. M., Sopory, S. K., and Veilleux, R. E. , eds.), Kluwer Academic, The Netherlands, pp. 411–423.
10. Murashige, T. and Skoog, F. (1962) A revised medium for rapid growth and bioassays with tobacco tissue cultures. *Physiol. Plant.* **15,** 473–497.
11. Müller, A. J. and Graffe, R. (1978) Isolation and characterization of cell lines of *Nicotiana tabacum* lacking nitrate reductase. *Mol. Gen. Genet.* **161,** 67–76.
12. Ohira, K., Ojima, K., and Fujiwara, A. (1973) Studies on the nutrition of rice cell culture. 1. A simple, defined medium for rapid growth in the suspension culture. *Plant Cell Physiol.* **14,** 1113–1121.
13. Chu, C. C., Wang, C. C., Sun, C. S., Hsu, C., Yin, K. C., and Chu, C. Y. (1975) Establishment of an efficient medium for anther culture of rice through comparative experiments on the nitrogen sources. *Sci. Sin.* **18,** 659–668.
14. Datta, S. K., Datta, K., and Potrykus, I. (1990) Embryogenesis and plant regeneration from microspores of both "indica" and "japonica" rice *(Oryza sativa). Plant Sci.* **67,** 83–88.
15. Datta, S. K. and Wenzel, G. (1987) Isolated microspore derived plant formation via embryogenesis in Triticum aestivum. *Plant Sci.* **48,** 49–54.

16. Shillito, R. D., Paszkowski, J., and Potrykus I. (1983) Agarose plating and a bead type culture technique enable and stimulate development of protoplast-derived colonies in a number of plant species. *Plant Cell Rep.* **2,** 244–247.
17. Smith, J. A., Green, C. E., and Grengenbach, B. G. (1984) Feeder layer support of low density populations of Zea mays suspension cells. *Plant Sci. Lett.* **36,** 67–72.
18. Yoshida, S., Forno, D. A., Cock, J. H., Gomez, K. A. (1976) Routine procedure for growing rice plant in culture solution. *Laboratory Manual for Physiological Studies of Rice.* The International Rice Research Institute. Los Baños, Philippines. pp. 61–66.
19. Alam, M. F., Datta, K., Vasqucz, A. R., Oliva, N., Khush, G. S., and Datta, S. K. (1996) Transformation of new plant type lines through biolistic and protoplast methods. *Rice Genet. Newsletter* **13,** 139–141.

31

Transformation of Wheat via Particle Bombardment

Indra K. Vasil and Vimla Vasil

1. Introduction

Wheat (*Triticum aestivum* L.) is the number-one food crop in the world based on acreage under cultivation and total production. It is also the most widely cultivated crop as a result of breeding for adaptation to a wide range of ecological conditions. Wheat was first domesticated nearly 10,000 years ago in the Fertile Crescent of the Tigris-Euphrates basin in southwestern Asia. As a major commodity in international agriculture, and an important source of nutrition and protein in the human diet, wheat has long played a central role in world food security. During 1965–1990, introduction of the Green Revolution high-input and high-yielding varieties led to a nearly threefold increase in world wheat production. However, increases in the productivity of whcat and other major food crops attained through breeding and selection have begun to decline. This is happening at a time when nearly a third (34%) of the wheat crop is lost to pests, pathogens, and weeds *(1)*, in addition to postharvest losses during storage. Introduction of single genes through genetic transformation into crops such as maize, soybean, potato, canola, and cotton, has shown that such losses cannot only be greatly reduced or even eliminated, but can also result in reduced use of pesticides and herbicides.

Wheat was the last of the major crops to be transformed *(2,3)*, because of technical difficulties and the long time needed to establish reliable protocols for high-efficiency regeneration and transformation. Owing to the inherent difficulties in the establishment of embryogenic suspension cultures and regeneration of plants from protoplasts, fertile transgenic wheat plants have so far been obtained only by the direct delivery of DNA into regenerable tissue explants by accelerated microprojectiles *(3–6)*. The following protocol

From: *Methods in Molecular Biology, Vol. 111: Plant Cell Culture Protocols*
Edited by: R. D. Hall © Humana Press Inc., Totowa, NJ

describes the genetic transformation of wheat by particle bombardment of immature embryos using the PDS-1000/He (Bio-Rad) system ***(3,4,6)***. Although genes for resistance to hygromycin ***(7)***, kanamycin ***(8)***, and glyphosate ***(9)*** have been used as selectable markers in wheat transformation, the best results have been obtained with the *bar* gene (phosphinothricin acetyltransferase, PAT; ***3–6,8,10,11***), which confers resistance to DL-phosphinothricin (PPT) or glufosinate, the active ingredient in the commercial herbicides Herbiace, Basta, and Bialaphos ***(12)***. The method described here is based on the use of the plasmid pAHC25 ***(4,13)***, containing the *uidA* (β-glucuronidase) and *bar* as reporter and selectable marker genes, respectively ***(4,6,14)***. Both genes are under the control of the maize ubiquitin promoter ***(4,13)***, which has been shown to provide high levels of gene expression in gramineous species ***(15)***. The stability of integration and expression of the transgenes has been shown for several generations ***(14,16,17)***.

2. Materials

2.1. Culture of Immature Embryos

1. Plant material: Wheat (*T. aestivum* L., cv Bobwhite) plants are grown either in the field, greenhouse, or in growth chambers (first 40 d at 15°C/12°C day/night temperature and 10-h photoperiod at 150 μmol/m^2/s, followed by maintenance at 20°C/16°C temperature and 16-h photoperiod; *see* **Note 1**).
2. Medium for culture of immature Embryos: Murashige and Skoog's ***(18)*** formulation was used as follows (MS + medium):

4.3 g	MS salts (Sigma #M5524).
20.0 g	Sucrose.
100.0 mg	Myo-inositol.
500.0 mg	Glutamine.
100.0 mg	Casein hydrolysate.
2.0 mL	2,4-dichlorophenoxyacetic acid (2,4-D) from stock (*see* **item 4** *below*).
2.5 g	Gelrite.

   Bring volume to 1 L with distilled water, and adjust pH to 5.8 with 4 *N* NaOH. After autoclaving, add 1 mL filter-sterilized vitamin stock (*see* **item 5** *below*). Pour medium in Petri dishes (100 × 15 mm, Fisher). The sterile media can be stored for 1 mo at room temperature.
3. Medium for osmotic treatment of embryos: Add 36.44 g sorbitol (0.2 *M*) and 36.44 g mannitol (0.2 *M*) to make up 1 L of MS+ medium. Pour in Petri dishes (60 x 15 mm) after autoclaving.
4. 2,4-D stock: Dissolve 100 mg 2,4-D in 40 mL 95% ethanol, and make up the volume to 100 mL with distilled water. Store in refrigerator.
5. Vitamin stock: Dissolve 10 mg thiamin, 50 mg nicotinic acid, 50 mg pyridoxine HCl, and 200 mg glycine in distilled water, and make up volume to 100 mL. Store at –20°C.

2.2. Preparation and Delivery of DNA-Coated Gold Particles

1. Sterile distilled water.
2. Gold particle stock is prepared essentially according to protocol provided by Bio-Rad with PDS/1000 He. Suspend 60-mg gold particles (ca. 1-μm diameter) in 1 mL 100% ethanol for 2 min in 1.5-mL Eppendorf tube, with occasional vortexing. Centrifuge for 1 min. Discard supernatant, and resuspend in 1 mL sterile distilled water by vortexing. Wash two more times, and finally resuspend in 1 mL sterile water. Aliquot 25 or 50 μL in 0.5-mL Eppendorf tubes, with vortexing between aliquots to ensure proper mixing. Store at –20°C.
3. $CaCl_2$ (2.5 *M*, filter-sterilized) in 1-mL aliquots, stored at –20°C.
4. Spermidine-free base (0.1 *M*, filter-sterilized) in 1-mL aliquots, stored at –20°C.
5. Ethanol (100%).
6. pAHC25 DNA at 1 μg/μL in TE (10 m*M* Tris-HCl, 1 m*M* EDTA, pH 8).
7. Helium gas cylinder.
8. Sonicator (optional).

2.3. Culture, Selection, and Regeneration

1. Medium for selection: MS+ medium without glutamine and casein hydrolysate, with 3 mg/L bialaphos (4 mg/mL stock) added after autoclaving. Pour in Petri dishes (100 × 15 mm).
2. Medium for regeneration: Selection medium without 2,4-D, and supplemented with 1–10 mg/L zeatin, added after autoclaving.
3. Medium for shoot elongation: MS salts (2.15 g), sucrose (15 g), myo-inositol (50 mg), and 2.5 g gelrite. Make volume to 1 L with distilled water, pH 5.8. After autoclaving, add 0.5 mL vitamin stock and 5 mg/L bialaphos. Pour in Petri dishes (100 × 20 mm).
4. Medium for root elongation: Shoot elongation medium (10 mL) with 4–5 mg/L bialaphos in culture tubes (100 × 25 mm). All media with bialaphos can be stored for 2 wk at room temperature.
5. Zeatin stock solution: Dissolve 40 mg zeatin in few drops of 1 *M* NaOH, and make volume up to 10 mL in distilled water. Filter-sterilize.
6. Bialaphos stock solution: Dissolve 40 mg of bialaphos in 5 mL distilled water, and make up volume to 10 mL. Filter-sterilize and store at –20°C.

2.4. Analysis of Transient and Stable Expression

2.4.1 Assay for GUS Activity

X-gluc (15-bromo-4-chloro-3-indolyl-b-glucuronide): 100 mL;

100 mg X-gluc dissolved in 40 mL distilled water (2 m*M*).
50 mL 0.2 *M* phosphate buffer, pH 7.0 (100 m*M*).
1 mL 0.5 *M* potassium ferricyanide (5 m*M*).
1 mL 0.5 *M* potassium ferrocyanide (5 m*M*).
1 mL 1 *M* EDTA, pH 7.0 (10 m*M*).

Bring volume to 100 mL and stir until dissolved. Add 100 µL Triton X-100, filter-sterilize (22 µm), and store in small aliquots at –20°C.

2.4.2. PAT Assay

1. PAT extraction buffer (EB buffer): 100 mL:
   5 mL 1 *M* Tris-HCl, pH 7.5 (50 m*M*).
   0.4 mL 0.5 *M* EDTA (2 m*M*).
   15 mg leupeptine (0.15 mg/mL).
   15 mg phenylmethylsulphonyl fluoride (PMSF, Sigma) (0.15 mg/mL).
   30 mg bovine albumin (BSA) (0.3 mg/mL).
   30 mg dithiothreitol (DTT) (0.3 mg/mL).
   Stir to dissolve, filter-sterilize, and store at –20°C.
2. AB buffer: 100 mL:
   5 mL 1 *M* Tris-HCl, pH 7.5 (50 m*M*).
   0.4 mL 0.5 *M* EDTA (2 m*M*).
   100 mg BSA (0.1%).
   Make up volume to 100 mL with distilled water, and store in refrigerator.
3. Saturated ammonium sulfate: 400 g $(NH_4)_2SO_4$ in 400 mL distilled water, pH 7.8. Autoclave and store at room temperature. Can be used for at least 3 mo.
4. Thin-layer chromatography (TLC) migration solvent: 500 mL:
   300 mL l-propanol.
   178.6 mL Ammonium hydroxide.
   Make up volume to 500 mL with water (saturate the chromatography tank overnight with 3-mm paper). Highly pungent; wear mask.
5. ^{14}C-labeled acetyl-Co-A: Wear gloves and work in area designated for radioactive work.
6. TLC plates.
7. Chromatography tank.
8. Kodak (X-Omat) AR X-ray film.
9. Bulk reaction mixture for PAT assay: Prepare the necessary volume for two reactions more than the total number of reactions to allow for pipeting error. For one reaction: 0.6 µL PPT (1 mg/mL), 1 µL ^{14}C acetyl Co-A (0.05 µCi/µL), 1.4 µL distilled water.

3. Methods

3.1. Culture of Immature Embryos

1. Collect spikes (11–14 d post-anthesis), and wrap in moist paper towels. Spikes can be used fresh or stored for up to 5 d, with cut ends immersed in water, in a refrigerator (*see* **Note 2**).
2. Remove caryopses from the middle half of each spike, and surface-sterilize with 70% ethanol for 2 min, followed by slow stirring for 15–20 min in 20% chlorox solution (1.05% sodium hypochlorite) containing 0.1% Tween 20. Remove chlorox by washing with four changes of sterile distilled water at 5-min intervals (*see* **Notes 3** and **4**).

3. Aseptically remove the immature embryos (0.8–1.5 mm) under a stereomicroscope, and place in Petri dishes containing MS+ medium, with the scutellum exposed and the embryo axis in contact with the medium (*see* **Note 5**). Twenty to 25 embryos can be cultured in one Petri dish. Seal the culture dish with Parafilm, and incubate in the dark at 27°C to induce proliferation of scutellar cells (*see* **Note 6**).
4. Transfer embryos 4–6 h prior to bombardment (4–6 d after culture, when cell proliferation is visible at the edges of the scutellum) to Petri dishes containing the medium for osmotic treatment. Arrange 30–40 embryos, with scutellum facing up, in a 2-cm diameter circle in the center of the dish (*see* **Note 7**).

3.2. Bombardment of Embryos

1. Sterilize the bombardment chamber of PDS 1000/He and the gas acceleration cylinder with 70% ethanol, and allow to dry.
2. Sterilize the rupture and macrocarrier disks by soaking in 100% ethanol for 5 min and air-drying in laminar flow hood.
3. Sterilize steel screens by autoclaving.
4. Add to two, 25-μL aliquots of washed sterile gold particles, an equal volume of sterile distilled water, and vortex for 5 s.
5. Add 5 μL plasmid DNA or 5 μL/construct for cotransformation to one tube of gold particles (the other tube of gold particles serves as control for bombardments). Vortex immediately for 30 s to ensure good mixing of DNA and gold particles.
6. Add, in quick succession, 50 μL of $CaCl_2$ and 20 μL of spermidine, and vortex for 2 min. Hold on ice for 5–10 min.
7. Centrifuge for 10 s, discard supernatant, and resuspend in 200 μL 100% ethanol (can sonicate briefly to disperse).
8. Centrifuge for 10 s, discard supernatant, and resuspend the pellet in 150–250 μL of 100% ethanol. Leave on ice, but use within 2 h.
9. Sonicate and quickly spread 5 μL of gold from control Eppendorf tube (without DNA) or from gold–DNA tube to the center of a macrocarrier disk placed in the holder (gold concentration 30–50 μg/shot). Air-dry. At least two disks should be prepared for control bombardments, and the required amount with gold–DNA based on the number of samples in the experiment. Five macrocarrier disks can be prepared at one time from each suspension (*see* **Note 8**).
10. With the helium tank on, set the delivery pressure to 1300 psi (200 psi above the desired rupture disk value).
11. Turn on the vacuum pump.
12. Place a rupture disk (1100 psi) in the holder, and screw tightly in place (*see* **Note 9**).
13. With the steel mesh in place, transfer the coated macrocarrier disk (coated side facing down) to the holder, and place the macrocarrier assembly unit in the chamber at level 2 from top.
14. Place a Petri dish containing the cultured embryos on the sample holder at level 4 from top, and close the door.
15. Follow manufacturer's directions for bombardment, and follow all indicated safety procedures.

3.3. Culture, Selection, and Regeneration

1. Sixteen to 20 h after bombardment, separate and transfer the embryos to Petri dishes containing either MS+ medium without glutamine and casein hydrolysate (delayed selection) or selection medium for callus proliferation (early selection), for up to 2 wk in the dark. Place 16–20 embryos (from each bombarded dish) in two dishes to avoid overcrowding and crossfeeding (*see* **Note 10**).
2. Transfer the embryogenic calli from callus proliferation medium to regeneration medium for shoot formation under 16-h photoperiod (40 μmol/m²/s) for 8–10 d. At the end of this period, green areas indicative of shoot formation are visible to the naked eye.
3. Transfer the green shoots, along with the callus, as a unit to shoot elongation medium and culture in the light for 1–2 cycles of 2 wk each. Keep each callus-shoot piece well separated from others (*see* **Note 11**).
4. Transfer green shoots (>2 cm in length) to root elongation medium in tubes. The smaller (<1 cm) shoots can be subcultured for one more cycle of shoot elongation in Petri dishes. Calli with no shoots or shoots smaller than 1 cm after two cycles on shoot elongation medium can be discarded (*see* **Note 12**).
5. Plants reaching to the top of the culture tubes and with well-developed roots can be tested for PAT activity and transferred to soil in 56–66 d (*see* **Fig. 1**). They are grown to maturity in growth chambers under conditions similar to those used for donor plants.

3.4. Analysis of Transgene Expression

3.4.1. Histochemical GUS Assay for Transient Expression

1. Remove a sample of embryos (2–4 from 4 dishes each of control and +DNA) 2 d after bombardment, and soak separately in X-gluc solution (50–100 μL/well of a microtiter plate).
2. Seal the plate with Parafilm, and incubate overnight at 37°C.
3. Examine embryos under a stereo dissecting microscope to visualize the blue, GUS-expressing units.

3.4.2. PAT Assay

This assay is based on the detection of ^{14}C-labeled acetylated PPT (nonradioactive PPT used as substrate) after separation by TLC.

1. Collect leaf samples on ice in 1.5-mL Eppendorf tubes (prior to transferring putative transformants to soil). Be sure to include ± control samples (optimum sample weight 20–30 mg; freeze in liquid nitrogen and store at –70°C if not to be used the same day).
2. Add about 2 mg Polyvinyl-Pyrrolidone (Sigma PVP40), some silicone powder and 100 μL of EB buffer to each tube. Grind for 30 s with a polytron and keep samples on ice.
3. Centrifuge 10 m at 4°C, and collect supernatant in a fresh tube. Repeat centrifugation.

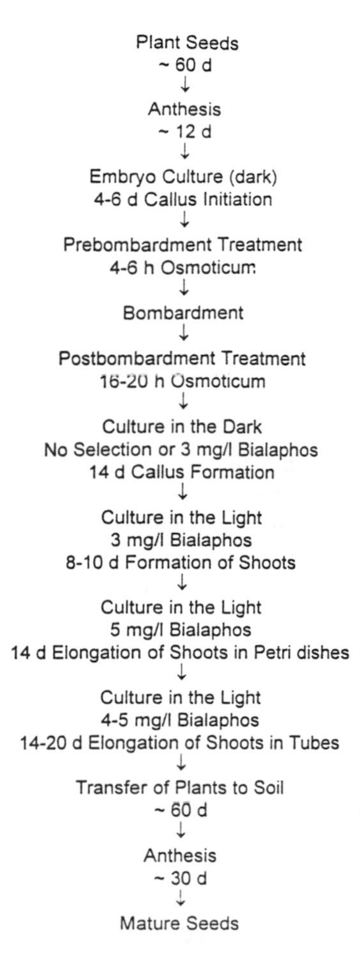

Fig. 1. Time frame for the production of transgenic wheat (cv Bobwhite) plants. Times shown are averages for experiments performed in 1995; those designated by ~ are approximate, and vary either with the batch of donor plants or the individual callus line or plant. Transgenic plants were transferred to soil 56–66 d after the initiation of cultures ***(6)***.

4. Measure protein concentration in the crude leaf extracts following the Bio-Rad protein assay relative to BSA as standard.
5. Adjust protein concentration to 2.5 (μg/μL for each sample with EB buffer. Assay can be stopped here overnight (store at –20°C). The background activity can be eliminated by precipitating the extracts with saturated $(NH_4)_2SO_4$.
6. Use 10 μL of extract for each sample for assay.
7. Aliquot 3 μL of bulk reaction mixture to 10 μL of each sample extract. Mix well, and incubate at 37°C for 1 h. Stop reaction on ice.

8. Spot 13 µL of each sample on a TLC plate (can be done by first spotting 6.5 µL, drying with a hair dryer, and spotting the rest to avoid spreading of mixture). Up to 15 samples can be spotted on one plate.
9. Transfer the TLC plate (after all the sample extracts have dried) to a saturated chromatography tank for 1–1.5 h.
10. Remove plate and let dry completely (at least 15 min).
11. Visualize the ^{14}C-labeled acetylated PPT by overnight exposure to X-ray film.
12. Plants testing positive for PAT are regarded as transformed (as the assay tests for expression of the enzyme encoded by the transgene *bar*) and can be transferred to soil. Characterization of the transgenes can be accomplished by DNA isolation and Southern hybridization using standard protocols.

4. Notes

1. Because the quality of donor plants directly affects the capacity of the immature embryos to produce embryogenic callus, it is important that the plants be grown under optimal conditions. Pesticide application from pollination to harvest time should be avoided.
2. For optimum response (especially when using field-grown material, which is often contaminated), use within 2 d after collection.
3. The size of the caryopsis is variable along the length of the spike. Therefore, a more uniform sample is obtained by collecting caryopses only from the middle half of each spike.
4. The duration of sterilization treatment depends on the quality of the donor plants. Caryopses from growth chamber grown plants tend to be clean and can be sterilized effectively in 10–15 min. Material from field-grown plants needs a longer (30-min) Clorox treatment.
5. Although embryos ranging in size from 0.5–1.5 mm are capable of producing embryogenic callus, the best postbombardment response is obtained from 1- to 1.2-mm embryos. Embryos at this stage of development are the easiest to dissect, since younger embryos are nearly transparent and difficult to locate, whereas the older ones are white owing to stored starch in the scutellum. Holding the caryopsis with forceps and making an incision at the base easily exposes the embryo (the outline of the embryo is visible before dissection).
6. Fifty to 60 immature embryos can be cultured in the same dish if the donor plants appear healthy and free of infection. Nonetheless, it is better to culture an average of 20 embryos/dish from field material owing to the probability of infection. The cultures should be examined on a daily basis to detect early signs of contamination. The contaminated embryos should be removed along with the medium around them to avoid infecting the adjacent embryos.
7. The highest frequencies of transformation and regeneration are obtained with 4–6 h of pre- and 16 h of postosmotic treatment.
8. The concentration of gold particles and DNA in gold/DNA mixture can vary. We recommend 30–50 µg gold/shot for a uniform spread on the macrocarrier disk, to obtain a more even and finer size of blue GUS units, and a consistent efficiency of transformation.

9. Rupture disks in a range of 650–1550 are used, with the 1100 psi disk being most common.
10. Among Basta, bialaphos, and PPT, bialaphos was found to be most reliable for selection of wheat transformants in the accelerated protocol ***(6)***. The time when selection is imposed, the number of explants/dish, and the concentration of the selective agent are critical variables that depend on the quality of the embryos cultured. Generally, early selection is recommended for high-quality embryos, and delayed selection for inferior embryos. Transformation frequencies of 0.1–2.5% were obtained with the protocol shown in **Fig. 1**.
11. Cultures should be examined frequently to assess the concentration of the selective agent required for the next step. For example, with high zeatin in the regeneration medium (10 mg/L), too many shoots are formed. With such cultures, it is safe to use 5 or 6 mg/L bialaphos during shoot elongation. With low (1 mg/L) or no zeatin, when only one or two green areas are seen in each explant, 4 or 5 mg/L bialaphos are sufficient.
12. During the period the explants and the regenerants are in Petri dishes, bialaphos can be used at 5 or 6 mg/L, since cross-protection is provided by the adjacent explants. However, if very few shoots emerge after the first cycle of shoot elongation, then 4–5 mg/L bialaphos are sufficient, because only a single explant is present and there is no crossprotection.

References

1. Oerke, E-C., Dehne, H-W., Schönbeck, F., and Weber, A. (1994) *Crop Production and Crop Protection: Estimated Losses in Major Food and Cash Crops.* Elsevier, Amsterdam.
2. Bialy, H. (1992) Transgenic wheat finally produced. *Bio/Technology* **10,** 675.
3. Vasil, V., Castillo, A. M., Fromm, M. E., and Vasil, I. K. (1992) Herbicide resistant fertile transgenic wheat plants obtained by microprojectile bombardment of regenerable embryogenic callus. *Bio/Technology* **10,** 667–674.
4. Vasil, V., Srivastava, V., Castillo, A. M., Fromm, M. E., and Vasil, I. K. (1993) Rapid production of transgenic wheat plants by direct bombardment of cultured immature embryos. *Bio/Technology* **11,** 1553–1558.
5. Weeks, J. T., Anderson O. D., and Blechl, A. E. (1993) Rapid production of multiple independent lines of fertile transgenic wheat *(Triticum aestivum). Plant Physiol.* **102,** 1077–1084.
6. Altpeter, F., Vasil, V., Srivastava, V., Stöger, E., and Vasil, I. K. (1996) Accelerated production of transgenic wheat (*Triticum aestivum* L.) plants. *Plant Cell Rep.* **16,** 12–17.
7. Ortiz, J. P. A., Reggiardo, M. I., Ravizzini, R. A., Altabe, S. G., Cervigni, G. D. L., Spitteler, M. A., et al. (1996) Hygromycin resistance as an efficient selectable marker for wheat stable transformation. *Plant Cell Rep.* **15,** 877–881.
8. Nehra, N. S., Chibbar, R. N., Leung, N., Caswell, K., Mallard, C., Steinhauer, L., et al. (1994) Self-fertile transgenic wheat plants regenerated from isolated scutellar tissues following microprojectile bombardment with two distinct gene constructs. *Plant J.* **5,** 285–297.

9. Zhou, H., Arrowsmith, J. W., Fromm, M. E., Hironaka, C. M., Taylor, M. L., Rodriguez, D., et al. (1995) Glyphosate-tolerant CP4 and GOX genes as a selectable marker in wheat transformation. *Plant Cell Rep.* **15,** 159–163.
10. Becker, D., Brettschneider, R., and Lörz, H. (1994) Fertile transgenic wheat from microprojectile bombardment of scutellar tissue. *Plant J.* **5,** 299–307.
11. Takumi, S. and Shimada, T. (1996) Production of transgenic wheat through particle bombardment of scutellar tissues: frequency is influenced by culture duration. *J. Plant Physiol.* **149,** 418–423.
12. Vasil, I. K. (1996) Phosphinothricin-resistant crops, in *Herbicide-Resistant Crops* (Duke, S. O., ed.), Lewis Publishers, Boca Raton, FL, pp. 85–91.
13. Christensen, A. H. and Quail, P. H. (1996) Ubiquitin promoter-based vectors for high-level expression of selectable and/or screenable marker genes in monocotyledonous plants. *Transg. Res.* **5,** 213–218.
14. Altpeter, F., Vasil, V., Srivastava, V., and Vasil, I. K. (1996) Integration and expression of the high-molecular-weight glutenin subunit 1Ax1 gene into wheat. *Nature Biotechnol.* **14,** 1155–1159.
15. Taylor, M. G., Vasil, V., and Vasil, I. K. (1993) Enhanced GUS gene expression in cereal/grass cell suspensions and immature embryos usng the maize ubiquitin-based plasmid pAHC25. *Plant Cell Rep.* **12,** 491–495.
16. Srivastava, V., Vasil, V., and Vasil, I. K. (1996) Molecular characterization of the fate of transgenes in transformed wheat (*Triticum aestivum* L.). *Theor. Appl. Genet.* **92,** 1031–1037.
17. Blechl, A. E. and Anderson, O. D. (1996) Expression of a novel high-molecular-weight glutenin subunit gene in transgenic wheat. *Nature Biotechnol.* **14,** 875–879.
18. Murashige, T. and Skoog, F. (1962) A revised medium for rapid growth and bioassays with tobacco tissue cultures. *Physiol. Plant.* **15,** 473–497.

32

Plant Transformation via Protoplast Electroporation

George W. Bates

1. Introduction

In electroporation, cells are permeabilized by the application of very short, high-voltage electric pulses. Molecules ranging in size from small organic metabolites and reporter dyes to large macromolecules—including antibodies and plasmids—can be introduced into cells by electroporation. Electroporation is effective on virtually any type of cell, and is now the method of choice for the genetic transformation of bacteria and certain animal cell lines. The primary application of electroporation to plants has been for DNA uptake for studies of transient gene expression and for stable transformation. However, electroporation has also been used to introduce RNAs *(1,2)*, antibodies *(3)*, and small molecules *(4)* into plant cells and isolated organelles *(5)*.

Because the thick plant cell wall restricts macromolecule movement, most work on plant cell electroporation utilizes protoplasts. This has limited the use of electroporation for stable transformation to species whose protoplasts are regenerable. As protoplast regeneration systems are improved, reports of new species of plants transformed by electroporation continue to appear (for example, **ref.** *6*). One advantage of electroporation over particle bombardment for stable transformation is that electroporation results predominantly in single-copy plasmid insertions (Bates, unpublished observations), whereas particle bombardment tends to introduce large plasmid concatemers. However, the main use of protoplast electroporation is in transient expression assays for studies of transcriptional regulation (for example, *7,8*). These studies do not require protoplast regeneration. A growing number of recent reports indicate that electroporation can be used to introduce DNA into walled plant cells and plant tissues *(9,10)*. Tissue electroporation does not work in all cases, and the parameters for successful plant tissue electroporation are not yet clear.

From: *Methods in Molecular Biology, Vol. 111: Plant Cell Culture Protocols*
Edited by: R. D. Hall © Humana Press Inc., Totowa, NJ

However, the success of tissue electroporation in crops, such as maize and soybean, reopens the use of electroporation for stable transformation in major crops.

This chapter provides a protocol for protoplast electroporation, a protocol for the selection of stable, kanamycin-resistant transformants, and notes on how to optimize these protocols for both stable transformation and transient expression. These protocols have been used for many years in the author's laboratory for the transformation of tobacco protoplasts, but they can be readily modified for use with protoplasts of other species and for the uptake of molecules other than DNA.

2. Materials

2.1. Protoplast Electroporation

1. Instrumentation: Electroporation equipment is available from a variety of commercial manufacturers and can also be homemade. Lists of commercial manufacturers and instrument specifications, as well as a general discussion of homemade equipment can be found in Chassy et al. ***(11)***.

   Two types of DC electrical pulses, square-wave pulses and capacitive discharges, may be used for electroporation. However, because the equipment is less expensive, most laboratories use capacitive-discharge electroporation systems. The equipment presently used in the author's laboratory is the Cell-Porator® Electroporation System I manufactured by Gibco BRL Life Technologies Inc. (Gaithersburg, MD). This capacitive-discharge instrument allows the pulse voltage to be adjusted from 0 to 400 V; pulse length can be varied by selecting one of eight different-sized capacitors (from 10–1980 μF) (*see* **Note 1**). The Gibco BRL Cell-Porator utilizes presterilized, disposable electroporation chambers to hold the cells during electroporation. For work with plant protoplasts, electroporation chambers should be selected that have a 0.4-cm electrode gap. Chambers with 0.1-cm electrode gaps can also be purchased, but are designed for electroporation of bacteria.
2. Protoplasts: Protoplasts isolated by standard procedures are suitable for electroporation. Protoplasts from a wide range of species, organs, and cell cultures have been successfully electroporated. However, it is important that the protoplasts be of high quality. Even in high-quality protoplast preparations, electroporation kills a substantial fraction of the protoplasts. Preparations of marginal or low-quality protoplasts are likely to be completely killed by the electric shocks.
3. DNA: The plasmid DNA used in electroporation does not have to be highly purified, but should be free of RNA and proteins (such as RNase and restriction enzymes) that might affect the viability of the protoplasts. The author's laboratory routinely uses plasmids isolated by alkaline lysis ***(12)***. RNA is removed from the plasmid preparations by RNaseA digestion followed by phenol extraction and ethanol precipitation (*see* **Note 2**).

   Before electroporation, the DNA must be sterilized. DNA that has been ethanol-precipitated and redissolved in autoclaved water or TE is probably sufficiently sterile. However, to be certain the DNA is sterile it can be passed through a 0.2-μm pore, low-binding, cellulose acetate syringe filter (e.g., Nalgene #190-2520). It is convenient to filter-sterilize the DNA after diluting it into electroporation medium at the start of an experiment.

For transient expression studies, linear and supercoiled plasmids are equally effective ***(13)***. However, for stable transformation, linear DNA is 3- to 10-fold more efficient than supercoiled ***(14)***.

4. Electroporation medium: Hepes-buffered saline (HBS): 150 m*M* KCl, 4 m*M* $CaCl_2$, 10 m*M* HEPES (pH 7.2), and enough mannitol to balance osmotically the protoplasts. For tobacco mesophyll protoplasts, 0.21 *M* mannitol is used. The HBS can be prepared in advance, sterilized by autoclaving, and stored at 4°C.

2.2. Selection of Stable Transformants

1. Media for protoplast culture and selection of stable transformants: Tobacco mesophyll protoplasts may be cultured in K_3G medium (K_3 salts, vitamins, and hormones ***[15]*** containing 0.4 *M* glucose as the carbon source and osmotic stabilizer). K_3G may be sterilized by autoclaving, and stored at 4°C.
2. Callus medium (CM): Murashige and Skoog salts and vitamins ***(16)*** plus 100 mg/L inositol, 3% sucrose, 1 mg/L benzyladenine, and 1 mg/L α-naphthaleneacetic acid. For selection of kanamycin-resistant stable transformants, CM is amended by addition of various amounts of mannitol, Sea plaque agarose (FMC Corp., Rockland, ME), and kanamycin as described below in the **Subheading 3.** CM and CM amended with mannitol and Sea plaque agarose can be prepared in advance, sterilized by autoclaving, and stored at 4°C. Kanamycin should not be autoclaved. A 1000X stock of kanamycin can be prepared by dissolving 100 mg/mL kanamycin sulfate in water; the stock should be filter-sterilized and stored at –20°C.

3. Methods

The following procedures must be carried out under sterile conditions, preferably in a laminar flow hood. All solutions should be at room temperature (*see* **Note 3**).

3.1. Protoplast Electroporation

1. Freshly isolated protoplasts should be used for electroporation: After washing the protoplasts free of the enzymes used for cell-wall digestion, pellet the protoplasts and resuspend them in 10 mL HBS + mannitol (*see* **Note 4**).
2. Determine the number of protoplasts/mL using a hemacytometer.
3. Place aliquots of 1×10^6 protoplasts into 15-mL conical centrifuge tubes, and pellet the protoplasts by centrifugation (50*g* for 5 min).
4. Discard the supernatant. Using a disposable plastic transfer pipet, gently resuspend each protoplast sample in 0.5 mL HBS + mannitol + DNA. The plasmid DNA concentration in the electroporation medium should be 10–100 μg/mL (*see* **Note 5**).

5. Use a disposable plastic transfer pipet to transfer the protoplast samples to electroporation chambers. Let stand for 5 min.
6. Resuspend the protoplasts by gentle agitation of the electroporation chamber, and then immediately apply a single electric pulse (325 μF, 300 V; *see* **Note 6**). If you are using the Gibco BRL Cell-Porator, make sure the instrument is on the low Ω setting.
7. Wait 2–3 min, and then resuspend the protoplasts by agitation of the electroporation chamber. Transfer the protoplasts to a conical centrifuge containing 5 mL of K_3G. Rinse the electroporation chamber with 0. 5 mL HBS + mannitol, and combine the rinse with the protoplast sample (*see* **Note 7**).
8. When all the protoplast samples have been electroporated, pellet the protoplasts, and resuspend each sample in 5 mL of culture medium.
9. Transfer the protoplast samples to a 60 × 15 mm Petri dish, and culture the protoplasts at 27°C.
10. Protoplasts may be sampled after 24 h for transient gene expression or maintained in culture for later selection of stable transformants.

3.2. Selection of Stable Transformants

The following procedure for selection of stable transformants is based on the agarose-bead culture technique of Shillito et al. ***(17)***. In this procedure, the protoplast culture is diluted and solidified by addition of low-melting-point agarose. When properly diluted, the embedded protoplasts are well separated from each other and grow into individual calli. Selection for transformants is carried out after embedding the cells. This procedure permits calculation of plating efficiency and transformation efficiency, and allows isolation of individual transformed clones. The method outlined here describes selection of kanamycin-resistant clones, but it can easily be adapted for use with other selectable markers or reporter genes. During selection, the culture is progressively diluted and the osmotic strength of the medium is reduced. The sequence of medium changes described here has been optimized for work with protoplasts of *Nicotiana tabacum,* but can be modified for work with other species.

1. Choose a plasmid containing a functional neomycin phosphotransferase II gene such as pMON200 ***(18)*** or pBI121 (Clonetech, Palo Alto, CA). Linearize the plasmid by digestion with a suitable restriction enzyme. It is convenient to cut enough of the plasmid for several experiments. Then phenol-extract and ethanol-precipitate the DNA, and resuspend it in water at a concentration of 1 μg/μL.
2. Electroporate the protoplasts in the presence of 10–100 μg/mL of linearized DNA, and culture them in 60 x 15 mm Petri plates in 5 mL of K_3G (*see* **Note 8**).
3. After 1 wk of culture, the protoplasts will have grown into small-cell clusters (*see* **Note 9**). At this point, the protoplast-derived colonies are immobilized by a 1:1 dilution of the culture with medium containing agarose. Prepare in advance CM medium + 0.23 *M* mannitol + 2.4% Sea plaque agarose, and sterilize it by

autoclaving. Just before the transfer, melt this medium, and let it cool to just above its gelling temperature. While the agarose is cooling, scrape any adhering protoplast colonies off of the Petri plate with a micropipet tip. Transfer half of the culture (2.5 mL) to a new 60 × 15 mm Petri plate. Now add 2.5 mL of the agarose-containing medium to both Petri plates of protoplasts. Mix thoroughly by swirling the plates. Place the Petri plates in the refrigerator for 15 min to solidify the agarose. Culture the protoplasts at 27°C.

4. At the end of the second week of culture, divide the solidified culture into wedges using a spatula, transfer the wedges to 100 x 15 mm Petri plates, and add 5 mL of liquid CM + 0.13 *M* mannitol supplemented with 100 µg/mL kanamycin (*see* **Note 10**). Culture at 27°C.
5. At the end of the third week, add 5 mL CM supplemented with 100 µg/mL kanamycin to each Petri plate.
6. Thereafter, at weekly intervals, remove 5 mL of liquid from each Petri plate, and replace it with 5 mL of fresh CM supplemented with 100 µg/mL kanamycin. Transformed colonies should be visible after 4–5 wk of culture.
7. When the transformed colonies are 2–3 mm in diameter (4–6 wk of culture), they can be picked out of the agarose using a spatula, and cultured on CM + 0.8% agar + 100 µg/mL kanamycin. After another 2 wk of growth, they should be large enough to be transferred to regeneration medium.

4. Notes

1. Pulse voltage, or more precisely electric field strength, and pulse length are two critical parameters in electroporation, and must be optimized for each species and cell type. Field strength depends on the voltage applied to the electroporation chamber and the distance between the electrodes in the chamber. The appropriate units for field strength are V/cm. Application of a 100 V pulse to a chamber with a 0.4-cm electrode gap results in a field strength of 250 V/cm.

   Pulse length is determined by the size of the capacitor and the resistance of the electroporation medium. Discharging a capacitor produces an exponentially decaying pulse. The length of such pulses is best described by their RC time constant, which is the time required for the pulse voltage to drop to 37% of its initial value.
2. Both RNA and proteins present in a crude plasmid preparation can be introduced into the protoplasts by electroporation along with the DNA and may affect the experimental outcome. For example, in some early experiments in the author's laboratory, protoplasts were electroporated in the presence of a plasmid that had been treated with RNase, but not phenol-extracted. No transient gene expression was observed until the RNase was removed from the plasmid preparations.
3. Some electroporation protocols call for chilling the cells in an ice bath during and immediately after electroporation. The pores that form in the plasma membrane owing to electric shocks have been shown to stay open longer if the cells are maintained at a temperature below the membrane's phase transition temperature. Chilling the cells would be expected to improve the efficiency of transformation

by electroporation, because it would allow more time for DNA uptake. However, several studies of transient and stable gene expression have shown that chilling does not improve electroporation efficiency. It turns out that the uptake of DNA during electroporation is electrophoretic (and not diffusive) and occurs during the electric pulse itself. Thus, chilling the cells is unnecessary if transformation is the goal of electroporation. Chilling may be useful, however, when electroporation is used to induce the uptake of molecules other than nucleic acids.

4. Because HBS is a nonphysiological, high-salt medium, it is advisable to limit the exposure of the protoplasts to this medium to 30 min. If a large number of samples are going to be electroporated, divide the protoplast preparation into two batches. Leave one batch in the protoplast wash or in culture medium, while the other batch is resuspended in HBS, divided into samples, and electroporated. Then when the first batch of protoplasts has been electroporated and diluted into culture medium (**Subheading 3.1., step 7**), pellet the second batch of protoplasts resuspend them in HBS and process them for electroporation.
5. The efficiency of both transient expression and stable transformation increases linearly with DNA concentration from 10 to 100 μg/mL. Transient expression and stable transformation are also increased by addition of "carrier DNA," such as salmon sperm DNA. For transient gene expression, the author's laboratory uses 10 μg/mL supercoiled plasmid DNA + 50 μg/mL salmon sperm DNA (sheared by sonication). Because the carrier DNA has been found to integrate along with the plasmid, carrier DNA should be avoided for stable transformation. When the goal is stable transformation, this laboratory uses 50 μg/mL of linearized plasmid DNA and no carrier.
6. The efficiency of electroporation depends on pulse length and voltage. A 325-μF capacitive discharge into 0.5 mL of HBS gives a pulse of about 10 ms (RC time constant). For tobacco mesophyll protoplasts, a 10-ms pulse of 300 V (750 V/cm field strength in the electroporation chamber) usually gives optimal results. The optimal setting can vary with species and cell type, and should be determined empirically in preliminary experiments. A quick way to begin to look for effective electroporation parameters is to find pulse settings that result in 50% protoplast death by 24 h after the shocks.

   Electroporation efficiency can also vary between batches of protoplasts because of batch-to-batch differences in protoplast viability. Batch-to-batch variability complicates transient expression studies. To handle this variability, replicates and appropriate controls must be included in every experiment. Some transient expression studies also include an internal control. The approach is to add a second plasmid carrying a reporter gene to each sample before electroporation as an internal control. Expression of the second plasmid can be used to normalize both batch-to-batch and sample-to-sample differences in transient gene expression.
7. Manufacturers of electroporation equipment intend for the electroporation chambers to be discarded after each use. However, the author finds that the chambers can be reused two or three times without affecting electroporation efficiency or

protoplast viability. Rinsing the chamber after electroporation with 0.5 mL HBS + mannitol not only helps remove all the electroporated protoplasts from the chamber, but also readies the chamber for the next sample.

8. High initial protoplast densities improve survival of the electric shocks. Up to 500 transformed clones can be recovered from a sample of 1 x 10^6 protoplasts. This would give an absolute transformation efficiency of 1 transformant for every 2000 electroporated protoplasts. However, the transformation efficiency is actually severalfold higher, because 50% of the protoplasts are killed directly by the electric shock and no more than half of the surviving protoplasts grow into calli.

   This laboratory uses 50 µg/mL linear DNA for stable transformation. Southern blotting shows that 50–75% of the transformants have a single copy of the plasmid integrated.
9. The timing of the media changes described here should not be adhered too rigidly, but should be modified depending on how fast the protoplasts grow after electroporation. For tobacco protoplasts, the first dilution of the cultures is done when the protoplasts have grown into microcolonies of 5–10 cells, which is sually after 6–7 d of culture. This first dilution may have to be delayed if the protoplasts are growing more slowly—as will happen if electroporation kills more than about 75% of the protoplasts. Too rapid a dilution of the culture results in death of the protoplast-derived colonies within 24 h of the dilution.
10. As described here, the selection pressure starts out at 50 µg/mL of kanamycin and increases to about 100 µg/mL over a period of a few weeks. Selection can be started earlier. For example, it can be conveniently started 1 wk after electroporation by adding kanamycin at the first dilution of the culture. This lab obtains the highest number of transformants when addition of kanamycin is delayed until the end of the second week. This may be because plating efficiency depends on cell density, so starting selection early reduces the cell density in the culture, and this in turn, inhibits the growth of some transformed clones. Selection for kanamycin resistance is very clean in tobacco. The author never observes untransformed clones growing in the presence of kanamycin.

References

1. Bailey-Serres, J. and Dawe, R. K. (1996) Both 5' and 3' sequences of maize adh1 mRNA are required for enhanced translation under low-oxygen conditions. *Plant Physiol.* **112,** 685–695.
2. Wintz, H. and Dietrich, A. (1996) Electroporation of small RNAs into plant protoplasts: mitochondrial uptake of transfer RNAs. *Biochem. Biophys. Res. Commun.* **223,** 204–210.
3. Maccarrone, M., Veldink, G. A., Finazzi Agro, A., and Vliegenthart, J. F. (1995) Lentil root protoplasts: a transient expression system suitable for coelectroporation of monocolonal antibodies and plasmid molecules. *Biochim. Biophys. Acta* **1243,** 136–142.
4. Jang, J.-C. and Sheen, J. (1994) Sugar sensing in higher plants. *Plant Cell* **6,** 1665–1679.

5. To, K. Y., Cheng, M. C., Chen, L. F., and Chen, S. C. (1996) Introduction and expression of foreign DNA in isolated spinach chloroplasts by electroporation. *Plant J.* **10,** 737–743.
6. Lee, L., Larmore, C. L., Day, P. R., and Tumer, N. E. (1996) Transformation and regeneration of creeping bentgrass (*Agrostis palustris* Huds) protoplasts. *Crop Sci.* **36,** 401–406.
7. Kao, C. Y., Cocciolone, S. M., Vasil, I. K., and McCarty, D. R. (1996) Localization and interaction of the cis-acting elements for abscisic acid VIVIPAROUS1, and light activation of the C1 gene of maize. *Plant Cell* **8,** 1171–1179.
8. Snowden, K. C., Buchholz, W. G., and Hall, T. C. (1996) Intron position affects expression from the tpi promoter in rice. *Plant Mol. Biol.* **31,** 689–692.
9. Laursen, C. M., Krzyzek, R. A., Flick, C. E., Anderson, P. C., and Spencer, T. M. (1994) Production of fertile transgenic maize by electroporation of suspension culture cells. *Plant Mol. Biol.* **24,** 51–61.
10. Chowrira, G. M., Akella, V., Fuerst, P. E., and Lurquin, P. F. (1996) Transgenic grain legumes obtained by in planta electroporation-mediated gene transfer. *Mol. Biotechnol.* **5,** 85–96.
11. Chassy, B. M., Saunders, J. A., and Sowers, A. E. (1992) Pulse generators for electrofusion and electroporation, in *Guide to Electroporation and Electrofusion* (Chang, D. C., Chassy, B. M., Suanders J. A., and Sowers, A. E., eds.), Academic, San Diego, pp. 555–569.
12. Sambrook, J., Fritsch, E. F., and Maniatis, T. (1989) *Molecular Cloning: A Laboratory Manual,* 2d ed. Cold Spring Harbor Laboratory Press, Cold Spring Harbor, NY.
13. Bates, G. W., Carle, S. A., and Piastuch, W. C. (1990) Linear DNA introduced into carrot protoplasts by electroporation undergoes ligation and recircularization. *Plant Mol. Biol.* **14,** 899–908.
14. Bates, G. W. (1994) Genetic transformation of plants by protoplast electroporation. *Mol. Biotechnol.* **2,** 135–145.
15. Nagy, J. I. and Maliga, P. (1976) Callus induction and plant regeneration from mesophyll protoplasts of *Nicotiana sylvestris. Z. Pflanzenphysiol.* **78,** 453–455.
16. Murashige, T. and Skoog, F. (1962) Revised medium for rapid growth and bioassays with tobacco tissue cultures. *Physiol. Plant.* **15,** 473–497.
17. Shillito, R. D., Paszkowski, J., and Potrykus, I. (1983) Agarose plating and a bead type culture technique enable and stimulate development of protoplast-derived colonies in a number of plant species. *Plant Cell Rep.* **2,** 244–247.
18. Rogers, S. G., Horsch, R. B., and Fraley, R. T. (1986) Gene transfer in plants: production of transformed plants using Ti plasmid vectors. *Methods Enzymol.* **118,** 627–641.

33

Transformation of Maize via Tissue Electroporation

Kathleen D'Halluin, Els Bonne,
Martien Bossut, and Rosita Le Page

1. Introduction

Introduction of DNA into plant protoplasts via electroporation is a well-known procedure. By giving electrical impulses of high field strength, the cell membrane is reversibly permeabilized, so that DNA molecules can be introduced into the cell *(1)*.

Originally, maize transformation efforts focused on electroporation of protoplasts *(2)*. However, transient transformation by tissue electroporation was first demonstrated by Dekeyser et al. *(3)*. Leaf bases of several monocot species, including maize, were shown to be amenable to electroporation-mediated DNA uptake based on the transient expression of reporter genes. Stable transformants via tissue electroporation of maize have been described for enzymatically treated or mechanically wounded immature embryos and type I callus *(4)*, for mechanically wounded immature zygotic embryos *(5)*, for enzymatically treated embryogenic maize suspension cells *(6)*, and for type II callus *(7)*. Furthermore, transient expression was observed after electroporation of intact immature maize embryos *(8)*. Transient and stable electrotransformations were obtained after electroporation of preplasmolyzed, intact black Mexican sweet maize suspension cells *(9)*.

The following schedule outlines the steps for the stable transformation of enzymatically treated, immature embryos and mechanically wounded type I callus of maize.

2. Materials

2.1. Electroporation

1. Embryo culture medium, Mah1VII: N6 basic medium *(10)* supplemented with 100 mg/L casein hydrolysate, 6 m*M* L-proline, 0.5 g/L MES, and 1 mg/L 2,4-D solidified with 2.5 g/L Gelrite (Duchefa), pH 5.8 (*see* **Notes 1** and **3**).

From: *Methods in Molecular Biology, Vol. 111: Plant Cell Culture Protocols*
Edited by: R. D. Hall © Humana Press Inc., Totowa, NJ

Mah1 VII substrate: M: N6 basic medium; a: 100 mg/L casein hydrolysate and 6 m*M* L-proline; h: 0.5 g/L MES; 1: 1 mg/L 2,4-D; VII: 2.5 g/L Gelrite (Duchefa).

2. Enzyme solution: 0.3% macerozyme (Kinki Yakult, Nishinomiya, Japan) in CPW salts ***(11)*** supplemented with 10% mannitol and 5 m*M* MES, pH 5.6, filter-sterilized (*see* **Note 2**).
3. Washing medium: N6 salt solution supplemented with 6 m*M* asparagine, 12 m*M* L-proline, 1 mg/L thiamin-HCl, 0.5 mg/L nicotinic acid, 100 mg/L casein hydrolysate, 100 mg/L inositol, 30 g/L sucrose, and 54 g/L mannitol; filter-sterilized.
4. EPM buffer: 80 m*M* KCl, 5 m*M* $CaCl_2$, 10 m*M* HEPES, and 0.425 *M* mannitol, pH 7.2. The same buffer without KCl is also needed.
5. Agarose substrate: water with 0.45% BRL agarose.
6. Electroporation apparatus, electrodes, and cuvets (*see* **Notes 4** and **6**).
7. DNA of an appropriate plasmid bearing the *bar* selectable marker gene (*see* **Note 5**).

2.2. Selection and Regeneration of Transformants

1. Selection medium Mahi1VII: Mah1VII substrate supplemented with 0.2 *M* mannitol and 2–5 mg/L filter-sterilized phosphinothricin (PPT) (*see* **Notes 1** and **3**).

   Mahi1 VII substrate: M: N6 basic medium; a: 100 mg/L casein hydrolysate and 6 m*M* L-proline; h: 0.5 g/L MES; i: 0.2 *M* mannitol; 1: 1 mg/L 2,4-D; VII: 2.5 g/L Gelrite (Duchefa).
2. Selection medium Mhi1VII: Mah1 VII substrate from which casein hydrolysate and L-proline are omitted, and supplemented with 0.2 *M* mannitol and 2–5 mg/L filter-sterilized PPT.

   Mhi1 VII substrate: M: N6 basic medium; h: 0.5 g/L MES; i: 0.2 *M* mannitol; 1: 1 mg/L 2,4-D; VII: 2.5 g/L Gelrite (Duchefa) (*see* **Notes 1** and **3**).
3. Selection medium Mh1VII: Mah1VII substrate from which casein hydrolysate and L-proline are omitted, and supplemented with 2–5 mg/L filter-sterilized PPT (*see* **Notes 1** and **3**).

   Mh1VII substrate: M: N6 basic medium; h: 0.5 g/L MES; 1: 1 mg/L 2,4-D; VII: 2.5 g/L Gelrite (Duchefa).
4. Selection medium Ahi1.5VII: corresponding substrate, of Mhi1VII substrate but MS (***12***, and *see* Appendix) basic medium ("A") instead of N6 basic medium ("M") supplemented with 0.2 *M* mannitol, 0.5 g/L MES, 1.5 mg/L 2.4-D, and 5 mg/L filter-sterilized PPT, solidified with 2.5 g/L Gelrite (Duchefa), pH 5.8 (*see* **Notes 1** and **3**).

   Ahi1.5VII substrate: A: MS medium; h: 0.5 g/L MES; i: 0.2 *M* mannitol; 1.5: 1.5 mg/L 2,4-D; VII: 2.5 g/L Gelrite (Duchefa).
5. Regeneration medium: MS medium supplemented with 5 mg/L 6-benzylaminopurine and 2 mg/L filter-sterilized PPT (*see* **Notes 1** and **3**).
6. Shoot development media (*see* **Notes 1** and **3**):

   MS6%: MS medium with 6% sucrose and supplemented with 2 mg/L filter-sterilized PPT.

   MS$^1/_2$: half-concentration of MS medium.
7. Basta 0.5%: 0.5% (v/v) of a Basta stock solution containing 200 g/L glufosinate-ammonium.

3. Methods

3.1. Electroporation

3.1.1. Immature Embryos

1. Maize plants of the public inbred line H99 are grown in the greenhouse in 20-L pots containing slow-release fertilizer. Growth conditions are at 25°C and 16-h light of ~20,000 lx (daylight supplemented by sodium vapor and mercury halide lamps); at night, temperature is reduced to 15–20°C.
2. Kernels from ears 9–14 d after pollination are surface-sterilized in 3.6% sodium hypochlorite for 15 min.
3. Immature zygotic embryos of the inbred line H99 (1–1.5 mm in length) are excised and plated on Mah1 VII substrate (*see* **Note 1**).
4. Freshly isolated immature embryos are collected from the Mah1VII substrate and enzymatically treated for 1–3 min at room temperature with the enzyme solution (*see* **Note 2**).
5. The embryos are then carefully, washed with filter-sterilized washing medium (*see* **Note 3**).
6. After washing, approx 50 embryos are transferred into a disposable microcuvet containing 100 µL maize electroporation buffer (EPM) (*see* **Note 4**).
7. Ten micrograms of linearized plasmid DNA are added per cuvet and coincubated with the enzyme-treated embryos (*see* **Note 5**).
8. After 1 h, the electroporation is carried out by discharging one pulse with a field strength of 375 V/cm from a 900-µF capacitor (*see* **Note 6**). The originally described *(4)* pre- or postelectroporation incubation on ice can be omitted.
9. After a few minutes, the embryos are transferred onto culture substrate.

3.1.2. Type I Callus

1. Immature embryos (1–1.5 mm in length) of the backcross (Pa91 × H99) × H99 are excised from ears 9–12 d after pollination and plated with their embryonic axis in contact with the Mah1 VII substrate (*see* **Note 7**).
2. Type I callus is initiated from immature embryos in the dark at 25°C.
3. Yellow, not too far differentiated, embryogenic tissue is microscopically dissected from developing, type I callus that has been cultured on Mah1VII substrate for a period of ~1.5–6 mo with subculture intervals of approx 3–4 wk. Tissues older than 6 mo are discarded, since the quality of the callus declines.
4. The embryogenic tissue is cut through the meristems, using a microscope, into pieces of approx 1–1.5 mm and collected in EPM buffer without KCl.
5. The finely chopped tissue is washed several times with EPM buffer without KCl to remove nucleases.
6. After 3 h of preplasmolysis in this buffer, the callus pieces are transferred to cuvets containing 100 µL of EPM supplemented with 80 m*M* KCl.
7. Ten micrograms of linearized plasmid DNA are added per cuvet, and subsequent conditions are as for electroporation of immature embryos.

3.2. Selection and Regeneration of Transformants

3.2.1. Immature Embryos

1. The embryos are transferred either immediately after electroporation to selective substrate Mahi1VII or to nonselective substrate for a few days (3–7 d) before transferring to selective substrate Mhi1VII (*see* **Note 8**). The quality of the embryos is the determining factor for the direct or nondirect transfer to selective substrate immediately after transformation. The omission of L-proline and casein hydrolysate from the selective substrate makes the selection more stringent. Good-quality embryos from maize plants grown under more optimal conditions are plated immediately on selective Mhi1VII substrate.
2. After 2–3 wk, the embryos are transferred to selective Mh1VII substrate.
3. After 3–4 wk, the embryogenic tissue is selected and subcultured for another 3–4 wk on selective Mh1VII substrate with 2–5 mg/L PPT.
4. Afterwards, the developing embryogenic tissue is isolated and transferred to regeneration medium at 25°C with a daylength of 16 h. Fluorescent lamps ("lumilux white" and "natural"; Osram, Munich, Germany) are used with a light intensity of 2000 l×.
5. After 2 wk, the embryogenic tissue is subcultured onto fresh substrate with the same composition.
6. Developing shoots are transferred to MS6% medium with 2 mg/L PPT from which hormones are omitted.
7. Further developing shoots are transferred to nonselective MS$^1/_2$ medium for further development into plantlets.
8. Plantlets are transferred to soil and grown to maturity in the greenhouse.

3.2.2. Type I Callus

1. Immediately after electroporation, the callus pieces are transferred to selective Ahi1.5VII substrate with 5 mg/L PPT.
2. After 3–4 wk, the proliferating callus pieces are subcultured onto Mh1VII substrate with 5 mg/L PPT for a period of approx 2 mo with a subculture interval of approx 3–4 wk.
3. Subsequent conditions for regeneration are as for immature embryos of H99, except that 5 mg/L zeatin are used instead of 6-benzylaminopurine for the induction of regeneration from the selected embryogenic tissue.

3.3. Analysis of Putative Transgenic Plants

3.3.1. Phosphinothricin Acetyltransferase (PAT) Activity

PAT activity is detected by an enzymatic assay using ^{14}C-labeled CoA. ^{14}C-labeled acetylated PPT is detected after separation by thin-layer chromatography *(13)*.

3.3.2. Dot Assay or Basta Spraying

Leaves of in vivo-grown plants are brushed with a 0.5% Basta® solution or sprayed with 100 mL/m^2 of a 0.5% Basta solution. Plants are assessed 8 d after Basta application. Plants expressing the *bar* gene have no symptoms, whereas leaves of nontransformed plants show necrosis.

3.3.3. Inheritance of the bar Gene

Most plants should develop normally and form a normal tassel and ear. Seed set is obtained by selfing or crosspollination. Transgenic lines can be used either as male or female parent in the crossing or as both. The segregation of the Basta resistance gene is then determined by using a Basta dot assay or by a Basta spraying, and scored for segregation of resistant and sensitive phenotypes. The majority of the lines obtained via electroporation show a Mendelian inheritance pattern for the *bar* gene locus (*see* **Note 9**).

4. Notes

1. All solid substrates are prepared in the following way: the sugar solution and the salt solution with MES and the solidifying agent are autoclaved separately. All other components are filter-sterilized and added after autoclaving.
2. The filter-sterilized enzyme solution is stored in the freezer at –20°C.
3. All liquid substrates and buffers are filter-sterilized and stored at room temperature in the dark, solutions of hormones and L-proline are stored in the freezer at –20°C, and vitamin and casein hydrolysate solutions are stored in the refrigerator.
4. Microcuvets (1938 PS; Kartell, Binasco, Italy) are sterilized in 70% ethanol for several minutes. Afterward microcuvets are filled with 1 mL agarose substrate (water with 0.45% BRL agarose), and after solidification 100 µL of the electroporation buffer (EPM) is pipeted on top of this agarose substrate. Parallel stainless-steel electrodes, 30 mm long, 2 mm thick, and 6 mm apart are inserted in such a way that by pushing the electrodes on the agarose substrate, the plant tissue is collected between the electrodes in the EPM buffer.
5. Plasmid DNA was purified on Qiagen (Qiagen Inc.) columns and resuspended in 10 m*M* Tris-HCl, pH 7.9, and 0.1 m*M* EDTA at a concentration of 1 mg/mL. The plasmid DNA, carrying a chimeric cauliflower mosaic virus (CaMV) 35S-*bar*-3'*nos* gene, was linearized prior to electroporation, using an appropriate enzyme.
6. A homemade electroporation unit consisting of a power supplier connected with an array of capacitors arranged in a circuit, as described by Fromm et al. *(1)*, was used.
7. Since we observed a higher transformation frequency from backcross material (Pa91 × H99) × H99 compared with the inbred lines H99 and Pa91, and since it was easier, under our climatic conditions, to produce good donor material from the backcross compared with the inbred lines, we continue to use the backcross in transformation experiments.

8. The mannitol is added in the first culture substrate immediately after electroporation to prevent an osmotic shock in the tissue that has been enzymatically treated or preplasmolyzed for 3 h in the EPM buffer with 0.425 *M* mannitol.
9. The majority of the lines obtained with electroporation show a Mendelian inheritance pattern for the *bar* gene locus in contrast with lines produced by particle bombardment. Transformation of maize based on electroporation or microprojectile bombardment of type I callus and immature embryos have been further compared. The level of transient expression as directed by anthocyanin biosynthetic regulatory genes revealed a significantly higher number of red spots on particle bombardment compared with electroporation.

   Both transformation procedures lead to the isolation of stable transformants derived from immature embryos and type I callus as targets for transformation using the *bar* gene as selectable marker. Upon electroporation, a clear selection was observed in tissue culture, but Basta-resistant plants could only be regenerated from a minor fraction of the selected calli. With particle bombardment, a significantly higher number of regenerants expressing the *bar* gene (up to ca. 40-fold) was obtained. However, segregation analysis of progenies from transgenic lines obtained from both transformation procedures revealed that in the majority of the electroporation-derived lines, the gene is inherited in normal Mendelian fashion, whereas a lower fraction (about 50%) of the *bar* lines obtained by microprojectile bombardment show a normal Mendelian segregation. About 30% of the lines produced by particle bombardment do not show an inheritance of the Basta resistance phenotype at all and yield only sensitive progenies. In reciprocal crosses of transformants produced by particle bombardment, segregation could be within the expected range in one direction and abnormal, even all-sensitive, in the other direction. Transformants should correctly express the introduced genes, if they are to be of practical value. A drawback of direct gene delivery is the unpredictable and often complex pattern of integration. On electroporation, about half of the transformants have a rather simple integration pattern, whereas the other half have a more complex integration pattern. With biolistic transformation, the majority of the transformants have integrated multiple copies of the transgene, often fragmented and rearranged. This could explain the high proportion of lines obtained with biolistics having an aberrant segregation, since the presence of multiple copies of transgenes can lead to instability of their expression ***(14)***.

   In conclusion, stably transformed maize lines using type I callus and immature embryos can be produced by both direct gene transfer methods. The efficiency of transformation is significantly higher with particle bombardment compared with electroporation. However, qualitatively better transformants are produced by electroporation, since a significant lower number of transformants have to be generated by electroporation compared with biolistics in order to obtain a transgenic line displaying a simple and "correct" integration pattern.

References

1. Fromm, M. E., Morrish, F., Armstrong, C. Williams, R., Thomas, J., and Klein, T. M. (1990) Inheritance and expression of chimeric genes in the progeny of transgenic maize plants. *Bio/Technology* **8,** 833–839.
2. Rhodes, C. A., Pierce, D. A., Mettler, I. J., Mascarenhas, D., and Detmer, J. J. (1988) Genetically transformed maize plants from protoplasts. *Science* **240,** 204–207.
3. Dekeyser, R. A., Claes, B., De Rijcke, R. M. U., Habets, M. E., Van Montagu, M. C., and Caplan, A. B. (1990) Transient gene expression in intact and organized rice tissues. *Plant Cell* **2,** 591–602.
4. D'Halluin, K., Bonne, E., Bossut, M., De Beuckeleer, M. and Leemans, J. (1992) Transgenic maize plants by tissue electroporation. *Plant Cell* **4,** 1495–1505.
5. Xiayi, K., Xiuwen, Z., Heping, S., and Baojian, L. (1996) Electroporation of immature maize zygotic embryos and regeneration of transgenic plants. *Transgenic Res.* **5,** 219–221.
6. Laursen, C. M., Krzyzek, R. A., Flick, C. E., Anderson, P. C., and Spencer, T. M. (1994) Production of fertile transgenic maize by electroporation of suspension culture cells. *Plant Mol. Biol.* **24,** 51–61.
7. Pescitelli, S. M. and Sukhapinda, K. (1995) Stable transformation via electroporation into maize Type II callus and regeneration of fertile transgenic plants. *Plant Cell Rep.* **14,** 712–716.
8. Songstad, D. D., Halaka, F. G., DeBoer, D. L., Armstrong, C. L., Hinchee, M. A. W., Ford-Santino, C. G., et al. (1993) Transient expression of GUS and anthocyanin constructs in intact maize immature embryos following electroporation. *Plant Cell, Tissue Organ Cult.* **33,** 195–201.
9. Sabri, N., Pelissier, B., and Teissie, J. (1996) Transient and stable electrotransformations of intact black Mexican sweet maize cells are obtained after preplasmolysis. *Plant Cell Rep.* **15,** 924–928.
10. Chu, C. C., Wang, C. C., Sun, C. S., Hsu, C., Yin, K. C., Chu, C. Y., et al. (1975) Establishment of an efficient medium for anther culture of rice through comparative experiments on the nitrogen sources. *Sci. Sin.* **18,** 659–668.
11. Frearson, E. M., Power, J. B., and Cocking, E. C. (1973) The isolation, culture and regeneration of *Petunia* leaf protoplasts. *Dev. Biol.* **33,** 130–137.
12. Murashige, T. and Skoog, F. (1962) A revised medium for rapid growth and bioassays with tobacco tissue cultures. *Physiol. Plant.* **15,** 473–497.
13. De Block, M., Botterman, J., Vandewiele, M., Dockx, J., Thoen, C., Gosselé, V., et al. (1987) Engineering herbicide resistance in plants by expression of a detoxifying enzyme. *EMBO J.* **6,** 2513–2518.
14. Matzke, M. A. and Matzke, A. J. M. (1995) How and why do plants inactivate homologous (trans)genes? *Plant Physiol.* **107,** 679–685.

34

Transformation of Maize Using Silicon Carbide Whiskers

Jim M. Dunwell

1. Introduction

As the most commercially valuable cereal grown worldwide and the best-characterized in genetic terms, maize was predictably the first target for transformation among the important crops. Indeed, the first attempt at transformation of any plant was conducted on maize *(1)*. These early efforts, however, were inevitably unsuccessful, since at that time, there were no reliable methods to permit the introduction of DNA into a cell, the expression of that DNA, and the identification of progeny derived from such a "transgenic" cell *(2)*. Almost 20 years later, these technologies were finally combined, and the first transgenic cereals were produced. In the last few years, methods have become increasingly efficient, and transgenic maize has now been produced from protoplasts as well as from *Agrobacterium*-mediated or "Biolistic" delivery to embryogenic tissue (for a general comparison of methods used for maize, the reader is referred to a recent review—ref. ***3***). The present chapter will describe probably the simplest of the available procedures, namely the delivery of DNA to the recipient cells by vortexing them in the presence of silicon carbide (SiC) whiskers (this name will be used in preference to the term "fiber," since it more correctly describes the single crystal nature of the material).

This needle-shaped material was selected as a means of perforating cells (thereby allowing the entry of DNA), since it was known to be one of the hardest of "man-made" products; it is used commercially as an abrasive and a component of saw blades. In fact, it is not a completely synthetic material, but is produced by carbonizing (at high temperature in carbon monoxide) the silicate found within the cells of rice husks. The chemically transformed product of one cereal is therefore being used to transform another cereal genetically.

From: *Methods in Molecular Biology, Vol. 111: Plant Cell Culture Protocols*
Edited by: R. D. Hall © Humana Press Inc., Totowa, NJ

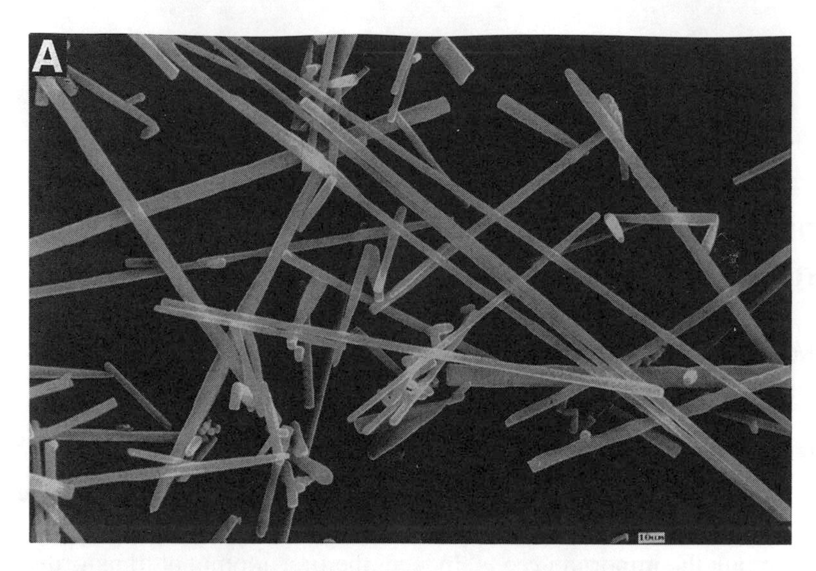

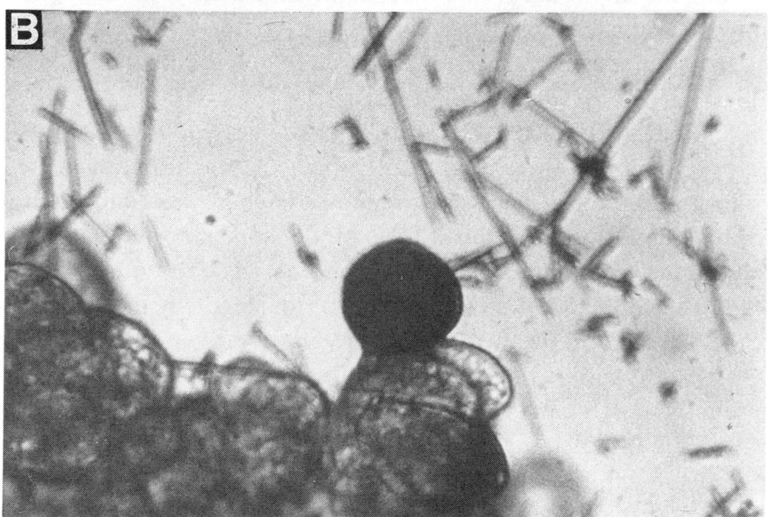

Fig. 1. **(A)** Scanning electronmicrograph of silicon carbide whiskers of the type used for transformation. **(B)** Group of maize cells after vortexing with silicon carbide whiskers, showing dark staining of a single-cell expressing the introduced GUS gene.

In essence, the transformation method described below involves the agitation of suspension-culture cells, with SiC whiskers and plasmid DNA (**Fig. 1**). The resultant transgenic cells are selected by growth on bialaphos and plants subsequently regenerated. Details of the procedure as developed for maize can be found in a series of recent papers and reviews ***(4–8)***. The scope of the method, however, is not restricted to maize, or indeed to cells of higher plants, as shown by successful delivery of DNA to cells of various algae ***(9)***, including *Chlamydomonas* ***(10)***, tobacco ***(11,12)***, *Agrostis* ***(13)***, and *Lolium* (Dalton, personal communication). Most recently, transient expression of the GUS marker was detected when dry wheat embryos were vortexed with DNA and whiskers for 10–30 min ***(14)***. It was also demonstrated that callus tissue induced from the treated embryos contained GUS-expressing sectors more than 1 mo after treatment. This very promising finding suggests that cells in culture are not necessarily the only possible target. Additionally, recent results obtained from node slices of American chestnut showed that treatment with whiskers increased the proportion of wounded tissue and thereby enhanced subsequent *Agrobacterium*-mediated gene delivery ***(15)***.

It should be noted that the method as developed for the production of transgenic maize is the subject of a granted US patent ***(16)*** and that other patents are pending. These do not of course prevent the use of the technique for research purposes.

2. Materials

2.1. Source of Whiskers

This is one of the most important factors in the process. The preferred type of whiskers is that designated "Silar SC-9" and obtained from Advanced Composites (1525 South Buncombe Road, Greer, SC 29651-9208). Other available types are TWS100 (Tokai Carbon Company, Tokyo, Japan) ***(13)*** and Alfa Aesar (Johnson Mathey, Ward Hill, MA). Most whisker preparations are highly heterogeneous, ranging in length from 5–500 µm. They can also differ in diameter (usual mean of 1 µm) and in the degree of hydrophobicity, both factors that are likely to affect their efficiency ***(8)***. SiC whiskers are known to be a respiratory hazard ***(17,18)*** and should therefore be treated with caution at all times, but especially when being weighed in the dry state. They should be disposed of as hazardous waste.

2.2. Donor Plants

Glass house-grown plants of the maize hybrid A188 × B73 are used as a source of embryos (*see* **Note 1**).

2.3. Callus Maintenance Medium

For 1 L of N6 medium ***(19)***, dissolve in distilled water: 4 g powdered N6 salts (Sigma), 30 g sucrose, 100 mg myo-inositol, 2 mg glycine, 1 mg thiamine, 0.5 mg pyridoxine HCl, 0.5 mg nicotinic acid, and 2 mg 2,4-D (from stock solution [1 mg/mL] made by dissolving 2,4-D in dilute KOH). Adjust to pH 6.0 with 1 *M* KOH, add 3 g Gelrite, and autoclave.

2.4. Callus Initiation Medium

Medium N6I is N6 ***(20)*** modified by adding 0.69 g/L proline (6 m*M*) and reducing the sucrose content to 20 g/L. Adjust pH to 6.0, add gelling agent as above, and autoclave.

2.5. Suspension Initiation Medium

For 1 L MS medium ***(21)***, dissolve in distilled water: 4.3 g powdered salts (Sigma), 2 mg glycine, 0.5 mg thiamine, 0.5 mg pyridoxine HCl, 0.05 mg nicotinic acid, and 0.0373 mg Na_2EDTA. Adjust the pH to 6.0 and autoclave.

2.6. Suspension Maintenance Medium

For 1 L of H9CP+, add to liquid MS: 30 g sucrose, 100 mg myo-inositol, 2 mg 2,4-D, 2 mg NAA, 0.69 g proline, 200 mg casein hydrolysate (both phytohormones added from predissolved stocks in dilute KOH). Adjust to pH 6.0, and autoclave. (Add 5% sterile coconut water [Gibco] before subculture.)

2.7. Pretreatment of Cells

Medium N6(S/M) used for osmotic treatment is liquid N6 with 45 g/L D-sorbitol, 45 g/L D-mannitol, and 30 g/L sucrose. Adjust pH to 6.0 and autoclave.

2.8. Selection of Transgenic Cells

1. Bialaphos is obtained by deformulation of the herbicide Herbiace (Meiji Seika, Japan).
2. Prepare N6(1B) from N6 and 1 mg/L bialaphos and N6(0.5B) with 0.5 mg/L bialaphos. Solidify with 0.3% gelrite, or for embedding, with 0.6% Sea plaque agarose (FMC Bioproducts). Add bialaphos by filter-sterilization after autoclaving the media.

2.9. Regeneration Media

1. Regen 1 medium is MS with (per L) 60 g sucrose, 1 mg NAA, and 1 g myo-inositol. Solidify with 0.3% Gelrite, pH 6.0.
2. Regen 2 medium is as above with 0.25 mg NAA and 30 g sucrose.
3. Regen 3 is half-strength MS with 30 g sucrose, solidified, and with pH as above.

2.10. DNA Preparation

Use a QIAGEN plasmid Maxi kit to purify plasmid DNA, and resuspend the DNA pellet in sterile Millipore-purified water at a final concentration of 1 μg/μL.

2.11. Additional Requirements

Bleach for sterilization, sterile water, forceps (curved tip), scalpel, Erlenmeyer flasks (250 mL), disposable pipets (10-mL wide-bore), and 20-mesh stainless-steel sieve are needed.

3. Methods

3.1. Establishment of Suspension Cultures

1. 10–12 d after pollination, sterilize husked ears with 50% commercial bleach (final concentration 2.6% available chlorine), rinse three times in water, remove immature zygotic embryos under aseptic conditions, and place scutellum surface upward on N6I medium.
2. Add 3 g of proliferating type II callus from a selected single embryo to 20 mL of H9CP+ medium.
3. Maintain in 125-mL Erlenmeyer flask at 28°C in darkness on a rotary shaker (125 rpm).
4. Subculture every 3.5 d by adding 3 mL packed cell volume (see **Subheading 3.2.**) and 7 mL old culture medium to 20 mL fresh medium.
5. If required, cell suspensions can be cryopreserved ***(22)***. To thaw, submerge tubes in a water bath at 45°C, and immediately use a 5-mL disposable pipet to plate cells onto a sterile filter paper in a 60 × 20 mm Petri dish. Remove excess cryoprotectant by repeated blotting with filter paper, place cells on solid N6 medium, and subculture weekly for 3 wk, after which the suspension can be reinitiated as above.

3.2. Pretreatment of Cells Prior to Transformation

1. One day after subculture, sieve suspension cells through a 20-mesh stainless sieve, and return to original Erlenmeyer flask in used medium.
2. Measure packed cell volume (PCV) by using 10 mL wide-bore disposable pipet and allowing cells to settle in lower half of pipet.
3. Replace H9CP+ medium with liquid N6(S/M) at rate of 3 mL to every 1 mL PCV.
4. Return flasks to rotary shaker for 45 min.
5. Resuspend cells evenly by pipeting cells and medium up and down repeatedly.
6. Using 10-mL disposable pipet, dispense 1-mL aliquots into 1.5-mL Eppendorf tubes.
7. When cells have settled to bottom of tube, remove 0.5 mL medium, leaving 1:1 proportions of PCV:medium.

3.3. Preparation of Whiskers

1. Transfer about 10–50 mg dry whiskers to preweighed 1.5-mL Eppendorf tubes in a fume hood.
2. Reweigh tubes, and calculate the weight of whiskers.
3. Pierce the lid to avoid its being displaced during heating, cover it in aluminium foil, and autoclave the tubes.

4. Store at room temperature prior to use. Whiskers should be prepared fresh in an aqueous suspension (5% w/v in sterile water) prior to use *(8)*.

3.4. Transformation Procedure

1. Add 40 μL of 5% whisker suspension and 25 μL plasmid DNA (1 μg/μL) to the pretreated cells in an Eppendorf; gently premix by tapping.
2. Place tube in multisample holder of Vortex Genie II, and mix at full speed for 60 s, or in the holder of a Mixomat dental amalgam mixer shaken at fixed speed for 1 s. (*See* **Notes 2**, **3**, and **4**).

3.5. Selection of Transgenic Tissue

1. Add 0.5 mL fresh N6(S/M) medium to the mixture in the tube.
2. Use a 200-μL wide-bore Pipetman tip to dispense the diluted contents onto a 55-mm Whatman no. 4 filter paper overlaying solid N6 medium in a 60 x 20 mm plastic Petri dish.
3. Wrap dishes in gas-permeable Urgopore tape (Sterilco, Brussels), and incubate at 28°C in darkness.
4. After 7 d, transfer the filter paper and cells to the surface of N6*(1B)* selection medium. Repeat after a further 7 d. (*see* **Note 5**)
5. Use a spatula to transfer cells from the filter paper into 5 mL N6(0.5B) medium containing 0.6% Sea plaque low-gelling-point agarose in a sterile test tube kept at 37°C.
6. Divide the suspension into two aliquots, and plate each evenly over 20 mL N6(0.5B) medium solidified with 0.3% Gelrite in a 100 × 25 mm Petri dish.
7. Seal with Nescofilm, and incubate at 28°C.
8. After 4–5 wk pick off potentially transformed calli, and transfer to the surface of fresh N6*(1B)* medium.
9. Subculture onto same medium every 2 wk.

3.6. Regeneration of Plants

1. Transfer type II callus onto Regen I medium, and incubate at 25°C in darkness.
2. After 2–3 wk, transfer opaque, mature somatic embryos onto Regen 2 medium. Incubate in light (~200 μmol/m^2/s with 16-h photoperiod).
3. After a further 2–3 wk, shoots and roots emerge, and small plantlets can subsequently be transferred to glass tubes containing 15 mL Regen 2 medium.
4. After 7 d, transfer to soil in a glass house for analysis (*see* **Note 6**). Grow to maturity in 3-gallon pots containing equal parts of peat:perlite:soil.

4. Notes

1. The genotype recommended in this study produces a form of dispersed suspension very suitable for the method described. Any other genotype with this characteristic is likely to be equally suitable. As mentioned in **Subheading 1.**, mature embryos have also been used successfully in wheat *(14)*.

2. Although these results were obtained using two particular forms of apparatus for mixing, it can be assumed that any other equivalent form of mixing device would give similar results. For example, the results on wheat *(14)* were achieved with a Micro tube mixer MT-360 (Tomy Seiko Co., Ltd., Japan).
3. To date, trials on a range of material have given the best results with SiC whiskers. There may, however, be other, as yet untested crystalline material of equal efficiency.
4. As an alternative to the use of a random-mixing process for cells and whiskers, a fixed array of microneedles has been developed *(23)*.
5. It can be expected that other selective agents could be used in place of bialaphos.
6. In an interesting comparative study *(24)*, it has been shown with *Chlamydomonas reinhardtii* that transformants produced by a glass bead method *(25)* give a higher frequency of random to homologous (1000:1) integration events than those produced by using particle bombardment (24:1) (ref. *26*). It may be that the form of the DNA (degree of condensation) is an important determinant of the type of integration pattern. If so, then the SiC method, in which the DNA is not precipitated, may be advantageous in those systems where random patterns are required.

References

1. Coe, E. H. and Sarkar, K. R. (1966) Preparation of nucleic acids and a genetic transformation attempt in maize. *Crop Sci.* **6,** 432–435.
2. Dunwell, J. M. (1995) Transgenic cereals. *Chem. Ind.* Sept. 18, 730–733.
3. Wilson, H. M., Bullock, W. P., Dunwell, J. M., Ellis, J. R., Frame, B., Register, J., et al. (1995) Maize, in *Transformation of Plants and Soil Microorganisms* (Wang, K., Herrera-Estrella, A., and Van Montagu, M., eds.), Cambridge University Press, pp. 65–80.
4. Frame, B. R., Drayton, P. R., Bagnell, S. V., Lewnau, C. J., Bullock, W. P., Wilson, H. M., et al. (1994) Production of fertile transgenic maize plants by silicon carbide whisker-mediated transformation. *Plant J.* **6,** 941–948.
5. Kaeppler, H. F. and Somers, D. A. (1994) DNA delivery to maize cell cultures using silicon carbide fibers, in *The Maize Handbook* (Freeling, M., and Walbot, V., eds.), Springer-Verlag, New York, pp. 610–613.
6. Thompson, J. A., Drayton, P. R., Frame, B. R., Wang, K., and Dunwell, J. M. (1995) Maize transformation utilizing silicon carbide whiskers: a review. *Euphytica* **85,** 75–80.
7. Wang, K., Frame, B. R., Drayton, P. R., and Thompson, J. A. (1995) Silicon-carbide whisker-mediated transformation: regeneration of transgenic maize plants, in *Gene Transfer to Plants* (Potrykus, I. and Spangenberg, G., eds.), Springer, Berlin, pp. 186–192.
8. Wang, K., Drayton, P. R., Dunwell, J. M., and Thompson, J. T. (1995) Whisker-mediated plant transformation: an alternative technology. *In Vitro Cell. Dev. Biol.* **31,** 101–104.
9. Dunahay, T. G., Adler, S. A., and Jarvik, J. W. (1997) Transformation of microalgae using silicon carbide whiskers. *Methods Mol. Biol.* **62,** 503–509.

10. Dunahay, T. G. (1993) Transformation of *Chlamydomonas reinhardtii* with silicon carbide whiskers. *BioTechniques* **15,** 452–460.
11. Kaeppler, H. F., Gu, W., Somers, D. A., Rines, H. W., and Cockburn, A. F. (1990) Silicon carbide fiber-mediated DNA delivery into plant cells. *Plant Cell Rep.* **9,** 415–418.
12. Kaeppler, H. F., Somers, D. A., Rines, H. W., and Cockburn, A. F. (1992) Silicon carbide fiber-mediated stable transformation of plant cells. *Theor. Appl. Genet.* **84,** 560–566.
13. Asano, Y., Otsuki, Y., and Ugaki, M. (1991) Electroporation-mediated and silicon carbide whisker-mediated DNA delivery in *Agrostis alba* L. (Redtop). *Plant Sci.* **9,** 247–252.
14. Serik, O., Ainur, I., Murat, K., Tetsuo, M., and Masaki, I. (1996) Silicon carbide fiber-mediated DNA delivery into cells of wheat (*Triticum aestivum* L.) mature embryos. *Plant Cell Rep.* **16,** 133–136.
15. Zing, Z., Powell, W. A., and Maynard, C. A. (1997) Using silicon carbide fibers to enhance Agrobacterium-mediated transformation of American chestnut. *In Vitro Cell. Dev. Biol.* **33,** 63A
16. US Patent 5302523, Transformation of Plant Cells. April 12, 1994.
17. Vaughan, G. L., Jordan, J., and Karr, S. (1991) The toxicity, in vitro, of silicon carbide whiskers. *Environ. Res.* **56,** 57–67.
18. Vaughan, G. L., Trently, S. A., and Wilson, R. B. (1993) Pulmonary response, *in vivo,* to silicon carbide whiskers. *Environ. Res.* **63,** 191–201.
19. Chu, C. C., Wang, C. C., Sun, C. S., Hsu, C., Yin, K. C., Chu, C. Y., et al. (1975) Establishment of an efficient medium for anther culture of rice through comparative experiments on the nitrogen sources. *Sci. Sin.* **18,** 659–668.
20. Armstrong, C. L. and Green, C. E. (1985) Establishment and maintenance of friable, embryogenic maize callus and the involvement of L-proline. *Planta* **164,** 207–214.
21. Murashige, T. and Skoog, F. (1962) A revised medium for rapid growth and bioassays with tobacco tissue cultures. *Physiol. Plant.* **15,** 473–497.
22. Shillito, R. D., Carswell, G. K., Johnson, C. M., Di Maio, J. J., and Harms, C. T. (1989) Regeneration of fertile plants from protoplasts of elite inbred maize. *Bio/Technology* **7,** 581–587.
23. US Patent 5457041, Needle Array and Method of Introducing Biological Substances into Living Cells using the Needle Array. October 10, 1995.
24. Sodeinde, O. A. and Kindle, K. L. (1993) Homologous recombination in the nuclear genome of *Chlamydomonas reinhardtii. Proc. Natl. Acad. Sci. USA* **90,** 9199–9203.
25. Kindle, K. L. (1990) High-frequency nuclear transformation of *Chlamydomonas reinhardtii. Proc. Natl. Acad. Sci. USA* **87,** 1228–1232.
26. Kindle, K. L., Schnell, R. A., Fernandez, E., and Lefebvre, P. A. (1990) Stable nuclear transformation of *Chlamydomonas* using the *Chlamydomonas* gene for nitrate reductase. *J. Cell Biol.* **109,** 2589–2601.

VII

Suspension Culture Initiation and the Accumulation of Metabolites

35

Directing Anthraquinone Accumulation via Manipulation of *Morinda* Suspension Cultures

Marc J. M. Hagendoorn, Diaan C. L. Jamar, and Linus H. W. van der Plas

1. Introduction

Secondary metabolite accumulation in cell suspensions is the result of an interplay between primary metabolism, supplying the machinery, the energy, and the precursors, and the secondary metabolic pathways. For this reason, a successful strategy for producing considerable amounts of secondary plant products can hardly be obtained without the knowledge of the interactions between primary and secondary metabolism. These interactions are best studied in cell suspensions that produce significant amounts of secondary metabolites.

In *Morinda citrifolia* cell suspensions, secondary metabolic pathways can easily be switched on, resulting in a considerable accumulation (more than 10% of the dry weight) of anthraquinones (AQs, *[1]*). Anthraquinones are dyes that can be treated to give a variety of colors, e.g., yellow, red, purple, brown, or black *(2)*.

Rubiaceae AQs are all derivatives from the basic structure as shown in **Fig. 1**. *M. citrifolia* is a tropical tree, inhabiting the islands on the edge of the Pacific Ocean *(2)*. In *M. citrifolia* plants, AQs are only accumulated in the roots, and then mostly in the bark. Most of the AQs of *M. citrifolia* cell suspensions are glycosylated, especially as *O*-glucosylxylosyl (therefore called primverosides *[3]*) and subsequently are stored in the vacuole.

In the Rubiaceae, anthraquinones are produced via the shikimate-mevalonate pathway. Intermediates from glycolysis and the pentose phosphate pathway are combined to form shikimate and subsequently chorismate. This compound can, via several steps, be converted into phenylalanine, tyrosine, tryptophan, and via isochorismate, into anthraquinones. Thus, the biosynthesis of anthraquinones

From: *Methods in Molecular Biology, Vol. 111: Plant Cell Culture Protocols*
Edited by: R. D. Hall © Humana Press Inc., Totowa, NJ

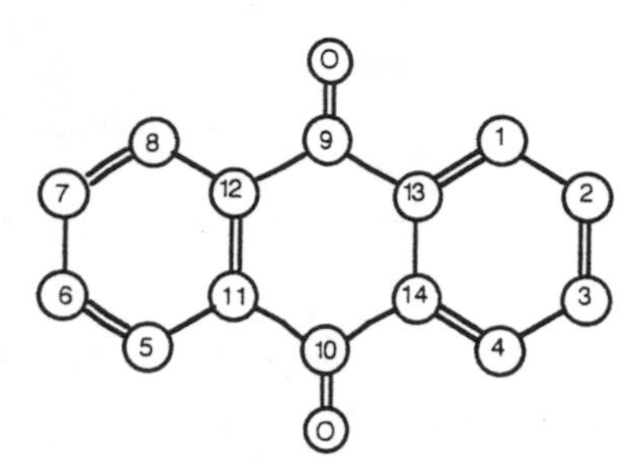

Fig. 1. Basic structure of anthraquinones found in the Rubiaceae. Substituents include methyl, hydroxyl, and methoxyl groups. Numbered atoms (1–14) are C-atoms. Os indicate the position of the 2-oxygen atoms.

shows interactions with protein synthesis, other secondary pathways (phenylpropanoid pathway via phenylalanine), and hormone synthesis (auxin synthesis via tryptophan).

We have studied the influence of several auxin analogs on anthraquinone production. For 2,4-dichlorophenoxy acetic acid (2,4-D) and 1-naphthalene acetic acid (NAA), we also have determined the effects on cell division. It appears that auxin analogs that are able to maintain cell division have a negative influence on AQ production ***(4)***. The auxin analog 2,4-D already stimulates cell division at a much lower concentration than NAA. In *Morinda* cells growing in the presence of 4.5 μ*M* NAA, the AQ production can be immediately blocked by the addition of 2,4-D (final concentration 4.5 μ*M*) ***(5)***. Therefore, *Morinda* cell suspensions can be used as a model system, in which AQ production can be easily switched on and off by changing the auxin concentrations.

Although there often is a negative correlation between cell division and anthraquinone accumulation (*see* **Fig. 2**), affecting specifically the cell division rate, this does not result in a change in AQ production. Possibly two distinct programs can be switched on under regulation of auxins: a program focused on fast cell division and one in which production prevails ***(4)***. These separate programs are also reflected in the cell composition. The slowly dividing cells show a lower respiration rate, a higher soluble sugar content, and contain less protein ***(5,6)***.

M. citrifolia cell suspensions can also be grown in fermenters, as so-called continuous cultures. After several weeks, a steady state is reached in which growth rate, cell composition and AQ content ***(4,7)*** are constant. Therefore, interactions between primary and secondary metabolism can be studied quantitatively over prolonged periods.

Another advantage of the *Morinda* system is that changes in AQ accumulation can easily be followed visually. A simple extraction and subsequent spec-

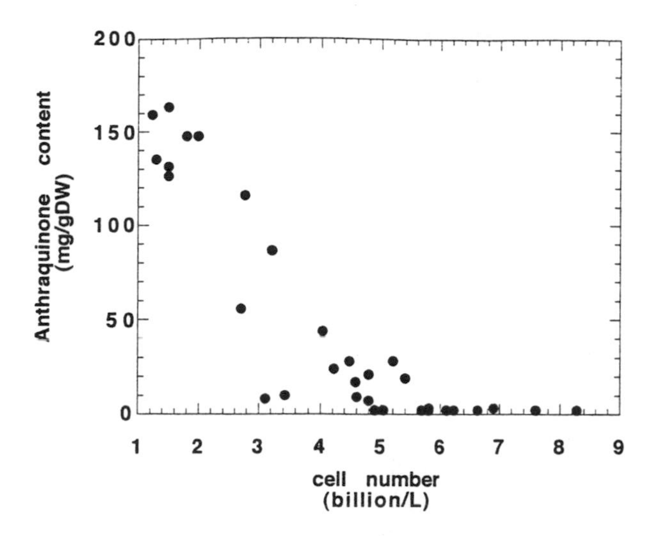

Fig. 2. Correlation between final cell number and AQ content in *M. citrifolia* cell suspensions at the end of a batch culture (14 d after inoculation). Data are used from several independent experiments. Each point is based on three AQ determinations and on 16 cell number determinations.

troscopic determination can give an indication of the AQ content. Methods are described for the growth and maintenance of batch and continuous *Morinda* cultures and for the measurement of the different specific anthraquinones (glycosides and/or aglucons).

2. Materials

2.1. Growth and Maintenance of Morinda Citrifolia Cultures

Batch cultures:

1. 250-mL Erlenmeyer flasks for the maintenance of the *M. citrifolia* L. cell line ***(1)***.
2. Growth medium: Gamborgs B5 medium (***8***, *see* Appendix), supplemented with sucrose (40 g/L), 0.2 mg/L kinetin (Sigma), and NAA or 2,4-D (Sigma) as auxin (analog), (*see also* **Note 1**).
3. Stock solutions of NAA and 2,4-D (1 g/L): dissolve in 1 *M* KOH (p.a.) and store at –20°C.

Continuous cultures

1. A standard 2-L bioreactor with round bottom vessel, stirred with an ADI 1012 motor controller (Applikon).
2. 0.2-µ*M* filters (Millipore) for aeration.
3. Gamborg's B5 medium supplemented with 5 or 10 g/L sucrose, 0.2 mg/L kinetin, and auxin (concentration and type depending on the experiment).

2.2. Measurement of Growth Parameters

1. A Fuchs-Rosenthal hemocytometer for cell counting.
2. Büchner funnel, filter papers (e.g., Schleicher and Schull).
3. Vacuum/water pump.
4. 60°C drying oven.
5. FDA solution: prepare a stock solution by dissolving FDA in acetone to give a concentration of 5 g/L. Dilute 100 times in growth medium before use. Make fresh each time.
6. Standard UV/visible microscope with FITC filter.

2.3. Anthraquinone Determinations

1. Ethanol (p.a.) was used.
2. Alizarin (BDH) as a standard (*see also* **Note 2**.)
3. Silica gel TLC (0.2 mm, Merck) plates. The mobile phase consists of isopropanol:water:ethyl ether:toluol (50:12.5:25:10; v/v/v/v).
4. Thymol-sulfuric acid is 0.5 g thymol in 95 mL ethanol, to which 5 mL 18 *N* sulfuric acid is carefully added.

3. Methods

3.1. Growth and Maintenance of M. Citrifolia Cultures

To start a new cell line, *Morinda* seeds are used.

1. Surface-sterilize seeds in a solution of 10 g/L Na hypochlorite and 50 mg/L Tween-20 for 10 min.
2. Rinse with sterile water (three times). Germinate the seeds in Petri dishes containing water-based agar.
3. When the first true leaves have appeared, transfer the obtained seedlings to liquid growth medium (*see* **Subheading 2.**) supplemented with 4.5 μ*M* 2,4-D.
4. Loose cells will detach from the seedlings within a few weeks. Use these cells as an inoculum, and initiate a new cell line.
5. Scaling up of such a newly formed line can be done by using the whole batch as inoculum for a new culture in a larger (6×) volume.

For culture maintenance, cells are grown in the presence of 4.5 μ*M* 2,4-D. Batch cultures should be subcultured every 14 d. Subculturing and/or the start of an experiment are as follows.

1. In a laminar flow cabinet, add 10 mL of a 14-d old suspension culture to 50 mL of fresh medium in a 250-mL Erlenmeyer flask.
2. Close the flasks with two layers of aluminum foil. After each handling, the foil should be replaced by a new, sterile one.
3. Shake the flasks on an orbital shaker (100 rpm), and culture at 25°C with a 16-h day period and an 8-h night period; irradiance 10–20 W/m^2 (Philips Fluorescent no. 84).
4. By repeating **steps 1–3** cultures can be maintained indefinitely.

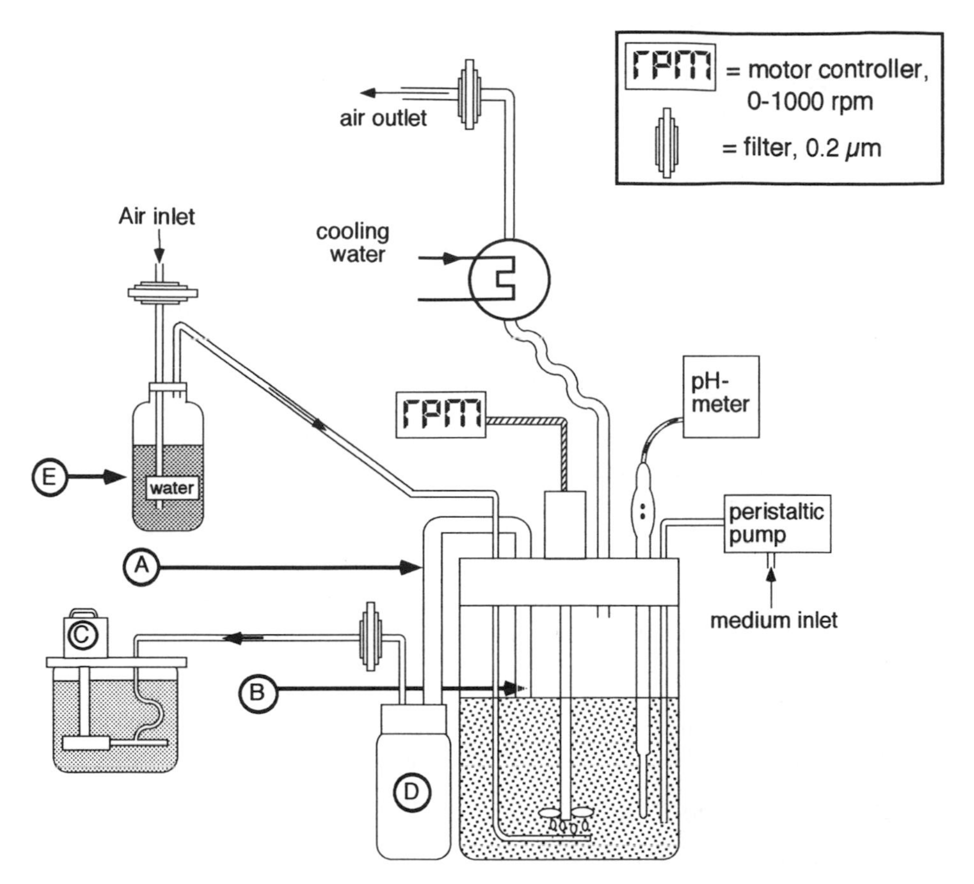

Fig. 3. Schematic drawing of the fermenter cultures. Fresh medium is pumped into the culture vessel at a constant rate. The overflow/sample is removed by a vacuum pump **(C)** through an enlarged tube **(A)** that is placed at a certain height to keep the culture volume constant **(B)**. The overflow is collected in a sample bottle **(D)**. The incoming air is moistened by leading it through a bottle filled with sterile water **(E)**.

3.1.1. Continuous Culture

Using the configuration as shown in **Fig. 3**, *Morinda* cells can be grown under sucrose or phosphate limitation. When cells are grown under phosphate limitation, 0.2 m*M* phosphate is used in combination with an otherwise standard Gamborg's B5 medium, supplemented with 10 g/L sucrose. For sucrose limitation, only 5 g/L sucrose is added to Gamborg's B5 medium. Air is continuously supplied at a rate of 20 L/h. The fermenter is stirred at 100 rpm. Cells are grown at a dilution rate of 0.1/d (i.e., 10% of the culture is replaced with fresh medium every day). Auxin type and concentration depend on the experiment.

Inoculation procedure:

1. Sterilize the entire apparatus in an autoclave for 15 min at 120°C and 1.4 bar.
2. In a laminar flow cabinet, add 300 mL cells from a 14-d old batch culture and 1.5-L fresh medium to a bottle that is subsequently connected to the sample port.
3. Pour the content of the bottle into the fermenter. The empty bottle can then be used for the first overflow/sample.
4. Connect the stirrer, the pH electrode, the thermostated water bath, and the medium pump, and the culture is started by the continuous addition of fresh medium.
5. The overflow is discontinuous; sample and overflow procedures are described in the legend to **Fig. 3** (*see also* **Note 3**).

3.1.2. Measurement of Growth Parameters

1. Cell number determination: dilute the cells in B5 medium (final number of cells per square should be between 10 and 30). Count two samples in a hemocytometer. The cells in 16 squares should easily be counted (*see also* **Note 4**).
2. Fresh weight and dry weight: collect a sample of the culture (up to 10 mL) with a sterile pipet (Greiner, wide-mouthed). Remove the medium using a paper filter on a Büchner funnel. Remove the cells from the paper filter to obtain the fresh weight (*see also* **Note 5**). To determine dry weight, cells are dried at 60°C for 24 h.
3. Viability of the cells: Mix 1 vol of cell suspension with 1 vol of 50 mg/L fluorescein diacetate solution. Using a light microscope, count the number of cells that show green fluorescence under UV illumination. Divide by the total number of cells (visible under white light). Multiply by 100 to give the percentage of viable cells.

3.2. Anthraquinone Determinations

3.2.1. Quantitative Estimation

Cells are harvested as described above. Cells are either directly used or stored at –20°C.

1. Add approx 0.05–0.2 g cells (fresh wt) to 5 mL 80% aqueous ethanol in a 15-mL centrifuge tube (*see* **Note 6**).
2. Heat the tubes in a water bath at 80°C for 10 min.
3. Collect the ethanolic extract by centrifuging the tubes at 1500 g for 5 min.
4. Add another 5 mL of 80% ethanol to the pellet, and repeat **steps 2** and **3**.
5. Pool the two supernatants, and make up to 10 mL with 80% ethanol.
6. Determine the absorption at 434 nm. When necessary, the extracts can be diluted to obtain an absorption between 0.1 and 1.0.
7. Calculate the AQ content by using a millimolar extinction coefficient of 5.5 and a mean mol-wt value of 580 (*see* **Notes 2** and **7**).

3.2.2. Qualitative Determination

To acquire a first impression of the anthraquinone composition, the extract described above can be analyzed using thin-layer chromatography (TLC) (*see also* **Note 9**). In this manner, the glycosides and the aglucons can be separated and the RF's compared with literature data (e.g., ***9***).

1. Apply a sample of the extract to the TLC plate, 1.5 cm from the bottom. The amount of extract that can be loaded onto a TLC plate depends on the concentration of the dissolved compounds. Spots should be clearly visible.
2. Develop the plate until the front of the mobile phase is 1 cm from the top of the plate in order to be able to determine the RFs (*see* **Note 8**).
3. Anthraquinone spots are readily visible. The glycosides can be identified after spraying with thymol-sulfuric acid and subsequent baking of the plate for 15–20 min at 120°C. In this way, the sugartails are colored dark pink.

3.3. Manipulation of Anthraquinone production

3.3.1. Upmanipulation by Auxin Deprivation

Growth in the presence of 2,4-D, down to a concentration of ca. 5×10^{-8} *M*, does not result in significant AQ accumulation ***(4)***.

1. Use cultures that have been grown in the presence of 5×10^{-8} *M* 2,4-D for several subcultures.
2. On d 0, transfer cells to auxin-free medium or to medium containing 4.5 μ*M* NAA.
3. Subculture every 14 d to auxin-free medium or to medium containing 4.5 μ*M* NAA.

Within a few days, AQ accumulation can be seen. When using NAA, the cells continue dividing, and the effects of AQ accumulation on cell growth can also be studied. After about four to five subcultures (ca. 60 d ***[6]***), the AQ content becomes toxic and the cells eventually die. In an auxin-free medium, *Morinda* cells produce AQ more rapidly, but they stop dividing and die within 2 wk.

3.3.2. Downmanipulation by Auxin Addition

1. Use cultures that have been grown in the presence of 4.5 μ*M* NAA. To study the kinetics of the inhibition of the AQ accumulation, cells should be used that contain ca. 50 mg AQ/g dry wt at the start of the batch culture.
2. Add 2,4-D (final concentration 4.5 μ*M*): the AQ accumulation is stopped within a day. When effects on growth are studied, 2,4-D can best be added in the early logarithmic growth phase (normally ca. 4–8 d after inoculation).
3. Effects on AQ accumulation can be measured after 1 d.

4. Notes

1. Culture medium can be prepared and stored in advance. Caution should be taken that some growth regulators are heat-labile (e.g., IAA) and that autoclaving at too high a temperature can lead to caramelization of the sucrose, the avoidance of which is especially important with medium for continuous cultures (prepared in larger quantities, 3 L).

2. The mixture of AQs extracted from *M. citrifolia* suspension cells shows a similar spectrum (400–600 nm) to the alizarin absorption spectrum. For quantitative determinations, a millimolar extinction coefficient of 5.5 should be used, also based on alizarin.
3. During continuous culture of plant cell suspensions, it is very important to have an efficient overflow and to be able to take representative samples. Because the dilution rate of plant cell continuous cultures is very slow, a continuous overflow would specifically remove only medium, because the upward movement of the culture through the overflow tube would take place at a lower velocity than the downward movement of the cells caused by gravity. Therefore, every few hours the excess of culture is removed at a high speed, with a vacuum pump (*see* **Fig. 3**). The same holds for sample taking. The diameter of the overflow/sample tube is rather large to prevent blockage by cell clumps.
4. Many plant cell suspension cultures contain clumps of cells, which makes counting the cells rather problematic. In these cell suspensions, the cell clumps have to be macerated in order to obtain a correct estimation of the cell number. In *M. citrifolia* suspensions, however, the cells are either loose or are attached to each other in chains. Both types of cells are easily counted. In order to avoid overestimation of the cell number, cells that cross the lines in a hemocytometer are only counted if they cross it on the upper or the left side of the square.
5. An exact estimation of the fresh weight content is hampered by two things. First, the layer of cells harvested on the paper filter will lose water. Therefore, one has to use a strict, reproducible time schedule. Second, in the course of growth of a batch culture, more cells are harvested and, therefore, the cell layer on the paper filter will become thicker; therefore, more intercellular water will be included in the FW determination at the later growth phases. One has to adjust the size of the sample to avoid this.
6. In 2,4-D cultures, 0.2 g FW is used because of the very low AQ content. So-called NAA cultures can produce so much AQ that the sample size must be reduced to 0.05 g FW or even less, in order to ensure that all AQs can be isolated from the cells. A good indication of a proper anthraquinone extraction is the color of the pellet after AQs have been extracted twice. The pellet should look whitish.
7. The mean molecular weight of the anthraquinones that occur in *M. citrifolia* is calculated, supposing that 95% occurs as glycosides and 5% as aglucons ***(1,9)***.
8. At least 30 min before the plates can be developed, the chromatography tank should be equilibrated. To do this, pour a 1-cm layer of mobile phase into the tank and close the lid.
9. To get more insight into the exact structure of the anthraquinones, other more elaborate methods have to be used such as HPLC, GCMS, and so forth ***(9)***.

References

1. Zenk, M. H., El Shagi, H., and Schulte, U. (1975) Anthraquinone production by cell suspension cultures of *Morinda citrifolia. Planta Med.* Suppl. 79–101.
2. Morton, J. F. (1992) The ocean-going Noni or Indian Mulberry (*Morinda citrifolia*, Rubiaceae) and some of its "colorful" relatives. *Econ. Bot.* **46,** 241–256.

3. Inoue, K., Nayeshiro, H., Inouye, H., and Zenk, M. H. (1981) Anthraquinones in cell suspension cultures of *Morinda citrifolia*. *Phytochemistry* **20,** 1693–1700.
4. Hagendoorn, M. J. M., Jamar, D. C. L., Meykamp, B., and Van der Plas, L. H. W. (1996) Cell division versus secondary metabolite production in *Morinda citrifolia* cell suspensions. *J. Plant Physiol.* **150,** 325–330.
5. Hagendoorn, M. J. M., van der Plas, L. H. W., and Segers G. J. (1994) Accumulation of anthraquinones in *Morinda citrifolia* cell suspensions. A model system for the study of the interaction between secondary and primary metabolism. *Plant Cell Tiss. Org. Cult.* **38,** 227–234.
6. Van der Plas, L. H. W., Eijkelboom, C., and Hagendoorn, M. J. M. (1995) Relation between primary and secondary metabolism in plant cell suspensions: competition between secondary metabolite production and growth in a model system (*Morinda citrifolia*). *Plant Cell Tiss. Organ. Cult.* **43,** 111–116.
7. De Gucht, L. P. E. and Van der Plas, L. H. W. (1995) Determination of experimental values for growth and maintenance parameters. *Biotechnol. Bioeng.* **47,** 42–52.
8. Gamborg, O. L., Miller, R. A., and Ojima, V. (1968) Nutrient requirements of suspension cultures of soybean root cells. *Exp. Cell. Res.* **50,** 151–158.
9. Wijnsma, R. and Verpoorte, R. (1986) Anthraquinones in the Rubiaceae. *Progress. Chem. Org. Nat. Prod.* **49,** 79–149.

36

Alkaloid Accumulation in *Catharanthus roseus* Suspension Cultures

Alan H. Scragg

1. Introduction

Plants have traditionally been the source of pharmaceuticals and fine chemicals, and at this time, plant material is still being screened for new compounds *(1)*. Initially, plant cell and tissue culture was developed in order to study the biochemistry and physiology of plants without the complication of the whole plant. However, plant cell or tissue cultures were also found to accumulate many of the compounds characteristic of the original plant *(2)*. Many plants are the source of plant-derived products, which include drugs, fragrances, colors, oils, and pesticides *(3)*. Most of these valuable plant products are defined as secondary metabolites, which are often complex in structure, the end product of long biochemical pathways, not essential for growth, but do often confer some advantage to the plant, such as reduction of predation *(3)*. A number of these secondary products are very valuable, such as taxol *(4)*, and plant cell culture may provide an additional source of supply if:

- There is a high demand for the product.
- The product is affected by disease.
- Supply is affected by the climate.
- Only low levels accumulate in plants.

It may also provide an alternative supply if:

- The plant is rare.
- The plant species is protected.
- No chemical synthesis is available.

From: *Methods in Molecular Biology, Vol. 111: Plant Cell Culture Protocols*
Edited by: R. D. Hall © Humana Press Inc., Totowa, NJ

These reasons, and in particular, the high values of some secondary products ***(2)*** have been the driving force in the attempts at producing them using plant cell cultures. Although high levels of some secondary products have been obtained, those of commercial value have remained stubbornly low. Therefore, numerous approaches have been taken in order to increase the yield of secondary products in plant cell cultures. These were:

- Selection of high-yielding lines.
- Changes in medium composition: carbon source quantity and quality, nitrogen, phosphate, growth regulators.
- Changes in culture conditions: pH, temperature, light.
- Cultivation strategies: immobilization, organ culture, hairy roots, two-stage cultivation.
- Specialized techniques: elicitation, product removal.

Much of this research has been carried out using cultures of *Catharanthus roseus* (L) G.Don, the Madagascan periwinkle. This plant has been extensively studied, since it contains the important antileukemics vincristine and vinblastine and the antihypertensives ajmalicine and serpentine ***(5,6)***.

Despite considerable efforts, the two dimeric alkaloids vincristine and vinblastine have not been detected in *C. roseus* cultures, even though a number of other alkaloids have been detected ***(6)***.

Although none of the approaches described above have developed a stable, high-yielding culture of *C. roseus,* the use of a two-stage process has proven to be the most successful. Secondary products are often accumulated after growth has ceased and a change in cultural conditions, which slows or stops growth, will often encourage the accumulation of secondary products. This led to the development of two-stage processes for the accumulation of ajmalicine and serpentine in *C. roseus* cultures ***(7,8)***. In the first stage, growth is encouraged, and in the second stage, secondary product accumulation is stimulated. In one *C. roseus* system, the change to the production medium was the replacement of the growth regulator 2,4-D by NAA ***(7,9)*** (*see* **Note 1**). An alternative two-stage system that has been used by other groups is based on L&S medium ***(10)***, and the production medium lacks nitrate, ammonium, and phosphate, combined with an increase in glucose from 20–80 g/L (*see* **Note 2**) ***(8,11)***. However, an alternative single-stage *C. roseus* system has been developed where the growth and production media have been combined and secondary product and growth are partially linked ***(12,13)***. All three *C. roseus* systems are capable of alkaloid accumulation despite their differences in cultural conditions, such as light/dark, use of sucrose or glucose, and growth regulators. Thus, it would appear that there are no specific conditions for secondary product accumulation, and these vary with the individual cell lines developed. **Table 1**

Table 1
Growth and Alkaloid Accumulation by *C. roseus* Suspension Cultures[a]

Medium	Growth regulators mg/L	Carbon source g/L	Growth rate μ/d⁻¹	Alkaloid accumulation mg/g dry wt: Serpentine	Alkaloid accumulation mg/g dry wt: Ajmalicine
B5	1.0 2,4-D 0.1 kinetin	20	0.21	0.1	0
Z[b]	0.17 IAA 1.13 BAP	50	0.12	3.0	1.2
8%[c]	Zero	80	0.01	1.2	0
Growth[e] (L&S)	2.0 NAA 0.2 kinetin	30[d]	—	0	0
Production	—	80[d]	—	0	12
APM[g]	0.17 IAA	20	0.25	0.023	0.001
APM	0.17 IAA	50	0.18	0.68	0.34
M3[h]	1.0 NAA	20	0.18	5.6	1.5

[a]Using production media in single and two-stage processes.
[b]Production medium based on L&S *(7)*.
[c]The medium contains only water and sucrose ***(8)***.
[d]The carbon source was glucose, whereas all other media used sucrose ***(11)***.
[e]Growth and production medium bases on L&S ***(11)***.
[f]L&S medium minus nitrate, ammonium, and phosphate ***(11)***.
[g]Based on B5 medium, single-stage culture ***(9)***.
[h]Based on M&S medium, single-stage culture ***(12)***.

gives details of the conditions encouraging alkaloid accumulation in *C. roseus* using both one- and two-stage processes. The last system (M3) is described in this chapter along with the extraction, separation, and identification of the alkaloids ajmalicine and serpentine.

2. Materials

2.1. Suspension Cultures and Growth Analysis

1. Suspension cultures of *C. roseus* can be developed from explants via callus using standard techniques ***(5,6)***. We use a selected line *C. roseus* ID1 ***(14)***.
2. Maintenance medium M3 is based on Murashige and Skoog salts (***15***, and *see* Appendix) (with the omission of KH_2PO_4) and contains 20 g/L sucrose, 0.1 mg/L kinetin, and 1.0 mg/L naphthaleneacetic acid (NAA).
3. 250-mL conical flasks containing 100 mL M3 medium. The flasks are closed with a double layer of aluminum foil.
4. Orbital shaker operating at 150 rpm.
5. A culture room at 25°C, and constant light at 50 μmol/m²/s.

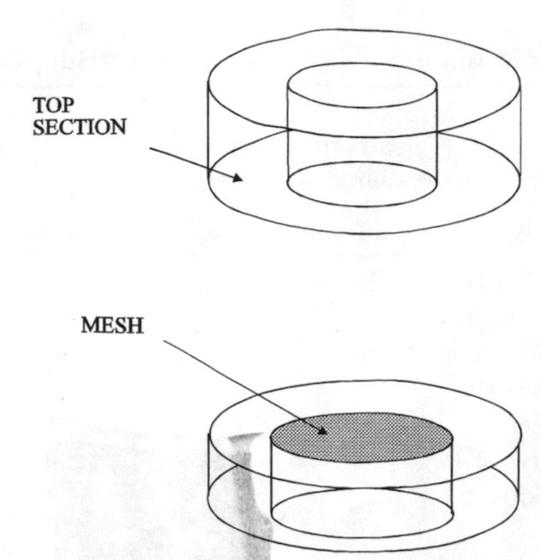

Fig. 1. A stainless-steel filter unit used for wet and dry weight determination. The filter is supported by the mesh on the lower half, and the weight of the unit holds the filter in place and the filter on the Büchner flask.

6. Whatman no. 1 filter paper disks (2.5-cm).
7. Oven at 60°C.
8. Stainless-steel filter bed (*see* **Fig. 1**).
9. 250-mL Büchner flask.

2.2. Alkaloid Extraction and Analysis

1. TLC plates: 20 × 20 cm, 0.25 mm silica 60_{f254} fluorescence (E. Merck, Darmstadt); activate by heating to 100°C for 1 h.
2. 9-cm Hartley funnel.
3. Miracloth (Calbiochem-Behring, La Jolla, CA).
4. 10 × 50 mm Whatman cellulose thimbles.
5. SEP-PAK C18 cartridges (Waters Associates).
6. Rotary evaporator (Rotavapor R110, BuchiLabs).
7. Sonicator bath (Decon Frequency Sweep bath, Decon Ultrasonic Ltd., Hove, UK).
8. Iodoplatinate stain: 5% w/v platinum chloride (10 mL), conc. HCl (5 mL), 2% w/v KI (240 mL).
9. Ceric ammonium sulfate stain: ceric ammonium sulfate (1 g) in 100 mL of 85% *o*-phosphoric acid.
10. 0.8 × 10 cm μm Bondpak C18 HPLC column.
11. HPLC-grade methanol.
12. *N*-heptane sulfonate.
13. Acetonitrile.
14. TLC tank suitable for 20 × 20 cm plates.

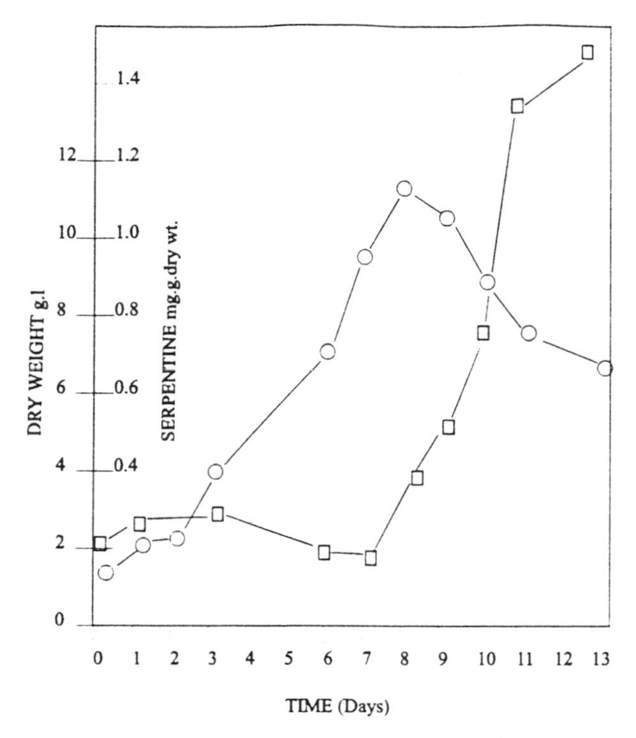

Fig. 2. The accumulation of serpentine in a single-stage system using *C. roseus* ID1 grown in M3 medium. (○) Dry weight g/L; (□) serpentine mg/g dry wt.

3. Methods

3.1. Growth and Growth Analysis

An example of the growth and alkaloid accumulation by a culture of *C. roseus* ID1 cultured in M3 medium is shown in **Fig. 2**. The accumulation of serpentine under these conditions can be increased by raising the concentration of sucrose in the medium (*see* **Table 2**). To avoid osmotically stressing the cells by adding high concentrations of sucrose at the start of the culture, the sucrose can be applied in a number of small aliquots, in a fed-batch process.

1. Subculture the suspension every 2 wk by adding 20 mL of culture to 100 mL of fresh medium in a 250 mL flask.
2. To measure wet and dry weight, place a supply of filter paper disks in the oven for at least 24 h before use.
3. Weigh the dried filter disk after removal from the oven using 1/sample and 3 samples/ measurement.
4. Place the filter paper on the filter unit seated on a bored rubber stopper on a Büchner flask under vacuum. The filter used is a stainless-steel unit as

Table 2
Improvements in the Alkaloid Accumulation in the One-Stage Process Using ID1 on M3 Medium

Sucrose %	Growth rate	Biomass g/L d^{-1}	Serpentine mg/g dry wt	Production mg/L/d
2	0.46	8.3	2.55	1.32
6	0.33	23.9	9.65	8.24
6 (fed batch)[a]	0.36	26.6	5.07	5.62

[a]Sucrose added in 3 × 2% batches during the growth of the culture.

shown in **Fig. 1**. Other filter units such as those supplied by Millepore can also be used.

5. Wet the filter with about 3 mL distilled water and retain under vacuum for a fixed time period (e.g., 10 s). Reweigh immediately to avoid loss owing to evaporation.
6. Replace the filter in the filter unit, and filter a known volume of culture (e.g., 3–5 mL). The culture should be well mixed before sampling to ensure a representative sample (*see* **Note 3**). When the cells appear "dry," continue the vacuum for the same fixed time period as above and weigh immediately.
7. Place the filter in a labeled Petri dish, and dry for at least 24 h in the oven before weighing the dry filter and cells.
8. The wet and dry weights of the cells are calculated as followed:

$$\text{Wet wt (g/L)} = [\text{wt of wet disk + cells (g)} - \text{wt of wet disk (g)}/\text{sample volume (mL)}] \times 1000 \quad (1)$$

$$\text{Dry wt (g/L)} = [\text{wt of dry disk + cells (g)} - \text{wt of dry disk (g)}/\text{sample vol (mL)}] \times 1000 \quad (2)$$

3.2. Alkaloid Extraction and Analysis

1. Samples of between 50 and 100 mg dry weight are required for alkaloid extraction. Thus, the sample will need to be of 0.5–1.0 g wet wt. For a fully grown culture, this will represent a volume of 5–20 mL culture (wet weights are 100–200 g/L) (*see* **Note 4**).
2. The samples are filtered using a 9-cm Hartley funnel fitted with a Miracloth filter. The cells are washed with 100 mL distilled water, placed in a small plastic bag, and rapidly frozen using liquid nitrogen (*see* **Note 5**). The frozen cells can be stored at –20°C, but are normally freeze-dried. Freeze-drying appears to disrupt the cell membrane and allows extraction of the alkaloids without using any other form of cell disruption (*see* **Note 6**).
3. Freeze-dried cells (0.05–0.1) are placed in the extraction thimble and plugged with nonadsorbent cotton wool. The actual weight to be extracted can be determined by weighing the thimble before and after adding the cells.

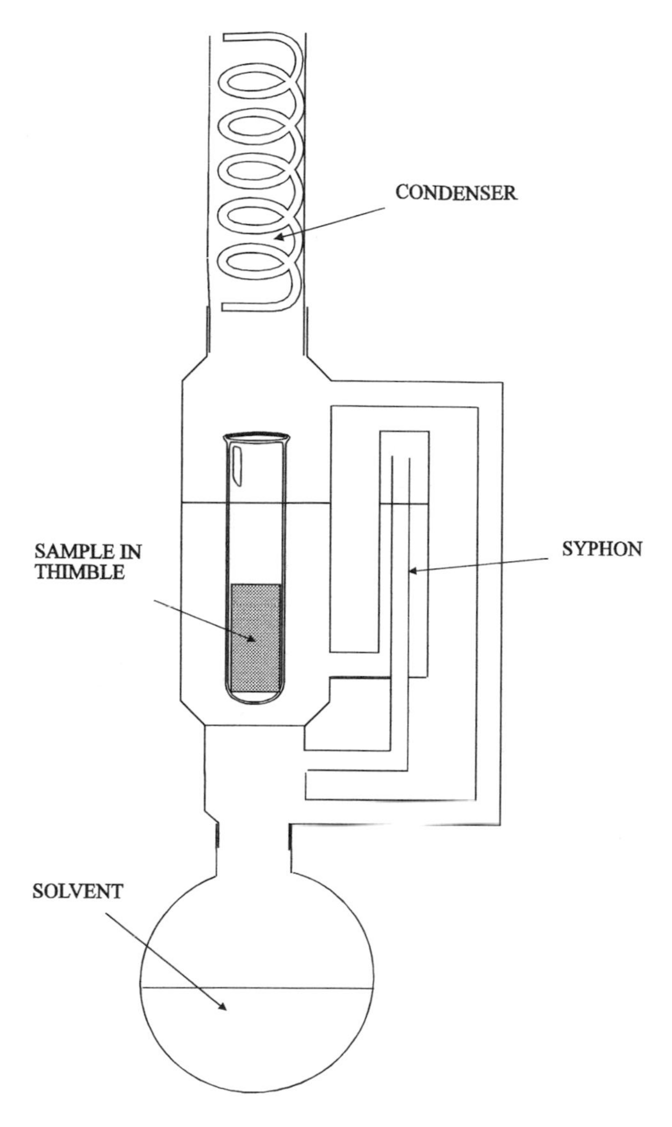

Fig. 3. A Soxhlet apparatus used to extract alkaloids from dried cells. The flask at the bottom contains 100 mL methanol, which is immersed in a water bath at 85°C. The methanol that boils off is condensed at the top, and the warm methanol bathes the sample in the thimble. Once the thimble unit is filled, the syphon drains the chamber and the process is repeated.

4. The thimble is placed in a small Soxhlet apparatus (**Fig. 3**) and refluxed with 100 mL methanol for 2 h at 85°C. The extract (in the round-bottomed flask) is reduced to dryness by rotary evaporation at 60°C under vacuum.

5. The residue is resuspended in 1 mL methanol using a 1-min exposure in a sonicator bath to help resuspension. Two further 1-mL methanol washes are used, and these are combined and stored in a sealed bottle at 4°C.
6. Alkaloid accumulation in the medium can be estimated by collecting 100 mL medium after filtration and extracting this with 50 mL chloroform. A more convenient method involves solid–liquid separation using a SEP-PAK C18 cartridge. The cartridges are preactivated with 4 mL of acetonitrile, followed by 4 mL of 95% methanol. The medium sample is centrifuged (10,000*g* for 10 min) or filtered to remove any cell debris, and 20–100 mL applied to the column. The alkaloids are eluted with 4 mL methanol and can be then identified by HPLC, and so forth.
7. For identification of the alkaloids in a routine system, HPLC is preferred, and if information on all alkaloids is required, UV detection at two wavelengths will give indications of peak purity. A diode array detection system will often resolve any purity problems.
8. For the identification of ajmalicine and serpentine, the following HPLC system can be used. The column is a 0.8 × 10 cm μm Bondpak C18 reverse phase, run isocratically at 55:45:5 methanol:water:*n*-heptane sulfonate (nHS) for 14 min followed by linear gradient to 70:30:5 methanol:water:nHS. The sample volume was 10–30 μL, with a flow rate of 1.5–2.0 mL/min, and the wavelengths of 254 and 280 nm are used for detection. The standard should be run every 6–10 runs, and the retention times updated, since these will vary owing to changes in run conditions, such as temperature.
9. Although diode array detection can give details of peak purity, a second method should be used to confirm the identity of a peak. This has been achieved by TLC ***(16,17)***, MS, or NMR ***(12)***. There are a number of methods that can be used for the analysis and detection of alkaloids by TLC, and the following are suitable for serpentine and ajmalicine.
10. Solvent system 1: chloroform, acetone, diethylamine 5:4:1
11. Solvent system 2: ethylacetate, absolute alcohol, NH_4OH 3:1:1
12. The plates are run in these solvents in normal TLC tanks for 45–50 min and dried at 60°C. The dried plates can be examined using the following techniques.
13. Under UV light at 254 nm, the alkaloids ajmalicine and serpentine fluoresce so that their R_f values can be compared with standards.
14. At a wavelength of 366 nm, the alkaloids will quench and thus form dark spots.
15. When sprayed with iodoplatinate, serpentine will give a purple color and the other alkaloids a red-purple or pink color.
16. When sprayed with ceric ammonium sulfate (CAS), ajmalicine gives a gray color and serpentine has no reaction.

4. Notes

1. The cell line *C. roseus* MCR17 ***(9)*** was maintained on Gamborgs B5 medium ***(18)*** containing 1.0 mg/L 2,4-D and 20 g/L sucrose. The pH was adjusted to 5.5 before sterilization. The cultures were subcultured every week by adding 20 mL

culture to 180 mL fresh medium in a 500-mL flask. The culture was incubated at 28°C under constant illumination and shaken at 150 rpm.

2. The cell line of *C. roseus* B1X was initiated from seed ***(11)*** and was maintained on L&S medium ***(10)***, containing 2.0 mg/L NAA, 0.2 mg/L kinetin, and 30 g/L glucose. The pH was adjusted to 5.8 prior to autoclaving. The cultures were subcultured every 2 wk by adding 35–165 mL fresh medium in a 1000-mL flask. The cultures were incubated at 25°C in the dark and shaken at 100 rpm.
3. Plant cell suspensions are generally aggregated, and thus settle rapidly and clog narrow pipet tips. A convenient method of avoiding problems with these aggregated cultures is to use a 5-mL automatic pipet fitted with tip, which has had the end 5 mm removed to give a very wide bore. These tips can be sterilized by autoclaving in beakers sealed with aluminum foil.
4. One to 3 samples can be removed from a single 120-mL culture in a 250-mL shake flask before the reduction in volume significantly alters conditions in the flask. However, if alkaloid accumulation is to be followed early in the culture when the cell density is low and duplicate samples are needed for accuracy, large-volume samples are needed (50–100 mL). Therefore, a large number of flasks will need to be initiated from a single inoculum, and the contents of a whole flask used in the early stages. The number of flasks used will depend on the frequency of sampling and the rate of growth of the culture.
5. The use of small plastic bags allows a large number of samples to be freeze-dried at one time using a large metal container and avoids the use of glass flasks. It is difficult to extract all the sample from a glass flask.
6. Most of the alkaloids can be exacted from plant material with organic solvents at alkaline pH. Thus, dichloromethane ***(19)***, ethyl acetate ***(20)***, and methanol have been used to extract alkaloids. The crude extract can be cleaned up and concentrated by using solid-phase extraction. In this case, the extract is passed through a cartridge of C18 reversed-phase ion-exchange groups as described for the analysis of the medium.

References

1. Cox, P. A. and Balick, M. J. (1994) The ethnobotanical approach to drug discovery. *Sci. Am.* **June,** 60–65.
2. Verpoorte, R., Van der Heijden, R., and Schripesma, J. (1993) Plant cell technology for the production of alkaloids: present status and prospects. *J. Nat. Prod.* **56,** 186–207.
3. Balandrin, M. F. and Klocke, J. A. (1988) Medicinal, aromatic and industrial materials from plants, in *Biotechnology in Agriculture and Forestry,* 4, *Medicinal and Aromatic Plants I* (Bajaj, Y. P. S., ed.), Springer-Verlag, Berlin, pp. 3–36.
4. Zhong, J.-J. (1995) Recent advances in cell cultures of Taxus spp. for production of the natural anticancer drug taxol. *Plant Tissue Biotechnol.* **1,** 75–80.
5. Van der Hijden, R., Verpoorte, R., and Ten Hoopen, H. J. G. (1989) Cell and tissue cultures of *Catharanthus roseus* (L) G.Don: a literature survey. *Plant Cell Tiss. Organ. Cult.* **18,** 231–280.

6. Moreno, P. R. H., Van der Hijden, R., and Verpoorte, R. (1995) Cell and tissue culture of *Catharanthus roseus*: a literature survey II updating from 1988 to 1993. *Plant Cell Tiss. Organ. Cult.* **42,** 1–25.
7. Zenk, M. H., El-Shagi, H., Arens, H., Stockigt, J., Weiler, E. W., and Deus, B. (1977) Formation of the indole alkaloids serpentine and ajmalicine in cell suspension cultures of Catharanthus roseus, in *Plant Tissue Cultures and Its Biotechnological Application* (Barz, W., Reinhard, E., and Zenk, M. H., eds.), Springer Verlag, Berlin, pp. 27–44.
8. Knobloch, K. H., Hansen, B., and Berlin, J. (1981) Medium-induced formation of indole alkaloids and concomitant changes of interrelated enzyme activities in cell suspension cultures of *Catharanthus roseus. Z. Naturforsch.* **36c,** 40–43.
9. Tom, R., Jardin, B., Chavarie, C., and Archambault, J. (1991) Effect of culture process on alkaloid production by *Catharanthus roseus. J. Biotechnol.* **21,** 1–20.
10. Linsmaier, E. M. and Skoog, F. (1965) Organic growth factor requirements of tobacco tissue cultures. *Physiol. Plant.* **18,** 100–127.
11. Moreno, P. R. H., Schlatmann, J. E., Van der Heijden, R., Van Gulik, W. M., Ten Hoopen, H. J. G., Verpoorte, R., et al. (1993) Induction of ajmalicine formation and related enzyme activies in *Catharanthus roseus* cells: effect of incoulum density. *Appl. Microbiol. Biotechnol.* **39,** 42–47.
12. Morris, P. (1985) Regulation of product synthesis in cell cultures of *Catharanthus roseus.* II Comparison of production media. *Planta Medica* **52,** 121–126.
13. Scragg, A. H. (1993) The problems associated with high biomass levels in plant cell suspensions. *Plant Cell Tiss. Org. Cult.* **43,** 163–170.
14. Cresswell, R. C. (1986) Selection studies on *Catharanthus roseus,* in *Secondary Metabolism in Plant Cell Cultures* (Morris, P., Scragg, A. H., Stafford, A., and Fowler, M. W., eds.), Cambridge University Press, Cambridge, pp. 231–236.
15. Murashige, T. and Skoog, F. (1962) A revised medium for rapid growth and bioassays with tobacco tissue cultures. *Physiol. Plant* **15,** 473–497.
16. Baerheim Svendsen, A. and Verpoorte, R. (1983) Chromatography of alkaloids, part A: thin layer chromataography. J. Chromatogr. Library **23a,** Elsevier, Amsterdam.
17. Verpoorte, R. and Baerheim Svendsen, A. (1984) Chromatography of alkaloids, part B: gas-liquid chromatagraphy and high-performance liquid chromatography. *J. Chromatogr.* Library **23b,** Elsevier, Amsterdam.
18. Gamborg, O. L., Miller, R. A., and Ojima, K. (1968) Nutritional requirements of suspension cultures of soybean root cells. *Exp. Cell Res.* **50,** 151–158.
19. Dos Santos, R. I., Scripsema, J., and Verpoorte, R. (1994) Ajmalicine metabolism in *Catharanthus roseus* cell cultures. *Phytochemistry* **35,** 677–681.
20. Facchini, P. J. and DiCosmo, F. (1991) Secondary metabolite biosynthesis in cultured cells of *Catharanthus roseus* (L) G.Don immobilized by adhesion to glass fibres. *Appl. Microbiol. Biotechnol.* **35,** 382–392.

37

Betalains

Their Accumulation and Release In Vitro

Christopher S. Hunter and Nigel J. Kilby

1. Introduction

Since the mid-1980s the diverse group of compounds known as plant secondary metabolites, or secondary products, have assumed even greater significance than previously as compounds of commercial (e.g., taxol, rosmarinic acid) and social (e.g., "natural" fragrances and flavors) significance. The commercial production of these secondary metabolites remained field-based until the production of shikonin from plant cells in vitro *(1)*. With few exceptions, the main world supply for most plant secondary metabolites remains the harvesting of field-grown plants followed by the extraction and/or direct utilization of the active principle(s). The proportion of the plant that constitutes the compound of interest is invariably only a few percent of the total plant dry weight. For example, the maximum quinine content of *Cinchona* bark is 6–14%, and the bark is only a small portion of the entire tree *(2)*. Some plants producing secondary metabolites are annuals or biennials, but many require long-term field-growth (*Cinchona,* 7–16 yr, *Lithospermum erythrorhizon,* 5–7 yr) before economic harvest. Field-grown crops are also subject to the vagaries of weather, pests, diseases, and nutrient availability. These, with other biotic and abiotic factors, continue to stimulate evaluation of alternative production systems for these compounds.

Recent developments in gene manipulation have given rise to transgenic organisms able to synthesize secondary products characteristic of the transgenic parent *(3)*. Although ethical and safety concerns dominate the statutory licensing and commercial development of such transgenic products, it is unlikely that their production in vivo will be superseded. Whether or not in vitro

From: *Methods in Molecular Biology, Vol. 111: Plant Cell Culture Protocols*
Edited by: R. D. Hall © Humana Press Inc., Totowa, NJ

production from nontransformed cell lines will be subject to comparable legislation remains unresolved, but is potentially less controversial.

The in vitro production systems investigated include *Agrobacterium*-transformed "hairy roots" ***(4)***, the growth of free cells and aggregates ***(5,6)*** suspended in stirred-tank chemostats and turbidostats, in air-lift bioreactors, in various immobilized systems, and as calluses, both in fluidized-bed bioreactors ***(7)*** and on agar-based media in flasks. To enhance product yield, media have often been especially developed. Depending on whether or not the product yield is substantially proportional to the biomass or whether the product is produced principally during stationary phase (that is, the period after which culture growth has essentially stopped), media may be designed to maximize cell growth or reduce potential cell growth, but enhance product yield. For shikonin production in Japan (Mitsui Petroleum Co.), cells were grown is stirred tanks. First, growth of mostly colorless cells was maximized for 9 d in medium MG-5. Then, cells were transferred to a medium (M9) for a further 14 d during which the red napthoquinone compound, shikonin, was produced.

The harvest of both shikonin, taxol/taxane and most other products of large-scale culture of plant cells depends on the destructive extraction of the biomass. With a cell doubling time ranging between 16 h and several days, the production of cells and secondary products is an expensive process ***(8)***. To produce secondary metabolites in a commercially sound way, it has been variously estimated that the compound of interest must have an end-user value >(US)$7500/kg. This figure is justified on the basis of the provision of aseptic culture conditions for several weeks, the (generally) low percentage yield of the secondary metabolite, and the costs of laboratory "overheads" and downstream processing. The long-term applicability of production in vitro thus depends on either product demand that can sustain the high price of the product on the market or the reduction of costs in production. Production costs can be reduced by a range of options, for example, increasing yields through the addition of elicitors ***(9)***, selecting higher-yielding clones, continuing to develop media that stimulate metabolite production ***(10)***, or permeabilizing the cells ***(11)*** to enable harvest of the secondary metabolite in a nondestructive way, thereby allowing the cells to produce second and subsequent metabolite crops ***(12,13)***.

A range of experimental techniques has been applied to cells to release the metabolites of interest. Permeabilization has three critical requirements:

1. Maintenance of asepsis.
2. Survival of a sufficient cell population to grow and produce subsequent crops.
3. Reversibility. Permeabilization using dimethyl sulfoxide, glycerol, media of low pH value, and other treatments has been of limited value only, principally because there has been either limited release of the metabolite, or cell death/lysis has occurred.

Permeabilization by either ultrasonication or electropermeabilization offers a robust system that satisfies the three critical requirements. Both techniques are amenable to small-scale operation in the laboratory, and ultrasonication has potential for use in pilot or large-scale production. This chapter provides details of a laboratory ultrasonic technique.

The ultrasonic process subjects cells to sonication for relatively short time periods; this ultrasonication is followed by short-term incubation of the cells, during which there is the gradual release of a proportion of the vacuole-located secondary metabolite into the medium during approx 30 min postsonication.

1.1. Red Beetroot: A Model System

Although of low financial value, the red pigment from *Beta vulgaris* (red beetroot) is a commercially important food colorant (E162) used in such products as salami and soups. Betalain-producing *B. vulgaris* has been used as a "model" system for studies on the production of secondary products in vitro. Its particular merit lies in the obvious color and the consequent simplicity of colorimetric quantification, in comparison with colorless compounds that are usually assayed by gas chromatography or HPLC. The principal pigments in beetroot are two red betacyanins (betanin and prebetanin) and two yellow betaxanthins (vulgaxanthin I and II) ***(14,15)***. In most of the recent studies on pigment production by *B. vulgaris* in vitro, attention has focused on cultures selected for their ability to produce the dominant red pigment betanin ***(16)***. The beetroot system has been used in hairy root ***(17)***, immobilization ***(18)***, and permeabilization ***(19)*** investigations.

The following protocol enables:

- Axenic beetroot cultures to be initiated.
- Callus to be grown on agar-based media.
- Suspension cultures to be initiated and subcultured.
- Pigment to be extracted using both destructive (whole-cell disruption) and non-destructive (ultrasonication) methods.

2. Materials

1. *B. vulgaris* cv. Boltardy: Seed is widely available from seed merchants or direct from Suttons Seeds plc, Torquay, Devon, UK. Surface-sterilants: 70% (v/v) ethanol, sodium hypochlorite (1 or 2% available chlorine), sterile distilled water.
2. Tissue-culture media: All media are based on that of Gamborg's B5 (***20***; and *see* Appendix). The basic medium without plant growth regulators, sucrose or agar can be purchased from Sigma, Poole, UK.
3. Suspension culture (S1) medium: 3.19 g Gamborg's B5 medium, 20 g sucrose, 0.1 mg kinetin (6-furfurylaminopurine), 0.02 mg 2,4-dichlorophenoxyactetic acid (2,4-D) (both hormones are taken from 1 mg/mL stocks dissolved in dilute KOH).

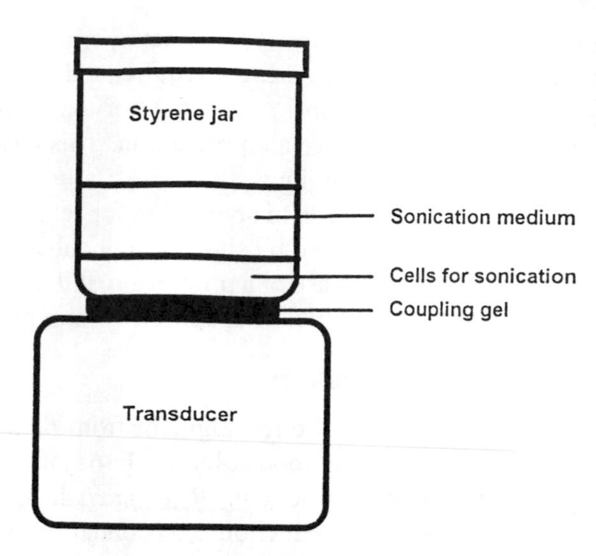

Fig. 1. Ultrasound transducer attached to the styrene jar prior to sonication.

Medium is made up to 1 L with single distilled water and then the pH adjusted to 5.5 prior to autoclaving at 106 kPa at 121°C for 15 min.

4. Agar-based medium (A2) for callus growth: as for S1, except that 6 g/L agar are added prior to pH adjustment.
5. Callus induction medium (A1): as for A2, except that the 2,4-D concentration is 1 mg/L.
6. Seed germination medium (A3): as for A2, except that it contains neither kinetin nor 2,4-D.
7. Erlenmeyer flasks (100 mL and/or 250 mL), with the necks closed by a double layer of aluminum foil and oven-sterilized (dry) at 160°C for 90 min, are used for seed germination, callus initiation, growth, and suspension cultures. Sterile styrene screw-capped jars (60 mL) (Sterilin, UK, type 125AP) are used to contain the cells and medium during the sonication process.
8. The ultrasound is produced from an Impulsaphon ultrasound therapy unit, type M55 with a 3.5-cm plane disk ceramic transducer (Ultrashall-Geräteban, Kronberg, Germany). The power output of the transducer is 3 W/cm^2. The frequency of the ultrasound is 1.02 MHz. The styrene jar is connected to the transducer by a few drops of a sound-conducting coupling gel (Electromedical Supplies, Wantage, UK) (*see* **Fig. 1**).
9. General equipment: wide-bore pipets, glass microfiber filter paper (type GF/C, Whatman, Maidstone, UK), spectrophotometer.
10. Culture incubation: all cultures are grown at 25°C in 16-h daylength in white fluorescent light (30 Wm^{-2}, but the intensity and spectral conditions are not critically important). Suspension cultures are incubated on an orbital shaker at 100 rpm.

3. Methods

3.1. Callus Initiation

Callus may be initiated from a range of sources—from seed, from petioles, or from flowering stems.

3.1.1. From Seed

1. Soak seeds in ethanol for 30 s. Then transfer them to a sterile flask containing an aqueous solution of sodium hypochlorite (2% available chlorine) plus 1–2 drops of Tween 20 (*see* **Note 1**).
2. Wash the seeds (4×) in sterile distilled water (SDW), and transfer them aseptically to the surface of A3 medium; use 5 seeds/250 mL flask (*see* **Note 2**).
3. When the seeds have germinated and seedlings grown to produce the second true leaf, aseptically remove individual seedlings, dissect petioles and hypocotyls into sections approx 15 mm long, and implant onto the surface of A1 medium.
4. Within 21 d, callus will have initiated and should grow into structures several millimeters in diameter. Excise these calluses, and transfer them to A2 medium.

3.1.2. From the Flowering Stems and Petioles of Mature Plants

1. Remove the flowering stem and/or the petioles, trim off any leaf lamina or flowers, and then surface-sterilize them in sodium hypochlorite (1% available chlorine + 1 drop Tween 20) for 20 min.
2. Wash the stems (4×) in SDW, and then cut aseptically into 20-mm lengths.
3. Bisect each 20-mm length longitudinally and place, cut sides down, onto A1 medium.
4. Incubate the flasks: within 21 d, callus should have grown and should be treated as described above for the seed-derived callus and subcultured to A2 medium.

3.2. Callus and Suspension Culture

1. When the calluses have grown to an approximate diameter of 20 mm, they should be subcultured by transfer with a sterile loop to fresh A2 medium. The quantity transferred at subculture is not critical, but a piece 5–10 mm in diameter is recommended. A fast-growing callus will require subculture at 3-wk intervals.
2. Transfer approx 5 g of callus to 50 mL of S1 medium, and incubate on the orbital shaker.
3. Subculture the suspension at 7-d intervals to fresh S1 medium (20% v/v inoculum volume). Incubate on the orbital shaker at 100 rpm.

3.3. Pigment Production Maximization and Analysis

Most calluses and suspension cultures of beetroot contain a mixture of cells that can visually be classified as "colorless, pink, or red." For the production of betanin, it is clear that at least in the short term, one must select for "red" cells.

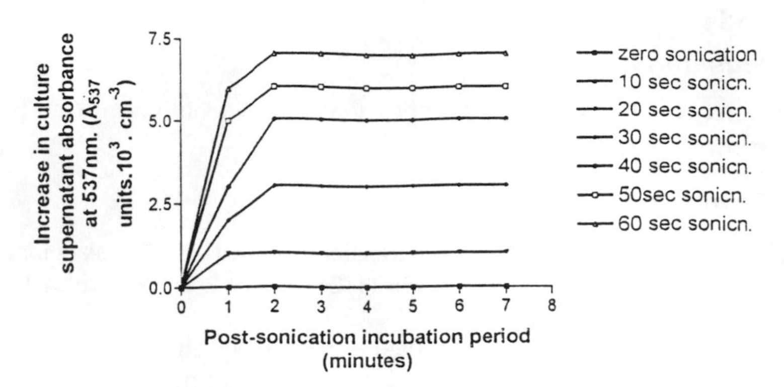

Fig. 2. Effect of duration of sonication pigment release.

1. To select for red cells, prepare several flasks of A2 medium, and onto the surface of each, pour approx 2 mL of suspension culture, swirl the suspension over the A2 surface to distribute the cells (single cells and aggregates), and incubate the flasks.
2. In each flask, many calluses will develop that can be assessed visually for their color, selected and subcultured to fresh A2 and later, after sufficient growth, to S1 medium for suspension culture. It is our experience that selected cell lines will require reselection after many subcultures. Also, we have found that it is prudent to have a stock of "backup" callus cultures in reserve to overcome the seemingly inevitable problems of a shaker power-supply failure or other such disaster!
3. To quantify the relative amount of betanin in a cell extract (*see* **Note 3**) or the medium surrounding cells postsonication, take a standard volume/extract dilution, filter it through glass microfiber paper, check and adjust the pH to 5.5, and measure the absorbance of the filtrate at 537 nm (betanin λ_{max}). Within the absorbance range 0–1.0, Beer's Law is observed. Thus, the quantity of betanin is proportional to the absorbance.

3.4. Ultrasonic Release of Pigments (see Fig. 2)

1. The suspension culture is removed from the shaker, and the cells allowed to settle. Using a 10-mL wide-bore pipet, modified by cutting off and discarding the tapered end, remove 4 mL of settled cells. Dispense the cells to the styrene jar, add 35 mL S1 medium, screw on the plastic lid, and place the styrene jar onto the transducer, using the coupling gel between the jar and the transducer.
2. Turn on the ultrasonicator for the desired period (*see* **Notes 4**, **5**, and **6**). After sonication, transfer the entire contents of the jar to a 250-mL Erlenmeyer flask containing 60 mL fresh S1 medium, and incubate on the orbital shaker. During the next 30–40 min the pigments will diffuse through the permeabilized membranes into the incubating medium: thereafter, no further pigment efflux is likely.

Remove samples for spectrophotometric analysis. If it is desired to continue to grow the cells for a further "crop" of pigment, ensure that all handling has been aseptic and that the cells are incubated with the appropriate volume of S1 medium.

4. Notes

1. The preparation of aseptic cultures from seed or from in vivo plants is the most difficult part of the entire protocol. The seed is irregular and rough-textured; often the plants in vivo appear to be contaminated with endophytic microorganisms. If external contamination is a problem, increase the NaOCl concentration for seed up to 5% available chlorine: endophytes will not be killed by surface sterilization, so different plant sources of beetroot should be used.
2. Normal aseptic techniques must be adopted for work with the cultures.
3. To harvest destructively all the pigments from a sample of cells, homogenize the cells in S1 medium at ca. 4°C, centrifuge the supernatant from the cells, retain the supernatant pigmented medium, and re-extract the cells in S1 medium. Repeat the process until no further pigment is extracted. Pool the supernatant solutions, adjust to pH 5.5, measure absorbance at 537 nm for betanin, or scan the extract between 350 and 700 nm to plot the absorption spectrum of the combined pigments.
4. Investigate a range of times of ultrasonic treatment times from 10 s to 5 min.
5. To observe the effect of dissolved gasses in the S1 medium on the ultrasonic release of pigments, try degassing the medium by bubbling helium into the medium for 5 min prior to sonication. Also try gassing the medium by bubbling laboratory air into the S1 medium prior to sonication. In both cases, a suitable air filter should be used to maintain the sterility of the cultures.
6. To observe the effect of temperature on the pigment efflux, cool or heat the medium (range 4–42°C) prior to sonication and during the postsonication incubation period.

References

1. Fujita, Y., Tabata, M., Nishi, A., and Yamada, Y. (1982) New medium and production of secondary compounds with the two-staged culture method, in *Proc. 5th Intl. Cong. Plant Tissue and Cell Culture* (Fujiwara, A., ed.), Maruzen Co., Tokyo, pp. 399–400.
2. McHale, D. (1986) The cinchona tree. *Biologist* **33(1),** 45–53.
3. Taya, M., Mine, K., Kino-Ora, M., Tone, S., and Ichi, T. (1992) Production and release of pigments by culture of transformed hairy root of red beet. *J. Fermentation Bioeng.* **73(1),** 31–36.
4. Flores, H. E. and Medina-Bolivar, F. (1995) Root cultures and plant natural products: "unearthing" the hidden half of plant metabolism. *Plant Tissue Cult. Biotechnol.* **1(2),** 59–74.
5. Scragg, A. H. (1995) The problems associated with high biomass levels in plant cell suspensions. *Plant Cell Tiss. Org. Cult.* **43,** 163–170.
6. Alfermann, A. W. and Peterson, M. (1995) Natural product formation by plant cell biotechnology. *Plant Cell Tiss. Org. Cult.* **43,** 199–205.

7. Khlebnikov, A., Dubuis, B., Kut, O. M., and Prenosil, J. E. (1995) Growth and productivity of *Beta vulgaris* cell culture in fluidized bed reactors. *Bioprocess Eng.* **14,** 51–56.
8. Ten Hoopen, H. J. G., van Gulik, W. M., Meijer, J. J., and Verpoorte, R. (1994) Economic feasibility of industrial plant cell biotechnology: the effect of various process options, in *Plant Cell, Tissue and Organ Cultures in Liquid Media* (Macek, T. and Vanek, T., eds.), Abstracts of symposium, July 8–11, Prague, CZ. pp. 22–30.
9. Sankawa, U., Hakamatsuka, T., Shinkai, K., Yoshida, M., Park, H-H., and Ebizuka, Y. (1995) Changes of secondary metabolism by elicitor treatment in *Pueraria lobatqa* cell cultures, in *Current Issues in Plant Molecular and Cellular Biology* (Terzi, M., Cella, R., and Falavigna, A., eds.), Kluwer Academic, Dordrecht, NL, pp. 595–604.
10. Taya, M., Yakura, K., Kino-Oka, M., and Tone, S. (1994) Influence of medium constituents on enhancement of pigment production by batch culture of red beet hairy roots. *J. Fermentation Bioeng.* **77,** 215–217.
11. DiIorio, A. A., Weathers, P. J., and Cheetham, R. D. (1993) Non-lethal secondary product release from transformed root cultures of *Beta vulgaris. Appl. Microbiol. Biotechnol.* **39,** 174–180.
12. Kilby, N. J. and Hunter, C. S. (1990) Towards optimisation of the use of 1.02-MHz ultrasound to harvest vacuole-located secondary product from *in vitro* grown plant cells. *Appl. Microbiol. Biotechnol.* **34,** 478–480.
13. Kilby, N. J. and Hunter, C. S. (1990) Repeated harvest of vacuole-located secondary product from *in vitro* grown plant cells using 1.02 MHz ultrasound. *Appl. Microbiol. Biotechnol.* **33,** 448–451.
14. Leathers, R. R., Davin, C., and Zryd, J. P. (1992) Betalain producing cell cultures of *Beta vulgaris* L. var. Biokores Monogerm (red beet). *In Vitro Cell. Dev. Biol.* **28P,** 39–45.
15. Böhm, H. and Rink, E. (1988) Betalains in, *Cell Culture and Somatic Cell Genetics of Plants,* vol. 5 (Vasil, I. K., ed.), Academic, pp. 449–463.
16. Kilby, N. J. (1987) An investigation of metabolite release from plant cells in vitro to their surrounding medium. PhD thesis, Bristol Polytechnic, UK.
17. Hamill, J. D., Parr, A. J., Robins, R. J., and Rhodes, M. J. C. (1986) Secondary product formation by cultures of *Beta vulgaris* and *Nicotiana rustica* transformed with *Agrobacterium rhizogenes. Plant Cell Rep.* **5,** 111–114.
18. Rhodes, M. J. C., Smith, J. I., and Robins, R. J. (1987) Factors affecting the immobilization of plant cells on reticulated polyurethane foam particles. *Appl. Microbiol. Biotechnol.* **26,** 28–35.
19. Kilby, N. J. and Hunter, C. S. (1986) Ultrasonic stimulation of betanin release from Beta vulgaris cells in vitro: a non-thermal, cavitation-mediated effect, in *Abstracts of VI International Congress of Plant Cell and Tissue Culture* (Somers, D. A., Gengenback, B. G., Biesboer, D. D., Hackett, W. P., and Green, C. E., eds.), University of Minnesota, MN, p. 352.
20. Gamborg, O. (1970) The effects of amino acids and ammonium on the growth of plant cells in suspension culture. *Plant Physiol.* **45,** 372–375.

APPENDIX

Widely Used Plant Cell Culture Media

The following represent the most widely used media for plant cell culture. Recipes for additional media used for specific applications are to be found in the relevant chapters and are listed in the index.

	Murashige and Skoog complete medium *(1)*	Murashige and Skoog plant salt mixture *(1)*	Linsmaier and Skoog medium *(2)*
Macroelements (mg/L)			
$CaCl_2$	332.020[a]	332.020[a]	332.020[a]
KH_2PO_4	170.000	170.000	170.000
KNO_3	1900.000	1900.000	1900.000
$MgSO_4$	180.540[b]	180.540[b]	180.540[b]
NH_4NO_3	1650.000	1650.000	1650.000
Microelements (mg/L)			
$CoCl_2 \cdot 6H_2O$	0.025	0.025	0.025
$CuSO_4 \cdot 5H_2O$	0.025	0.025	0.025
FeNaEDTA[c]	36.700	36.700	36.700
H_3BO_3	6.200	6.200	6.200
KI	0.830	0.830	0.830
$MnSO_4 \cdot H_2O$	16.900	16.900	16.900
$Na_2MoO_4 \cdot 2H_2O$	0.250	0.250	0.250
$ZnSO_4 \cdot 7H_2O$	8.600	8.600	8.600
Organics (mg/L)			
Glycine	2.000		
Myo-inositol	100.000		100.000
Nicotinic acid	0.500		
Pyridoxine-HCl	0.500		
Thiamine-HCl	0.100		0.400

[a] 440 mg/L $CaCl_2 \cdot 2H_2O$
[b] 370 mg/L $MgSO_4 \cdot 7H_2O$
[c] Original recipe uses 5 mL/L of a stock: 5.57 g $FeSO_4 \cdot 7H_2O$, 7.45 g Na_2EDTA dissolved in 1 L.

From: *Methods in Molecular Biology, Vol. 111: Plant Cell Culture Protocols*
Edited by: R. D. Hall © Humana Press Inc., Totowa, NJ

	Gamborg's B5 medium *(3)*	Schenk and Hildebrandt medium *(4)*	Nitsch's medium *(5)*
Macroelements (mg/L)			
$CaCl_2$	113.230[a]	151.000[b]	166.000[c]
KH_2PO_4			68.000
KNO_3	2500.000	2500.000	950.000
$MgSO_4$	121.560[d]	195.050[e]	90.27[f]
NH_4NO_3			720.000
NaH_2PO_4	130.440		
$(NH_4)_2SO_4$	134.000		
$(NH_4)H_2PO_4$		300.000	
Microelements (mg/L)			
$CoCl_2 \cdot 6H_2O$	0.025	0.010	
$CuSO_4 \cdot 5H_2O$	0.025	0.020	0.025
FeNaEDTA	36.700	19.800	36.700
H_3BO_3	3.000	5.000	10.000
KI	0.750	1.000	
$MnSO_4 \cdot H_2O$	10.000	10.000	18.940
$Na_2MoO_4 \cdot 2H_2O$	0.250	0.100	0.250
$ZnSO_4 \cdot 7H_2O$	2.000	1.000	10.000
Organics (mg/L)			
Biotin			0.050
Folic acid			0.500
Glycine			2.000
Myo-inositol	100.000	1000.000	100.000
Nicotinic acid	1.000	5.000	5.000
Pyridoxine-HCl	1.000	0.500	0.500
Thiamine-HCl	10.000	5.000	0.500

[a]150 mg/L $CaCl_2 \cdot 2H_2O$.
[b]200 mg/L $CaCl_2 \cdot 2H_2O$.
[c]220 mg/L $CaCl_2 \cdot 2H_2O$.
[d]250 mg/L $MgSO_4 \cdot 7H_2O$
[e]400 mg/L $MgSO_4 \cdot 7H_2O$.
[f]185 mg/L $MgSO_4 \cdot 7H_2O$.

References

1. Murashige, T. and Skoog, F. (1962) A revised medium for rapid growth and bioassays with tobacco tissue cultures. *Physiol. Plant.* **15,** 473–479.
2. Linsmaier, E. M. and Skoog, F. (1965) Organic growth factor requirements of tobacco tissue cultures. *Physiol. Plant.* **18,** 100–127.
3. Gamborg, O. L., Miller, R. A., and Ojima, K. (1968) Nutrient requirements of suspension cultures of soybean root cells. *Exp. Cell. Res.* **50,** 151–157.

4. Schenk, R. U. and Hildebrandt, A. C. (1972) Medium and techniques for induction and growth of monocotyledonous and dicotyledonous plant cell cultures. *Can. J. Bot.* **50,** 199–204.
5. Nitsch, J. P. and Nitsch, C. (1969) Haploid plants from pollen. *Science* **163,** 85–87.

Index

G

H

I

K

L